Defenders of the Text

Defenders *of the* Text

THE TRADITIONS OF SCHOLARSHIP IN AN
AGE OF SCIENCE, 1450–1800

Anthony Grafton

HARVARD UNIVERSITY PRESS
Cambridge, Massachusetts
London, England
1991

Copyright © 1991 by the President and Fellows of Harvard College
All rights reserved
Printed in the United States of America
10 9 8 7 6 5 4 3 2 1

This book is printed on acid-free paper, and its binding materials
have been chosen for strength and durability.

Library of Congress Cataloging-in-Publication Data
Grafton, Anthony.
Defenders of the text : the traditions of scholarship in an age
of science, 1450–1800 / Anthony Grafton.
p. cm.
Includes bibliographical references and index.
ISBN 0-674-19544-2 (alk. paper)
1. Renaissance   2. Learning and scholarship—History.
3. Europe—Intellectual life.   I. Title.
CB361.G69   1991
001.1—dc20
90-37895
CIP

For J. B. Trapp

CONTENTS

Introduction: The Humanists Reassessed   *1*
1   Renaissance Readers and Ancient Texts   *23*
2   The Scholarship of Poliziano and Its Context   *47*
3   Traditions of Invention and Inventions of Tradition in Renaissance Italy: Annius of Viterbo   *76*
4   Scaliger's Chronology: Philology, Astronomy, World History   *104*
5   Protestant versus Prophet: Isaac Casaubon on Hermes Trismegistus   *145*
6   The Strange Deaths of Hermes and the Sibyls   *162*
7   Humanism and Science in Rudolphine Prague: Kepler in Context   *178*
8   Isaac La Peyrère and the Old Testament   *204*
9   Prolegomena to Friedrich August Wolf   *214*

Notes   *247*
Acknowledgments   *321*
Index   *323*

Defenders of the Text

INTRODUCTION

The Humanists Reassessed

THE ESSAYS that follow were written at different times and for different audiences. But they attack a single, general dogma—one formulated in the late sixteenth and early seventeenth centuries by such advocates of the New Science as Francis Bacon and René Descartes. These men enhanced the authority of their enterprise by denigrating the history of Western culture before their time. In fact, they devised a cultural history of the West that proved as influential as it was polemical. In it they demoted classical culture to the level of the merely literary; for all their eloquence and wit, they argued, the Greeks had not been able to produce the sort of knowledge that gave men power over nature. They dismissed scholasticism as sterile. And they treated with particular condescension the most recent stage in the development of Western thought—the tradition of Renaissance humanism, which they had encountered in the course of their expensive and excellent educations. They admitted that the humanists had driven the barbarous scholastics out of the liberal arts and raised the standard of intellectual and literary discourse. But they had wasted their time and that of their students on literary rather than scientific questions. They had cared only for morality and eloquence, not for rigor and power. Only a new science that could prove itself by direct interference in nature's operations could attain intellectual certainty and serve human needs. This only Bacon and Descartes could provide.¹

The prophets of the new intellectual world looked with particular disdain on their predecessors' method. The humanists had explicated other men's texts instead of investigating the world around them. This textual approach, which limited its purview to what other men had already thought, was necessarily sterile; trying to wring new knowledge from the used-up components of the classics was like trying to make fresh tea from the bitter leaves left in an old pot. "Down with antiquities," wrote Bacon, "and citations or supporting testimonies from texts; down with debates and controversies and divergent opinions; down with everything philological."² Whatever their disagreements on other points, almost all the manifesto-writers of the New Science agreed about the sterility of exegesis.³ Descartes pleasantly mocked the study of past texts, which could reveal only that their authors had disagreed because they lacked a philosophical method as rigorous as his. Pascal insisted that texts could inform the reader only about brute facts or theological doctrine; in either case one must have recourse only to the texts themselves, not to the tradition of interpretation which could only obscure the original message. In science, of course, one must stick not to reading but to reasoning.⁴ Even Galileo, who took a deep interest in problems of exegesis and drew on the work of humanists and neo-Platonists, mocked those who tried to decode the book of Nature with the methods of humanist philology rather than observation and computation. "This kind of man," he wrote to Kepler, "thinks that philosophy is a sort of book like the *Aeneid* and *Odyssey,* and that truth is to be found not in the world or in nature but in the collation of texts (I use their terminology). I wish I could spend a good long time laughing with you."⁵

These arguments had an enormous impact in the seventeenth century. They induced many prominent intellectuals to devote themselves to direct study of mathematics and the natural world. They convinced many polemicists that the humanists' scholarship and education were as sterile as the scholasticism they had replaced. And they left adherents of the classical languages and classical learning in the depressed state natural to votaries of an obsolete, arteriosclerotic divinity. Even a proficient philologist like Ismaël Boulliau felt by 1657 that he had to admit that, as the excellent

Gronovius had advised him, "the age of criticism and philology has passed and one of philosophy and mathematics has taken its place." "The first blow of misfortune," Isaac Casaubon's son Meric reflected in 1668,

> came from the Cartesian philosophy, which knocked the good books out of the hands of the young, who are foolish and desire novelty. Then one passed on to "Experiments," where all learning and all wisdom are now located. These clever men call themselves "Realists," an attractive title; the rest, whatever sort of literature they are distinguished for, are dismissed as "Verbal" and "Idealist" . . .[6]

Casaubon links Descartes's method and Bacon's empiricism—in some respects an odd couple—as jointly responsible for the decline of humanism.

Despite the achievements of late-seventeenth-century scholarship, as practiced by the likes of Richard Simon and Richard Bentley, modern historians have accepted Bacon's and Descartes's polemics and Boulliau's and Casaubon's laments as an accurate description of the past. They have treated Renaissance humanism as an influential but transitory effort to renew Western culture by reviving a classical literary education and applying the tools of philology to ancient texts. They have agreed that newer men with newer scientific brooms swept the humanists from the center stage of Western thought after 1600. And they have argued that humanistic scholarship did not regain intellectual profundity or popular appeal until it was transformed, after the French Revolution, by the rise of a new German hermeneutics and historiography which learned from the radical changes of its own time to see the past as a genuinely foreign country. These views dominate the teaching of Western civilization and intellectual history. They provide the central thesis of so well-informed a survey as Robert Mandrou's *From Humanism to Science* (1973). More surprisingly, they inspire the moving final pages of Robert Bolgar's still standard history *The Classical Heritage and Its Beneficiaries* (1954). Yet recent research in fields as varied as historiography and the history of science itself has shaken their foundations.

This book uses case studies to challenge the orthodox history of Western culture in both implicit and explicit ways. In the first place, it argues that humanism was neither simple nor impractical. The humanists studied a vast range of texts, issues, and problems. They forged many of the technical methods still applied by the supposedly revolutionary German philology of the late eighteenth century. But they also found practical lessons in the classics—lessons applicable to warfare and administration as well as oratory and epic poetry. And they continued to find formal Latin eloquence a supple and expressive tool for both technical and literary purposes. Each major humanist was a complex figure, each local or national scholarly tradition different from the rest. Humanism remained a rich and vital—though also a varied and embattled—tradition for at least two centuries after the end of the Renaissance. Italian and German historians have long known that the disciplines of humanism survived into the eighteenth century, that history and philology formed a powerful part of the critical weaponry of the Italian and German Enlightenments.[7] But the essays collected here study the afterlife—and demonstrate the continued impact—of humanism in new geographical and intellectual areas.

Secondly, the book shows that the method of the humanists was classical in more than one sense. The humanists not only found models for epic poetry, moral philosophy, and historiography in the classics they revived; they also turned up what they took as examples of lower and higher, textual and historical criticism. Science, as Noel Swerdlow, Vivian Nutton, Lynn Joy, and Peter Dear have shown, was transformed in the sixteenth and early seventeenth centuries by the systematic use of newly discovered classical texts and methods.[8] The history of humanism, these essays show, closely resembles that of science. It too transformed itself by systematic efforts to return to ancient methods and models. Indeed, in scholarship the classical substrate remained vital long after the scientists had abandoned their classical ancestors to the dustbin of history.[9] The claims of Renaissance Italian and German Enlightenment humanists to have revolutionized the world of historical thought need to be reconsidered in the light of the long-term continuities and complex genealogies traced here.

Relations between science and humanism were more complex than Bacon and Descartes suggested in another respect as well.

The humanists did not confine themselves to strictly literary areas of study. After 1450 they often analyzed scientific texts and produced results of interest to specialists in medicine and astronomy as well as to general readers. The scientists, for their part, often did work of great penetration and originality in the humanist fields of textual exegesis and cultural history. The two cultures, in short, were not locked in the battle that the pamphleteers of the New Philosophy called for; they coexisted and often collaborated, and sometimes the scientists proved to be better readers of texts than their scholarly friends.

An Innocent Abroad

This view of the humanist tradition, as I have suggested, is hardly standard. I arrived at it by a long and sometimes painful journey, an account of which may both illuminate the argument just sketched and explain why it needs to be advanced. In the hot July of 1973 I arrived in London ready to begin work on a doctoral dissertation about Joseph Scaliger, the most famous classical scholar of the French Renaissance. I had been trained at the University of Chicago in Renaissance intellectual history, classics, and the history of science. I had read secondary literature in English, French, and German. And I knew precisely the coordinates that Scaliger occupied on my mental map of Renaissance culture. His work on the history of classical texts and calendars formed part of the "historical revolution" of the sixteenth and seventeenth centuries. It reflected, that is, the development of a wider historical consciousness and the effort to frame precise critical rules for evaluating sources that preoccupied better-known intellectuals of his time and place, like Louis Le Roy and Jean Bodin. My task, I thought, was to draw from Scaliger's work a more technical and detailed account of this revolution than previous scholars had constructed from Bodin & co.

My convictions were not irrational; they were not even wholly wrong. But they stemmed from the wrong source: from modern historiography rather than Renaissance texts and their sources. Intellectual historians in the English-speaking world had identified a number of smaller topics within the history of Renaissance thought that deserved the historian's attention and needed monographic research. These included the civic humanism ascribed by Hans Baron to early Renaissance Florence, the connections teased out by Michael Baxandall and others between humanism and the arts, and the revolution in historical thought. A plausible dissertation topic had to be anchored to one of these firm and prominent markers; otherwise the author might float off into the unpathed waters of antiquarianism or suffer shipwreck on the shoals of obscurity—might, indeed, waste his life on a set of texts and issues not included in the canon of important topics. To this fate, I thought, Scaliger would not condemn me. His name had recognition value. His work had not been studied in depth for more than a century. And he seemed directly connected to one of the canonical topics of my generation.

Powered by these notions, not weighed down by learning, and kept afloat by a welcome but exiguous Fulbright fellowship, I steered my fragile craft into what my teachers at Chicago and the Fulbright authorities agreed was the only appropriate harbor: the long, dark, narrow office, crammed with apparently unrelated books *de omni re scibili et quibusdam aliis,* of Arnaldo Momigliano, then in his last years as Professor of Ancient History at University College London. He had not yet left for his customary summer in Italy when I came knocking at his door. In an hour's impromptu conversation he gave me many small surprises and a large new outlook on the definition of my project.

Momigliano asked if I had read much of the scholarly work of Angelo Poliziano. I knew the name—a small point in my favor. Poliziano was familiar to me as a brilliant Latin and Italian poet, a client of Lorenzo de' Medici and a friend of Pico della Mirandola. But it had never occurred to me that a study of a modern figure like Scaliger, a sixteenth-century Frenchman who wound up his career at the innovative University of Leiden in republican, mercantile Holland, should begin from or stop at the equivocal Florence of the Medici and the Platonic Academy—a world of courtly elegance and ritual magic that inspired more discomfort than interest in most American historians.[10] Fortunately, ignorance never surprised Momigliano, and he had a partial remedy at hand. He advised me to begin with Poliziano. He also told me that I could do so easily around the corner from his office, in the library of the Warburg Institute.

With most of July still to come, I entered that austere modernist building, obtained a reader's card, and found shelves burdened with Poliziano's Latin works and modern Italian studies of them. Within a few days it became clear to me that Poliziano's work had marked a watershed in the history of classical scholarship: a break between an older, rhetorical style of humanism and a newer, technical philology. The summer passed in a blur of sixteenth- and twentieth-century type faces. It soon emerged that if Scaliger's wide historical interests were shaped by his original French milieu, in the technical details of his work he responded not only to the historians and philologists of his own time and place but even more to the precedents Poliziano had offered and to later French and Italian adaptations of these. The larger point of Momigliano's first suggestion was also clear. I must study not a single individual who had been identified as important by the modern academic culture that I came from, but the whole scholarly tradition that Scaliger himself had belonged to—one that began in the fifteenth century, though its most distant roots ran all the way back to antiquity, and that survived, in a sense, in the European scholarship of my own day.

American scholars knew in the 1960s and 1970s that the Renaissance humanists had devised a new, historical approach to reading texts. This approach first became visible in the works of fourteenth-century notaries and lawyers in northern Italy, sometimes called "prehumanists." In Petrarch's scholarship, a little later, it grew into a more systematic effort to revive classical Latin literature as a whole. In the work of Valla and Erasmus it gave rise to sharp criticism of accepted beliefs and practices in the pre-Reformation church. And in the thought of the mid-sixteenth-century French lawyers, like Bodin, it spawned general rules for reading and thinking historically. Finally, thanks to the discovery of a new world that the ancients had not known and the rise of a New Philosophy that they had not anticipated, the humanist enterprise became obsolete.

If humanism seemed a vitally important movement, it also seemed narrowly limited. It sought to revive the grammatical scholarship and rhetorical skills of ancient Rome, not to reform the Western intellectual tradition. It challenged textual errors, not the supremacy of scholasticism. Within humanism, by an odd contrast, modern historians' attention was claimed by the few individuals like Valla, Erasmus, and Bodin, who came to quotable general conclusions about ancient history, early Christianity, or modern theology. Only broad and polemical theses—not the long and difficult journeys by which they were arrived at—seemed to matter. Paradoxically, those who practiced the technical disciplines of humanism, like textual criticism, at the highest level but did not draw general morals from the texts seemed far less innovative and interesting. This was the category to which Poliziano, for example, seemed to belong.

The Warburg library offered a different prospect. Italian scholars, whose books and articles lay heavy on the shelves, held that the historical method of the humanists implied nothing less than a new view of the world, a systematic and profound alternative to the methods of traditional Western philosophy. The humanists saw knowledge as being concerned above all with what earlier men had thought and written. But understanding of earlier thought must rest on textual exegesis—on an exegesis the exponents of which read the texts in the light of the fears and aspirations of their authors and the prejudices and assumptions of their first readers. The scholastics had read their texts as structures, systems of interlocking propositions that they tested for coherence as an engineer tests the load-bearing parts of a building. The humanists read theirs as clouded windows which proper treatment could restore to transparency, revealing the individuals who had written them. The scholar could thus come to know dead men as they had really been, and the serious effort to obtain that sort of knowledge became the first characteristically modern form of intellectual life. Such knowledge, moreover, lurked most richly and vividly in the details of the ancient texts. When a humanist explicated a particular passage in the light of the author's circumstances, he revealed a novel view of the world in its most sharp and distinct form. At this central humanist enterprise Poliziano had excelled all his rivals.

The Italian vision of Poliziano and his context was eloquently expressed in general essays by senior Italian intellectual and literary historians, like Eugenio Garin and Vittore Branca.¹¹ But even more exciting insight was afforded by the monographic studies of his individual technical works by Sebastiano Timpanaro, Alessandro Perosa, Manlio Pastore Stocchi, and many others. The modernity of Poliziano's mind emerged with startling clarity from his solutions to basic and difficult philological problems. Timpanaro, for example, showed that Poliziano had seen the problem of assessing manuscript evidence as one to be settled not by seat-of-the-pants navigation but by strict historical reasoning. In all cases he considered later manuscripts less authoritative, because farther removed from the ultimate, ancient source, than older ones. In some favored cases, moreover, he could show that one surviving manuscript was the parent of all the rest. This alone should serve as the basis for textual scholarship. The demonstration that Poliziano had arrived, long before Scaliger, at a fundamentally historical approach to textual criticism, was dizzyingly exciting. But Timpanaro's book was only the most shatteringly novel of a series of vastly informative studies. These included brilliant articles and monographs by Timpanaro, Carlo Dionisotti, and John Dunston, which drew the fascinating moral of Poliziano's relationship with a scholar of a more traditional stripe, Domizio Calderini; by Konrad Krautter and Maria Teresa Casella, who described in patient and appreciative detail the normal science of the old-fashioned commentators that Poliziano abused so heatedly; by Brugnoli and Perosa, who followed the humanists through some complex Suetonian nooks and crannies, to emerge with new insight into their historical methods and conclusions; and by many others.¹²

I could have stayed with Poliziano forever, rather like the character in David Lodge's *The British Museum Is Falling Down* whose study of sanitation in the Victorian novel never emerges from its first background chapter on the sewer in prehistory. But I doubted my ability to compete with the Italians as a manuscript hunter—a doubt amply confirmed by the brilliant detective and editorial work that has appeared since the early 1970s, thanks to Lucia Cesarini Martinelli, Vincenzo Fera, the late Roberto Ribuoli, and many others. And in any event I still wanted to explore the afterlife of high Italian scholarship in the North. So I continued to read Scaliger as well as his Italian—and, as it turned out, French—sources.

I went to Leiden, where his books and papers were preserved, and spent happy weeks summoned to the university library by the bells that seemed always to be ringing, randomly, the quarter hour. I became a habitué of some of the other pleasant places where Scaliger's published works and private letters were stored, gaining familiarity with the worn Victorian elegance of the British Museum and the crisp Enlightenment splendor of the Burgerbibliothek in Bern as well as the melancholy charm of Leiden. I bought copies of Scaliger's works in dusty side-street shops, accumulated a vast store of wrong and outdated solutions to the problems he had attacked in classical texts, and even shook his hand when his bones were briefly disinterred. And I wrote about his work.[13]

But even the effort to stick to one subject and the need to write one book did not simplify my interests unduly. In fact, as I pursued Scaliger and his sources around Europe, I found in each country another case of the phenomenon first met at the Warburg: a strong national version of the history of humanist scholarship, independent of the rest in inspiration and methods and usually complementary to them in conclusions. Holland nourished a tradition of research into the historical erudition, biblical criticism, and textual scholarship that had been the glory of the Dutch universities when they were the glory of European letters. Germany harbored a set of Renaissance scholars who demonstrated at length that classical texts had retained a vital practical value, especially for administrators and military men, deep into the seventeenth century. The universities and academies that stuck to the classical curriculum after 1600, they argued, were not sterile bastions of classical reaction but practical institutions serving well-defined needs. A second set of German scholars investigated the humanism of the late eighteenth and early nineteenth centuries, the time when German universities came to the intellectual forefront of Europe. They powerfully emphasized the interdisciplinary interests and historical insight of eighteenth-century humanist savants like Christian Gottlob Heyne, even if most of them underscored even more heavily the gap that separated such men from the heroes of nineteenth-century historicism, like Wilhelm von Humboldt. French scholars relished the tradition of humanist rhetoric, and showed in detail how Jesuit theorists and teachers had not abandoned humanist oratory but updated it to fill seventeenth-century needs. American literary scholars, finally, made their own powerful contribution, one that cast some doubt on the strain of Italian historiography from which I began. They showed—as I learned while teaching collaboratively with critics at Princeton—that Renaissance humanists across Europe had not always applied a historical method to the ancient classics. Indeed, they had often read the ancients allegorically, making them not human figures in a vividly realized historical landscape but atemporal vehicles of an eternal wisdom—and finding good classical precedents in the ancient neo-Platonic tradition and elsewhere for doing so.[14]

This diet proved both rich and indigestible. Each tradition filled lacunae in the rest. But they did not seem to add up to a single coherent account—much less to the normal picture of European intellectual history, in which humanism ceased to figure substantially on the scene after the first third of the sixteenth century. General treatments existed. But they were straitjacketed by doxographical tradition, fragmented by national prejudice, or artificially limited by disciplinary boundaries. The standard histories of classical scholarship arranged the field into a neatly teleological system of ages, each dominated by one nation: the Italians in the fifteeenth century, the French in the sixteenth, the Dutch in the seventeenth, and so on. The neatness of this classification revealed its spuriousness at once. But knowing that did not help me find a new general account to replace it, especially since the nature of humanist scholarship was in effect sharply contested ground. Was its method historical or allegorical? Were its goals pragmatic or idealistic? Powerful accounts based on rich sources could be cited for either position in either case. Even the chronology was uncertain. American historians tended to believe that the humanist tradition succumbed in the seventeenth century to a form of death by sclerosis. Many European scholars, however, stated or implied that humanist rhetoric and scholarship retained their vitality a century later. And if one accepted that humanism was not replaced by, but coexisted with, the New Philosophy, the New Science, and even the Enlightenment, what forms did their relations take? Collision, collaboration, conflation?

I could not hope to answer these questions by making my own

the methods and interests of any single national tradition. By doing so I might or might not gain that second identity which social historians of my generation sought; but I would certainly find myself constrained by the same limitations of viewpoint that had created so many impervious boundaries and deep fissures in the existing literature. I decided, accordingly, to carry on a wide-angled series of case studies in the later life of humanist scholarship across early modern Europe as well as a sharply defined program of research into Scaliger's life and work. The essays that resulted from this decision try to make a virtue of the American's natural problem as a scholar: his distance from the other shore, which certainly induces ignorance but may also lend perspective. They try to juxtapose, where possible to fuse, and sometimes simply to contrast the divergent traditions of research I have encountered. They apply these to a wide range of mostly understudied primary sources. And they argue that the humanists remained intellectually powerful—both in the content of their work and in the audience it reached—long after we have thought they did.

Scholarship and Its History: The Address of Bentley's "Letter to Mill"

For all their variety of subject matter, the essays collected here converge in method as well as in thesis. They assume that feats of scholarship are just as complex—and require just as rich and flexible a set of interpretative techniques—as feats of philosophical or scientific work. The scholar reasoning about a difficult text works within a set of contexts. Personal needs and circumstances, professional customs and institutions, long-standing intellectual and technical traditions, and recent polemics all shape his methods and help to dictate his conclusions. He is the prisoner of his own tastes and obsessions, interests and insensitivities. His deceptively modern-sounding arguments often address now-forgotten and unlikely issues or follow from now-obscure and alien premises. Hence no early work of classical scholarship—however austerely technical and modern it may seem—can simply be read off like a modern journal article (not that these lack their own subterranean politics of allusion and quotation, often imperceptible to the outsider). Only systematic comparison between a given work under analysis and many earlier and contemporary texts can make the modern reader familiar with the inherited technical language a past scholar used; without that familiarity one cannot distinguish between the novel and the traditional, the original and the tralatitious. Only careful study of the responses that the work in question evoked from contemporary and later scholars, finally, can enable the modern reader to uncover its original agenda of personal and technical polemics. And only an inquiry that gives due attention to each of these factors can do historical justice to a complex work of scholarship.

These principles may seem so obvious as not to need formal statement. But one example will suggest how completely they have been ignored by those who actually write about the technical content of early scholarship. No philological masterpiece occupies a grander niche in the canon than Richard Bentley's *Epistola ad Millium* of 1691. This was a short essay in Latin, written while Bentley served as chaplain and tutor in the household of Edward Stillingfleet, onetime Dean of St. Paul's and now Bishop of Worcester. It took off from the Greek text of the world chronicle written by John Malalas of Antioch in the sixth century. But its concerns were not Byzantine. Bentley set out to make historical sense of the chapters of ancient Greek literary history that Malalas had mangled and to make classical Greek out of the many fragments that he and similar writers had misquoted. Despite the obscurity of its occasion and the difficulty of its contents, the *Epistola* made Bentley a European celebrity. It showed that he had surveyed in magisterial detail those favorite quarries for the Hellenists of his day, the Byzantine lexicographers and scholiasts (Meric Casaubon described the *Lexicon* of Hesychius as "the greatest treasure of humane learning now extant").[15] More strikingly, it revealed that he now understood the meters of Greek poetry more thoroughly than any other modern scholar—including Scaliger and Grotius, whose errors he corrected with characteristic self-confidence.[16]

Modern scholars have described, praised, and even reprinted the *Epistola*. They have relished Bentley's rhetoric—especially, perhaps, the passages in which he pretends to address Malalas as he would later address the Fellows of Trinity, rebuking him when he nods and slapping him when his attention strays. And they have listed and explained with meticulous care the many hurdles that Bentley leaped where all others had faltered or fallen: the many corrupt passages, that is, that he definitively restored.[17]

They have had little to say, by contrast, about any larger purposes or consistent themes that might unify the *Epistola*. Usually, in fact, they characterize it as a miscellany, a random set of observations only loosely connected by the thread of Malalas's text. They treat the *Epistola* rather as historians of science once treated the Queries in Newton's *Optics*: as a wholly novel, wholly iconoclastic masterpiece that departs from more primitive conventions of argument and method, a work not determined by or rooted in context and convention. And this view is at least partly problematic.

Every admirer of the *Epistola* has called attention to the passage in which Bentley analyzes a supposed fragment of Sophocles. Here the tragedian expresses the convictions of a monotheist in the language, more or less, of Attic tragedy:

> One in truth, there is one God,
> Who made the heavens and the broad earth,
> The gray ocean and the powerful winds.
> But many of us mortals err in our heart,
> Setting up as consolation for our woes
> Images of the gods in stone or bronze,
> Or figures worked in gold and ivory.
> Honoring these with sacrifices and useless feasts
> We think that we thus act most piously.[18]

Bentley dissected this unattractive little specimen with infectious zest and meticulous attention to detail. He pointed out that only Christian writers, not pagans, had quoted the fragment. How, he asked, could the learned and diligent Plutarch, Porphyry, and Stobaeus have missed so striking a text? The fragment had linguistic features not found in Greek tragedy. And the content revealed the spuriousness of the whole even more powerfully than the style. Sophocles' tragedies had formed part of a central religious ritual of a pagan society. They could hardly have included attacks on pagan observances or polytheism: "How could Sophocles have been bold enough to bring the holidays and solemn shows into contempt on the very holiday on which the shows were performed?"[19] A text that the Greek Fathers, the beneficiaries in some cases of a splendid literary education, had considered a jewel was revealed by a mere modern to be paste.

Modern students of Bentley have made these points. But they have not pressed beyond them to the further and historically more vital one: that his attack on ps.-Sophocles was neither isolated nor unpremeditated. It appears bracketed by scathing remarks about other supposedly pagan sources which ancient and modern Christians had cited to show that the wisest Greeks had rejected polytheism and idolatry. Bentley ridiculed the verses ascribed to Orpheus and the modern scholars who offered interpretations of their profound messages "based on the foolish trifles of the Cabalists."[20] He mercilessly mocked those "men of elegant judgment who revere the oracles commonly ascribed to the Sibyl as the real effusions of that prophetic old lady, Noah's daughter."[21] He thus turned his formidable guns on the prime pagan witness to monotheism and the prime pagan prophecy of the coming of Christ, the Sibylline oracles. Evidently Bentley set out not only to discredit one bit of fake Sophocles but to destroy the whole armory of dubiously ancient texts with which Christian scholars had tried to show that the best of the Greeks agreed with them. The attack on Sophocles was only part of a well-aimed broadside.

Bentley stuck to his guns, moreover, in the face of well-informed opposition from an early reader whom he respected—always good evidence that an author has invested an apparently straightforward theory with a high emotional charge. Edward Bernard, Savilian professor of astronomy at Oxford, read the proofs of the *Epistola* at Bentley's request. He found Bentley's critical arguments depressing, even shocking. He urged Bentley to emend "those splendid iambs" rather than to reject a testimony cited in good faith by "our Justin and other orators of Christ" as "unworthy of the buskined Sophocles." But Bentley refused, politely but firmly, to modify his conclusions, even though he agreed to mitigate the sharpness of his tone—a concession Bernard thought unnecessary in the end.[22]

The correspondence was stately, detailed, and unfailingly polite. It remains a model of scholarly discourse which many budding controversialists could read with profit. But it did not make Bentley shift his stance in any important respect. Instead, it provoked him to amass new evidence for his views in an appendix to the original text. He offered his best single piece of linguistic evidence—the close parallelism between a phrase in ps.-Sophocles, "many of us mortals err in our heart," and the Septuagint text of Psalm 95.10—in a letter and in his addenda.[23] And in the end it led him to sharpen the point of his original thesis as well as to broaden the base of data on which it rested. He argued that all the Fathers who quoted the Sophoclean fragment had derived it, ultimately or directly, from one suspect source, which they ascribed to Hecataeus of Abdera. This now lost work the great ancient scholar Herennius Philo of Byblos (Origen *Contra Celsum* 1.15) and the great modern scholar Scaliger had rightly condemned as a forgery by a Jew (or an uncritical admirer of the Jews). Bentley concluded that ps.-Hecataeus and the author of the third Sibylline oracle had both been Jews "trained up in the same school" to show from forged evidence that the best of the Greeks had been monotheists too.[24]

Once illuminated by Bentley's correspondence, the *Epistola* no longer seems simply a random series of technical arguments. It takes a firm theological and historical position. And that has a larger context of its own in the intellectual history of the sixteenth and seventeenth centuries. Bentley knew that the exposure of Jewish and Christian forgeries had preoccupied the great Hellenists of the 1580s and 1590s, Scaliger and Casaubon—the very men whose errors in meter he had revealed with such pleasure and aplomb. Both the tone and the content of Bentley's remarks ranged him on their side. Indeed, in this area—unlike that of meter—he quoted them and their ancient forerunner, Philo, as authorities. He compared his treatment of ps.-Hecataeus with that already meted out to ps.-Aristeas, another Jewish text, which "Scaliger and a great many scholars after him have shown to be spurious."[25] He thus self-consciously represented his own work as one link in a solid existing chain. The *Epistola,* then, did not simply break with the traditions of humanist scholarship; in some respects it continued them.

Contexts, of course, are personal and social as well as technical and intellectual. And the attack on these forgeries had a special resonance and urgency in the Restoration Church of England where Bentley moved and had his living. From the 1640s onward, the Cambridge Platonists and their allies, Henry More and Anne Conway, revived and amplified the Florentine neo-Platonist tradition with its insistence that the pagans had enjoyed a separate, powerful monotheist revelation. They did so in opposition to the Puritans, trying to prove that classical culture had its uses for Christians. But in their assiduous defense of classical studies they ignored or denied the conclusions of the sophisticated classical scholars of their day. They overstated the similarities between classical and Christian professions of monotheism. And they lent credence to particular texts, like the Hermetic Corpus and the Sibylline oracles, which had been exposed as fakes.[26]

Just after the English revolution ended, more orthodox divines like Theophilus Gale and Stillingfleet attacked what they saw as the Platonists' excessive tolerance in large-scale polemical treatises. Stillingfleet's *Origines Sacrae* argued that "there is no credibility in any of those Heathen Histories, which pretend to give an account of ancient times"; the Old Testament was the only reliable source, and Jewish writers the only reliable witnesses, to the early history of mankind. The pagans had been foolish men sunk in ignorance, not monotheist sages.[27] Gale argued, drawing at paralyzing length on the even more learned French theologian Samuel Bochart, that any apparently valid elements in ancient culture or mythology were actually feeble and derivative. They were pale reflections of the biblical history and theology of the Jews, borrowed and corrupted by Greeks and others.[28]

These criticisms provoked a reply from the most learned of all the devotees of the ancient theology. Ralph Cudworth, head of a house in Cambridge and master of far more original learning than Gale or Stillingfleet, defended the Platonist tradition at vast and absorbing length in his *True Intellectual System of the Universe* of 1678. This great book not only restated the historical thesis that Gale and Stillingfleet had denied but also argued the literary and philosophical case of the ancient sages on a level that their attackers had not touched. Cudworth subtly defended the notorious ancient forgeries, like the Hermetic Corpus, arguing that they must have contained some genuine elements even if the transmitted texts he knew were late. Otherwise contemporaries would have seen through them at once and exposed them. More dangerously still, he amassed heaps of evidence from classical literary works for the monotheist piety of the Greeks. He neatly combined an appearance of critical rigor with his assertion of an unhistorical thesis. And he did so with particular deftness in the case of the Sophoclean fragment that Bentley took such pains to attack only ten years later. Cudworth quoted three lines from it and remarked:

> That though this be such as might well become a Christian, and be no where now to be found in those extant Tragedies of this Poet (many whereof have been lost) yet the sincerity thereof, cannot reasonably be at all suspected by us, it having been cited by so many of the Ancient Fathers in their Writings against the Pagans, as particularly, *Justin Martyr, Athenagoras, Clemens Alexandrinus, Justin Martyr, Eusebius, Cyril,* and *Theodoret;* of which number, *Clemens* tells us, that it was attested likewise, by that ancient Pagan Historiographer *Hecataeus.*[29]

It would be hard to disarm potential adversaries more neatly than Cudworth did here by his opening admission. But it was Cudworth's misfortune to encounter an adversary whom he could neither win over by his learning nor frighten off by his preemptive critical argument. Bentley found in Cudworth both the Sophoclean fragment and the arguments that made it dangerous. He dissected the *True Intellectual System,* as well as ps.-Sophocles and his brethren, in the *Epistola.*

The *True Intellectual System* and its author, however, were not Bentley's only targets. Behind both of them lurked the more disturbing figure of More. He too had argued for the priority and purity of pagan revelations. But he had done so far less scrupulously than Cudworth, seizing any scrap of evidence that came to his hand. As he wrote to Anne Conway about his *Conjectura Cabbalistica,* a numerological interpretation of Genesis 1–3, "Though the Conceptions in the Cabbala be most what my own, yett I do what I can in my Defense to gette Godfathers all along to these births of my own braine, and so to lessen the odium of these inventions by alledgeing the Authority of Aunceint Philosophers and Fathers."[30] More had fused Cabalism and the *Orphica*—two branches of literature Bentley found especially detestable—long before Cudworth. His arbitrary appropriation of ancient texts, genuine and spurious, early and late, to support his doctrine on the "Pythagorick mysteries of numbers" in the Bible represented an intellectual threat to the Restoration order being forged by Stillingfleet and others.[31] These men used history and philology to prove the uniqueness of Christianity and the soundness of the Bible, not to conflate all texts and dispensations. They disliked More's dissolution of historical boundaries as much as Spinoza's dissolution of Revelation. And they felt they had far less common ground with him than with Cudworth.[32]

In fact, Stillingfleet and Bentley probably saw More as a political as well as a scholarly threat to their ecclesiastical and scholarly enterprises. He had written against Enthusiasm and called on the magistrate to suppress the sects. And he had reviled such spiritualists as the Familist prophet H.N. (Henry Niclaes), whom he described as "a meer Mock-Prophet, so mad and phrantick are his allusions to and interpretations of the Scripture."[33] But he had also dabbled in learned magic and spiritual healing, encountered spirits in dreams, and populated the cosmos with rank on rank of angels and daemons.[34] He had popularized and debated with Descartes and helped to create the infinite cosmos of the New Science. But he had also eagerly collected stories of

> the Spirit of a murdered Boy in Sheffield in Yorkshire, which has appeared to a man nigh twenty times in pursuance of the revenge of his own murder ... And another also, of a woman bewitched, as it would seem by some circumstances, that was brought to bed of a [fiend?] Catt with her fore feet and hinder feet cutt off as in Rabbits ... A fortnight after she brought forth the legges of that Catt, and she complained of a Catt in the bed

with her which was sensible to the graspe of the by-standers but when they look'd under the cloaths it vanish'd . . .[35]

By the 1680s More seemed an enthusiast rather than an enemy of Enthusiasm. His cosmos had to be replaced by Newton's, which Bentley would expound so lucidly in his Boyle Lectures. And his party had to be eliminated from the church—or at least clearly distinguished from the party of Stillingfleet, the members of which also insisted on the value of pagan culture and the study of the past. The attack on disorderly and eclectic use of the past thus had a sharp and practical bearing on the present.[36]

Bentley carried on Stillingfleet's argument, taking on dangerous new and old opponents with his own shiny new technical weapons. He used the finely honed criticism of the late humanist tradition not only to purify the classical canon but also to please his patron by repairing a live theological wire; he addressed his *Epistola,* in effect, as much to More and Cudworth as to Mill. Bentley's own tastes and beliefs undoubtedly resembled Stillingfleet's; but his choice of enemies also undoubtedly helped him to prove his value to the church and thus to win preferment in a career that was hardly academic. Short-term ecclesiastical controversies thus did much to define the tone and direction of Bentley's work, which became in part a bid for further patronage—a move in the game of clerical politics as well as in that of historical theology. The personal and the professional, prejudice and philology, learned tradition and iconoclastic innovation are inextricably interwoven in the fabric of Bentley's little book.[37]

Yet in the end this analysis is most revealing in another, more internal respect. It shows that Bentley's tastes and opinions are as fully embodied in his aggressively technical *Epistola* as in the more discursive Boyle lectures and Phalaris pamphlets. And the technical and rebarbative *Epistola* in turn sheds light on the origins of the more obviously significant arguments that Bentley advanced in those more public contexts. When Bentley compared the universe to the *Aeneid* in the Boyle lectures, for example, he did so only because he saw in both the traces of a rational creator at work. He certainly did not mean to suggest that one could read Nature like a text—or explicate it in the light of any number of texts. Indeed,

Introduction: The Humanists Reassessed 21

he took this further opportunity to attack the tradition of More, insisting that one could speculate about such deep questions in natural philosophy as the plurality of inhabited worlds without even worrying about the testimony of Scripture, which did not forbid the philosopher

> to suppose as great a Multitude of Systems and as much inhabited, as he pleases. 'Tis true; there is no mention in *Moses's* Narrative of the Creation, of any People in other Planets. But it plainly appears, that the Sacred Historian doth only treat of the Origins of Terrestrial Animals: he hath given us no account of God's creating the Angels; and yet the same Author in the ensuing parts of the Pentateuch makes not unfrequent mention of the *Angels of God*.[38]

Bentley's sense of history and analysis of nature converged in his insistence that even the most authoritative of all texts, the Bible, did not contain all knowledge. And Bentley's position here rested not only on the new science but also on the older humanist tradition that he had assimilated and reworked in the *Epistola,* with its attack on Cabala and *prisca philosophia.* Here as elsewhere the history of scholarship cannot profitably be separated from general intellectual history, of which it forms a discouragingly complex, but still an organic, part. This simple moral will find repeated confirmation from other instances in the body of the book.

Postscript

The essays that follow apply the interdisciplinary method sketched here in a variety of ways and to a variety of sources. One final point that unites them—and that does not apply to Bentley, that firm believer in English common sense—is an interest in the unexpected and the paradoxical. In a number of cases—notably Chapters 3 and 8—the work of the humanists studied is shown to have had curious, and even counterintuitive, elements and effects. Chapter 3 argues that a forger devised basic principles of early modern historical criticism. Chapters 5 and 6 show that one of the preeminent triumphs of historical criticism in early modern Europe, Isaac Casaubon's dissection of the Hermetic Corpus, was far less modern and somewhat less critical than most historians have claimed. Chapter 8 follows a scholarly amateur, La Peyrère, as he misunderstands and misappropriates the real discoveries of Scaliger and Saumaise—only to shape in his turn the scholarship of the next generation of professionals. Perhaps this emphasis will occasion surprise; yet it certainly fits the present context. Any student of the humanist tradition should be able not only to savor learning and wisdom but also to praise innocence and folly.

1

Renaissance Readers and Ancient Texts

IN THE AUTUMN OF 1465, two humanists agreed to have a battle. Lorenzo Guidetti was a disciple and friend of Cristoforo Landino, the most prominent teacher of rhetoric in the Florentine Studio. Buonaccorso Massari was a pupil of a less famous master, Giovanni Pietro of Lucca. Naturally, Massari began the fight, with all the pretended friendliness and genuine desire to insert knives in backs that characterize the intellectual at his deadliest. Massari had heard that Landino would lecture on Cicero's *Epistolae ad Familiares* during the academic year 1465–66. "I am an eager student," he wrote to Guidetti on 14 September, "and I would be grateful to know what Landino said when he explained the first letter. For in my view it is quite hard because of the historical questions involved, and few have the ability to tackle so great an enterprise."[1]

Guidetti answered with civility. Landino, he explained, had just finished a course of lectures "on the precepts of poetics and rhetoric"; he would use Cicero's letters to illustrate these. The historical difficulties that the letters posed seemed unimportant to Guidetti: "I don't see why someone who wishes to attain style and elegance in writing letters should work very hard at this. My primary interest—and Landino's teaching confirms me in this—is to learn what style, what constructions, what 'flowers' and what 'sobriety' we should use when writing letters."[2]

Matters like these had a far stronger claim on the humanist's attention than petty details of the history of Cicero's Rome, "which hardly anyone knew about even when they happened"; these "seem the province of a pedantic and trivial mind rather than of one which diligently seeks important forms of knowledge."[3]

Massari thanked Guidetti for his frank answer but pressed him on the issue of history. He dismissed as arbitrary and vague the technical rhetorical terms that Guidetti had used to characterize his own interest in the letters: "How and out of what words and tropes one constructs 'flowers' and 'sobriety' I do not understand, nor have I ever read or heard a satisfactory account."[4] And he argued that Cicero's letters, permeated as they were with the events of the Roman revolution that their author had lived through, were unintelligible to anyone who lacked a firm grounding in Roman republican history. To highlight the difference in their approaches Massari asked Guidetti to explain a specific text as Landino had: "I would like you to tell me the meaning of the sentence that reads: 'senatus religionis calumniam non religione, sed malivolentia et illius regiae largitionis invidia comprobat' [*Fam.* 1.1.1]."[5]

Guidetti replied less calmly than before. He admitted that one needed a minimum of historical knowledge, "enough to make the argument of the letter clear." But one should not make the hunt for such details an end in itself:

> For when a good teacher undertakes to explicate any passage, the object is to train his pupils to speak eloquently and to live virtuously. If an obscure phrase crops up which serves neither of these ends but is readily explicable, then I am in favor of his explaining it. If its sense is not immediately obvious, I will not consider him negligent if he fails to explicate it. But if he insists on digging out trivia which require much time and effort to be expended in their explication, I shall call him merely pedantic.[6]

As to the particular passage that Massari had called into question, Guidetti continued:

> In my view the teacher's duty is first of all to show from history that the ancients took no public decisions in the Senate without first consulting the gods. If they found that the gods opposed their action, they said that it was forbidden 'by religion' [cf. the first two words of the Cicero text]. If it is appropriate to the

seniority of the students and to the occasion, he may add an account of the different ways that they determined this, one natural and one artificial . . . But if someone is unpleasantly insistent on asking what *kind* of religion was talked about in the Senate, I would tell him if I knew; if not I would admit that I didn't, and I still would not be afraid that I had failed in the duties of a teacher . . . After that I would come back to the meaning of the individual words. I would explain what the Senate is, what the word is derived from, who founded it at Rome, who enlarged its powers, how great its worth and authority were within the state, how many sorts of senator there were . . . I would also deal with the definition of "religio" and whether the word comes from "religare" or "relegare" or "relegere"; I would deal with "calumnia" in the same way . . .[7]

Massari found no satisfaction in all this verbiage. "I asked you, in a friendly way," he plaintively wrote, "for an explanation of the passage, not the duty of an exegete and the office of a teacher. I too have read Macrobius, Valerius Maximus, Aulus Gellius, and Cicero *On Divination,* but I still don't know what Cicero means in that letter. Please explain to me the phrase 'religionis calumniam'; I don't care so much whether 'religio' comes from 'religare' or 'relegere' "[8] For Massari, in short, the task of the interpreter working his way through a passage is to decipher, phrase by phrase, what it meant to its author and its original readers; for Guidetti, the task of the interpreter is to amass around the individual words of the passage general information useful to the modern student.

These exchanges are more than just another paper confrontation between two minor gladiators of the Republic of Letters. They reveal two different notions of classical scholarship in conflict. On the one hand, Guidetti (and presumably Landino) saw the purpose of scholarship as pedagogical: to produce well-behaved young men who could write classical Latin. The teacher should equip his pupils with the tools of rhetorical analysis and a broad—if shallow— grounding in the Latin language and in Roman *Realien.* So trained, the pupil would be able to extract from his text—an ideal thing outside of any particular time, space, or individual experience—a central core of moral and literary instruction. On the other hand, Massari saw the purpose of scholarship as scientific: to offer exact knowledge about minute details of ancient culture and to transmit sophisticated techniques for resolving difficulties in the ancient sources. The scholar should be able to extract from his text—a deposit of a specific age, author, audience, and context—original answers to technical problems. Guidetti views his texts as classics, as ideal and unproblematic objects for imitation in the present. Massari views his texts as artifacts, as human and difficult products of an irrecoverable past. The ideal vehicle for Guidetti's views was the fifteenth-century school commentary, with its vast mass of paraphrase, rhetorical analysis, and elementary lexical and historical detail. The ideal vehicle for Massari's views was the late fifteenth-century selective commentary or scholarly monograph, with its tightly organized series of quotations marshaled to solve a single difficulty.[9]

This debate is exemplary for the clarity with which its proponents stated the issues that divided them. But it was far from unique, or even unusual, in the Renaissance. If Raphael had lived long enough and had had another wall to cover in the Stanza della Segnatura, he could have produced a School of Hermeneutics well populated by stately pairs of humanists arguing about how to read their texts. One could imagine, in the center, Machiavelli insisting that the only way to reform political life is to have constant recourse to the Roman historians, and Guicciardini replying "How wrong it is to cite the Romans at every turn."[10] On the left, one might find the Lyons humanists who printed the late-medieval moralization of Ovid by Pierre Bersuire and produced an updated one as well, being upbraided by that other Lyons humanist Rabelais,[11] who began his *Gargantua* with an attack on that "true bacon-picker," the Dominican allegorist Petrus Lavinius. On the right, one might find those preeminent Leiden scholars Justus Lipsius and Joseph Scaliger, the former insisting that direct study of the ancient sources would provide the modern statesman with an ideal training in everything from morality to military tactics—and the latter telling his students that Lipsius "is no teacher of politics, nor can he achieve anything in government. Pedants can do nothing in these matters."[12] We have passed from Florence to Leiden, from politics to poetry and back, and yet we still encounter the same contradictory maxims and methods. One set of humanists seeks to make the ancient world live again, assuming its undimmed relevance and unproblematic accessibility; another set seeks to put the ancient texts back into their own time, admitting that reconstruction of the past is difficult and that success may reveal the irrelevance of ancient experience and precept to modern problems.

A rich body of research has taught us much about the historical reading of the ancients. Remigio Sabbadini and Pierre de Nolhac almost a century ago, B.L. Ullman, Roberto Weiss, Giuseppe Billanovich, Sebastiano Timpanaro, and Silvia Rizzo more recently have shown that the humanists did indeed create a new mode of experiencing old texts.[13] They saw the ancients as inhabiting a world different from theirs and devised what we would now call the methods of historical philology in order to bring themselves closer to it. Petrarch's exemplary work on assembling and correcting the text of Livy, Valla's historical scrutiny of the *Donation of Constantine* (and of the Latin language as a whole, which he saw as the fullest deposit of and truest key to Roman history), Poliziano's devising of the principle of *eliminatio codicum descriptorum,* and Erasmus's application of the principle of *lectio difficilior potior*—these triumphs of humanist philology have been described in several recent works that concentrate on the great men who won glittering prizes.[14]

Naturally, not every humanist carried off a gold medal. As historians have turned away from the towering brilliant icebergs like Poliziano and Scaliger and poked their heads below the waterline to examine the vast submerged bulk of humanist scholarship, we have learned just how untypical the heroes were. The work of Martin Sicherl and Martin Lowry on Aldo Manuzio, for example, has shown that he (and his editors) used their manuscript authorities with far less discrimination than a Poliziano. In Aldo's famous shop a creative editor tried to change a verse in the Greek Anthology to read not "Drink to me only with thine eyes" but "Drink to me only with thy lips."[15] In Aldo's shop, too, someone misread the word *paidōn* (children) in an early line in Aeschylus's *Agamemnon* as *podōn* (feet). So emended, the matchless simile in the first chorus described Menelaus and Agamemnon raging for all-out war "like vultures who in terrible pain for their feet wheel high above their nests." And as Monique Mund-Dopchie has shown, this striking image of vultures suffering from corns was preserved by some editors and translators even after the Greek text was corrected in print in 1552.[16]

But we should not be hypercritical. If Letizia Panizza's elegant essay on the reception of Seneca's first Letter in Renaissance Italy portrays a series of humanists making little headway with either their author's historical context or his Latin style, most recent work has borne out the more positive results of two pioneering scholars, John Dunston and Timpanaro. Donatella Coppini has portrayed the little-known Antonio Volsco unriddling a hard distich in Propertius exactly as the modern commentator Max Rothstein would, in ignorance of his predecessor. Jill Kraye has shown us the industrious German Hellenist Hieronymus Wolf explaining a hard point of Greek philosophical lexicography which had baffled cleverer men, notably Poliziano. And Carlotta Dionisotti, by a startling feat of detective work, has breathed life into the forgotten bones of the Flemish Hellenist Jean Strazel, showing that he offered his Parisian students around 1540 penetrating guidance into both the linguistic mysteries and the political teachings of a very difficult Hellenistic author, the historian Polybius.[17] Moreover, the humanists' successes were not confined to the explication of individual passages. Two books on the study and editing of Aeschylus during the Renaissance have recently appeared. They diverge on many details but converge in showing that Pier Vettori and Henri Estienne, working together, solved the basic editorial problems in a way that still commands assent—and so provided a prominent exception to the norms of editorial practice described in E. J. Kenney's influential *The Classical Text: Aspects of Editing in the Age of the Printed Book* (Berkeley, 1974).[18]

The humanists, moreover, not only cracked specific textual problems but developed generally valid methods for textual analysis. In the 1560s, for example, Jean Bodin elaborated in print a set of rules for assessing the credibility of statements made by ancient (and later) historians. The very existence of such rules implied a novel point—that ancient history had to be reconstructed from, not simply found in, the ancient historians. And the criteria Bodin employed seem as novel as his general enterprise. Bodin considers the ideal historian "less the indefatigable on-the-spot inquirer than the somewhat more detached historian who has diligently consulted records":[19] "The narratives of those who have only what they have heard from others, and have not seen public records, ... deserve less approval. Therefore the best writers, to win greater authority for their works, say that they have gathered their material from public records."[20] In evaluating the ancient historians less as writers than as researchers, Bodin seems closer to the German scholars of the nineteenth century than to the critics of his own day, who saw history chiefly as *magistra vitae* (and, thus, as *opus oratorium*).[21]

The temptation to stress Bodin's modernity does need some resistance. As will appear in Chapter 3, Bodin drew his belief in the crucial importance of a historian's use of sources less from his mother wit than from the work of an earlier Renaissance scholar. Giovanni Nanni—or Annius—of Viterbo published his *Commentaria* in 1498. This gallimaufry presented to the learned world a set of complementary forged histories of ancient nations, attributed to Berosus, Manetho, "Metasthenes," and other gentlemen. Annius discussed the rules of historical criticism because he wished to prove the superiority of his novelties to the works of Herodotus and Thucydides. Some principles he expounded in the forgeries themselves. Metasthenes warns the reader not to follow those earlier writers who have their facts "auditu et opinione," but only those ancient priests (like Metasthenes) whose annals enjoyed "publica et probata fides." But Annius offers his most explicit guidance to the modern reader in his annotations. Glossing Metasthenes, he makes clear that the priestly annalists deserved credence because their work rested upon public records: "they were in older times the public recorders (*notarii*) of events and dates; either they were present or they made copies from older ones."[22] The unclassical notion that the best historians had worked not from hearsay but by archival research was formulated, in short, not by the sixteenth-century scholar but by the fifteenth-century forger. And even Annius's rules were more original in formulation than in content. For as Bernard Guenée has shown, they amounted in part to little more than reformulations of critical principles long applied to documents in ecclesiastical controversies.[23]

Yet it would be wrong to underestimate the novelty of Renaissance readings. Offering general rules like those of Annius and Bodin was a bold innovation in itself. And Renaissance readers certainly developed a form of historical awareness that had few precedents in their immediate environment. The humanists realized that any classical text was not just the product of a single intellect but a subordinate part of an organic cultural and historical whole. Accordingly, they set out to eliminate from the corpus of genuine antiques all works that used a vocabulary, employed concepts, or referred to events that their supposed authors could not have known. Naturally, the humanists made mistakes and expressed personal and religious prejudices in their work of criticism. But they also demolished the claims to deep antiquity of such long-established (and long-convincing) fakes as the letter of Aristeas, the *Testaments of the Twelve Patriarchs,* and the *Corpus Hermeticum*.[24] And in doing so they framed, drawing on classical precedent, the first clear general rules for testing the external form, internal consistency, and vocabulary of documents—and thus laid the foundations on which later scholars, challenged and stimulated by the New Philosophy, would rear the baroque edifices of palaeography, diplomatics, and source-criticism.[25]

At the same time, the humanists set out to show precisely how the genuine ancient works they studied fitted into the larger development of ancient thought and writing. Sometimes they did so simply and directly—as Henri Estienne did when he put the surviving works attributed to the ancient bards Orpheus and Musaeus at the end, not the beginning, of his corpus of Greek hexameter poetry.[26] Sometimes their historical awareness expressed itself in more ambitious forms. Francis Bacon, for example, called for the creation of a new kind of history of human culture from the earliest times: a "literary history." This should not only deal with the great authors and their opinions but explain the causes of each development in the arts and sciences: "it should include the nature of regions and peoples; their disposition, whether suited or unsuited to the various disciplines; the accidental qualities of the period which were harmful or favorable to the sciences; rivalries between and minglings of religions; the ill-will and the benevolence of the laws; and finally the outstanding virtues and ability of certain individuals for promoting letters." In the end, Bacon held, this history should transcend rational analysis to arrive at an intuitive perception of the unity of each period. By direct examination of the primary sources, it would "evoke, as by an incantation from the dead, the literary spirit of each age."[27] This program, as Erich Hassinger and Ulrich Muhlack have shown, had already been adumbrated by such earlier thinkers as Petrarch and Erasmus, and found more elaborate expression in the works of Barclay and other seventeenth-century thinkers than it did in Bacon.[28] And it remained the model for cultural history down to the late eighteenth century, when Bacon was still cited as an authority on what a "literary history" of human culture should be—and when Christian Gottlob Heyne recreated in a masterly essay the "spirit of the age of the Ptolemies" and the geographical, climatic, and political conditions that had shaped it.[29] Humanist philology, in short, eminently deserves the attention it has received from modern classicists and historians. Its proponents not only filled in the gaps and cleaned off the stains of the classical corpus but rethought the nature of reading itself.

At the same time, a second grasp of scholars—mostly literary—have shed new light on the methods by which some humanists made their texts yield a meaning directly useful to modern readers. The great surveys by Jean Seznec and Don Cameron Allen vividly portray generations of Renaissance scholars building around their texts a vast wedding cake of interpretation, mingling ancient, medieval, and modern ingredients in the hope of disguising the awkward pagan features and apparent errors of the texts. Ida Maïer and Alice Levine have shown that even Poliziano seized upon a late Byzantine physical allegory as a potential key to Homer. Ann Moss has shown how successive generations of Ovidian allegorists modernized their methods and their use of evidence without ever fully abandoning the assumption that Ovid must be reconciled with Christianity—and thus kept the *Metamorphoses* fresh and instructive for very different audiences. Michael Allen has analyzed in subtle and revealing detail the ways in which Marsilio Ficino fused earlier Neoplatonists' readings and misreadings of Plato with his own. And Michael Murrin and David Quint have traced the impact of allegorical modes of reading, ancient and modern, on Renaissance poets' modes of writing.[30]

Recent scholarship has demonstrated that allegorical exegesis persisted longer—and influenced more fields of thought—than anyone would have imagined a generation or more ago. Charles Lemmi, Paolo Rossi, and others have traced the paths by which Bacon came to believe that he would find support for his iconoclastic, modern natural philosophy in the myths transmitted by the ancient poets.[31] He seems to have turned to myths opportunistically, hoping to find a modish and inoffensive cloak for theories that could provoke resistance if stated baldly. But the enterprise eventually captivated him. By the end of his life he gave it as his considered opinion that "a good many of the myths of the ancient poets were instilled with some mystery."[32] And the up-to-date physical allegories of the *De sapientia veterum* were, we now know, by no means the last venture down a shaft stripped of ore. A century later Newton still groped for anticipations of his theories in the myth of the pipes of Pan.[33]

Of forms of classicism other than the allegorical we know less. Yet the varied efforts of the humanists to select a truly up-to-date canon of literary classics for imitation have been studied by Morris Croll long ago, and by John D'Amico, Marc Fumaroli, and Wilhelm Kühlmann more recently. We are beginning to see how broad a spectrum of possibilities confronted humanists looking for antique models for discussion of such apparently nonantique subjects as scholastic theology and Machiavellian political theory. And we are at least aware that the stylistic decisions of the humanists normally expressed broad political and cultural choices rather than a technical preference for Cicero or Seneca.[34]

Yet if recent work has continued to reveal the progress of humanist scholarship and the varieties of humanist *imitatio*, it has done less to explain how a single set of scholars could employ two apparently opposite sets of assumptions and methods. In fact, scholars have tended to seize upon only one of the two approaches I have described as the one really characteristic of humanism. Eugenio Garin, for example, has insisted that the essence of the humanist movement lay in its call for a direct, historical contact with the ancient masters. He soundly chided Seznec, in a famous article, for overemphasizing the traditional and unhistorical elements in humanist scholarship.[35] Yet another student of Florentine humanism, Roberto Cardini, has almost reversed the terms of Garin's argument. He has claimed that only allegorical and rhetorical exegesis of the sort Landino practiced made the classics usable by modern writers. By contrast the new historical philology of a Massari or a Poliziano could lead only to the isolation of a merely historical past from modern needs and problems.[36]

Most scholars seem to find the current of thought that Garin emphasizes easier to accept as novel and important. He and other critics and historians offer a vision far more gripping than the mock-Raphael I described before: a group of heroic humanists energetically wipe the fog from a vast window, behind which appears the ancient world *as it really was*—or at least as Mantegna portrayed it, with meticulous attention to archaeological detail and perspective. Yet difficulties arise when we test this vision against the sources.

In the first place, few of the humanists actually reveal on close inspection that commitment to a strictly historical approach which most secondary sources lead one to expect. True, one finds occasional efforts to treat late or early products of the Greeks and Romans as not inferior to the canonical classics (some of which were not *our* classics, of course—for example Lucian). Poliziano argued in the introduction to his course on Quintilian and Statius that "We should not simply dismiss as inferior everything that is different."[37] Scaliger divided the history of Greek literature into four natural seasons in a famous letter to Salmasius, and clearly described the late poetry of the Hellenistic age as different from, but not worse than, what had gone before: "Autumnus ab aestate non degenerans."[38] But such assertions are rare. Even the most sharply historical thinkers tended to set a particular canon of classics aside as those in which the Greek and Latin languages—and, so some held, Greek and Roman thought—had attained a unique height. Andrea Alciato, for example, took great pride in that expert command of late-antique sources like the *Scriptores Historiae Augustae* which enabled him to give the first modern commentary on the last three books of Justinian's *Code*. He lamented his contemporaries' ignorance of "the events of later times." But he also admitted that the late sources which he knew so well had put most other scholars off because they showed unmistakable signs of literary degeneration: "In these [constitutions] the style is not absolutely correct, since the elegance of Latin had begun to decay." What was late, in short, could be important, could have "non parva . . . utilitas"; but it was not in any sense as good as what preceded it.³⁹

Furthermore, if we allow the noble marmoreal figures to climb down from our imaginary School of Hermeneutics and examine them in the round, we will see that many humanists read their texts both as classical and as historical—that is, that many Renaissance intellectuals managed to take their stand on both sides of the gap that separates a Massari from a Guidetti, a Guicciardini from a Machiavelli. This balancing act required cool nerves and considerable boldness, yet many engaged in it.

Consider Bacon. We have already seen him playing the part of historian and that of allegorist. In the set-piece preface to *De sapientia veterum* he tried to combine these irreconcilable characters. Borrowing from Cicero, he ridicules the allegories of Chrysippus, "[who] used to assign the opinions of the Stoics to the oldest poets, like a sort of dream-interpreter." Borrowing from Proclus and Julian, he argues that the myth-makers had signposted the deeper meanings of their stories, by including in them palpable absurdities that "reveal the parable even at a distance, and as it were cry it aloud." He admits that he may be deceiving himself, "captus veneratione prisci seculi." He argues that the ancient poets had not created, but only transmitted, the myths, which must have been created in order to accommodate the conclusions of the philosophers to the "rude intellects" of their contemporaries. And he rounds off this Hamlet-like soliloquy by confessing his inability to prove the historical validity of his allegorical views—and simultaneously insisting on their correctness:

> The wisdom of those early times was either great or fortunate; great, if this figure or trope was consciously devised, fortunate, if men who had something quite different in view provided the matter and the stimulation for meditations of such value. But in either case I will consider my work successful, so long as it proves useful at all. For I will illuminate either antiquity or nature.⁴⁰

Other witnesses provide less explicit, but equally revealing, testimony. Consider, for example, Poliziano, whose practice as an interpreter is richly documented by his lectures of 1480–81 on the *Silvae* of Statius, recently—and meticulously—edited by Lucia

Cesarini Martinelli. The Poliziano of these lectures is hardly the rigorous advocate of a purely historical philology that one might expect from some recent descriptions of his work. Certainly he studies the text detail by detail, illustrating every allusion with lashings of unpublished Greek and striking out with his customary zeal at the errors of the previous commentator, Domizio Calderini (of whom Poliziano once said: "I point Domizio out to scholars as I would point out an unexpected ditch to travelers").[41] We can watch him tear the text apart like an eager terrier after a bone in order to prove a new historical point: that Statius married Lucan's widow, Polla Argentaria. Statius calls her at one point "rarissima uxorum." Poliziano comments:

> Smell these words out one by one. You'll clearly see that they are too familiar to fit another man's wife. He says "rarest of wives"—"wives," not "women"—"wives," because she both venerates the memory of her dead husband and sweetly loves her living one. "Cum hunc forte diem consideraremus" ["When we by chance considered this day"]—both the adverb "by chance" and the plural number of the verb "we considered" clearly connote a certain familiarity . . . Nor should the fact that Statius calls his wife Claudia in a letter worry you . . .[42]

The argument is wrong, as the last, give-away sentence shows. But the attitude behind it is sharply historical. Poliziano sees Statius as a historical figure, a man who belonged to the same circle that included Martial and Lucan. And he uses this perception in his explanation of important features of Statius's work—above all, his repulsive flattery of Domitian, which Poliziano explained as a necessity for a poet whose patron genuinely believed himself a god.[43]

Yet much of the commentary is less historical than rhetorical. Poliziano is as concerned with the literary genres of the *Silvae* as with their historical occasions. More important, he treats some of them not as historically conditioned products of human arts but as absolutely perfect examples of externally valid principles. Analysing 2.1, he quotes the precepts of Menander and "almost all the other rhetors" on what any *consolatio* should be:

> A monody ought to deal with praise of the dead [person's] family, disposition, education, learning, studies, deeds, in such a way that

no clear order is followed. Thus it seems that the speaker's grief has made him lose control of his material. If the dead person was young, one should elicit emotion on the grounds that he died prematurely, that he deprived relatives and friends of their hopes, that he was no common person . . .[44]

And he goes on, for five pages and more of the modern edition, to show that Statius's *consolatio* employs every conceivably relevant artifice and figure of speech from exclamation and repetition to hyperbole and homoeopathia.[45] This analysis begins from a historical vision, to be sure: Statius did consciously write within genre conventions and deploy all the formal tricks of the rhetorician's trade. But by the end Poliziano has elevated Statius's poem from the work of a historical individual to an atemporal ideal that can be imitated by any modern writer needing to mourn or console— "a garland to be dedicated to the Muses from the most excellent flowers," as he sums it up.

Bacon and Poliziano—admittedly an odd couple—were not isolated cases. Petrarch and Salutati, as is well known, combined a commitment to philological research and a remarkable imaginative ability to work themselves into the circumstances of ancient writers with what seems to have been an equally powerful belief in the allegorical interpretation of some ancient poetry.[46] And even in the late sixteenth century, the gravest and most learned of philologists still used allegorical keys to open certain doors. Isaac Casaubon fiercely denied, as we will see below, that the author of the Hermetic corpus could have enjoyed a special revelation of Christian truths clearer than the one granted to the Jews. But when he wished to emphasize the piety and seriousness of his beloved Aeschylus, he treated even a small verbal detail (a reference to "God" rather than "gods") as evidence that the ancients "by an instinct of nature" had realized the truth of monotheism.[47]

Even more startling is the case of Casaubon's friend and correspondent Joseph Scaliger. He never read Christian morals into pagan texts, and he bitterly criticized his old friend and former teacher, Jean Dorat, for "seeking the whole Bible in Homer."[48] Yet he too found one ancient allegorical method exciting. Challenged at the very end of his life by the Heidelberg theologian David Pareus, who considered the pagans entirely mendacious and cited their myths as evidence in favor of this view, Scaliger responded by arguing—as Euhemerus and Varro had in antiquity—that the myths were not veiled accounts of philosophical doctrines but confused accounts of historical events. He claimed that simple common sense sufficed to find the facts beneath the myths: "For it is equally certain that Hercules existed, and that the Hydra, continually reborn with its innumerable heads, did not exist." And he offered to transform the ancient mythological manual of Apollodorus, "opus sane ingeniosissimum et elegantissimum," into a "certum Chronicon" organized by generations.[49] No criticism of allegorical principles, in short, ever did away with allegorical practice—even on the part of the critic himself. At the end of the seventeenth century Jean Le Clerc still included scathing attacks on unhistorical readings of classical texts and ventures into allegory and modernization of his own devising in his great manual of hermeneutics and textual criticism, the *Ars critica*.[50]

True, some Renaissance intellectuals held that their job was not to devise a single, absolutely valid interpretation of a text but to collect all remotely plausible ones. Filippo Beroaldo, the influential commentator on Apuleius, Suetonius, and Propertius, explicitly took this position, citing Saint Jerome as his authority. He at least could read some texts in apparently self-contradictory ways without feeling undue strain.[51]

But this simple explanation cannot account for all cases of interpretative schizophrenia. No one favored the historical mode of exegesis more strongly than Erasmus. He ridiculed the *Ovidius moralizatus* as a "crassly stupid" work which "gives a Christian adaption—distortion, rather—of all the myths in Ovid."[52] He abused those inept early commentators on Seneca who had thought him a Christian.[53] Yet he too often allegorized in his treatment of ancient texts and myths. "But as divine Scripture bears no great fruit if you stick obstinately to the letter," he warned in the *Enchiridion*, "so the poetry of Homer and Virgil can be quite useful if you bear in mind that it is entirely allegorical."[54] "What difference does it make," he asked, "if you read the books of Kings or Judges or the history of Livy, if you take no account of the allegory in either?"[55] And at least once he made clear that he saw allegorical readings as necessary in some contexts even though they might distort the literal sense. In the *De ratione studii* Erasmus instructs his reader on how to teach a classical text. No doubt deliberately, he takes as his example the worst line—from a Christian point of view—in the entire Virgilian corpus: "Formosum pastor Corydon ardebat Alexim" ("Corydon the shepherd was hot for pretty Alexis"), the beginning of the second Eclogue. Here are his comments:

> If the teacher is clever, even when something crops up that could corrupt the young, it not only does no harm to their characters but contributes something useful to them. For their attention is partly turned to taking notes and partly raised to higher levels of thought. If the teacher is to lecture on the second Eclogue, let him prepare his students' minds with an appropriate introduction, or rather fortify them, thus: Let him say that friendship cannot be established save between those who are alike, that similarity breeds mutual benevolence while dissimilarity gives rise to hatred and discord. In proportion as the similarity is greater, truer, and firmer, the friendship will be more intimate. That after all is the upshot of a great many classic proverbs: "The good come to gatherings of the good even without an invitation"; "like rejoices in like"; "one equal pleases another"; "seek a wife who is your equal" . . .[56]

Examples of bad friendships—Romulus and Remus, Cain and Abel—would drive home by contrast the virtues of true *amicitia*. The story of Narcissus would prove the vital role of similitude between lover and beloved. Plato's two Venuses should be invoked to distinguish between sacred and profane love. And the eclogue itself could be explicated as a moral "image, so to speak, of a friendship too weak to hold" between Corydon the *rusticus* and Alexis the *urbanus*.[57] "If the teacher gives this preface," Erasmus concludes, ". . . I think that nothing shameful can occur to a listener who is not already corrupt."[58] Erasmus is certainly right. The students, buried under a flood of adages and examples, would never suspect that Virgil had described a passion Christians could not acknowledge—even though Erasmus knew perfectly well that he had done so. Here Erasmus plays a role unusual for him, not Proteus but Procrustes—and he was only one of many scholars who lopped off anything imperfect in their perfect sources.

Recent work by Kühlmann and others suggests that the late sixteenth century saw serious efforts to bridge the gap between the humanists' two sets of ideals and methods. In particular, Marc-Antoine Muret and Justus Lipsius, Muret's clever plagiarist and pupil, argued in words that soon won notoriety that the modern scholar must select for study and imitation those aspects of antiquity which were strictly comparable—and relevant—to his situation.[59] Tacitus, for instance, deserved close study not for his stylistic and intellectual virtues but because he cast a special illumination on the chiaroscuro world of revolt and repression in which late-sixteenth-century scholars lived:

> Tacitus [so Lipsius wrote in a famous preface in 1581] does not present you with showy wars or triumphs, which serve no purpose except the reader's pleasure; with rebellions or speeches of the tribunes, with agrarian or frumentary laws, which are quite irrelevant in our time. Behold instead kings and rulers, and—so to speak—a theater of our modern life. I see a ruler rising up against the laws in one passage, subjects rising up against a ruler elsewhere. I find the devices that make the destruction of liberty possible and the unsuccessful effort to regain it. I read of tyrants overthrown in their turn, and of power, ever unfaithful to those who abuse it. And there are also the evils that accompany liberty regained: chaos, rivalry between equals, greed, looting, wealth pursued from, not on behalf of, the community. Good God, he is such a great and useful writer! And those who govern should certainly have him at hand at all times.[60]

Lipsius prided himself on the power and the success of his effort to ground imitation in historicism. In a letter to Johannes Woverius, he described his entire life as a coherent effort to make classical studies serve practical ends. He treats his systematic manuals, the *Politica* and the *De constantia,* as the culmination of his whole career as editor and exegete. And he sums up his intellectual ambitions in one climactic boast: "I was the first or the only one in my time to make literary scholarship serve true wisdom. I made philology into philosophy."[61] No wonder that this cogent intellectual program revolutionized the study of history and politics, impressed statesmen as well as academics, and found imitators and parallels across the intellectual spectrum, from military affairs to medicine.[62]

Yet the letter raises its own interpretative problems. Lipsius—as his readers were meant to know—quoted Seneca's letter 108 when he spoke of converting philology into philosophy. In that letter Seneca had described the reverse process: the degeneration of philosophy into philology in his own time. He had blamed this partly on the teachers "who train us in argument, not for life"; partly on the students "who come to their teachers to be made not wise but clever" (108.23). And he had argued that the root of the problem lay in bad exegesis of the classics, the natural result of the proliferation of philological studies. "The future grammarian studying Virgil does not read that incomparable line 'fugit inreparabile tempus' with the thought in mind that 'We must be on our guard; unless we hasten we will be left behind; swift time drives us onward, and is driven . . . ' but in order to observe that whenever Virgil describes the speed of time's passage he uses the verb *fugio*."[63] Seneca supported the productive, "philosophical" interpreter who found a moral message in his texts against the passive, "grammatical" interpreter who found merely a poet's habitual usage. Glossed by Seneca, as it was meant to be, Lipsius's letter shows that his philosophical philology did not reconcile the opposing methods of his predecessors. His commitment to historically valid readings was superficial. Like Seneca, he asserted the primacy of the modern scholar who chooses the texts for his canon over the texts, the primacy of modern, practical needs over textual details, and the primacy of the modern context in which a work is used over the original context in which it was composed. All of these assertions were calculated to provoke the anger of hard-nosed philologists like Joseph Scaliger (who filled his copies of Lipsius's work with critical marginalia).[64] None of them offered a defense against the argument that if utility was the sole criterion for choice of study, modern history and travel should be preferred to classic texts.[65] And none of them could repel the increasingly sophisticated attacks of skeptics who—like Montaigne—denied that any classical text or situation resembled modern circumstances closely enough to serve as an adequate explanation or model for imitation:

As no event and no shape is entirely like another, so none is entirely different from another. An ingenious mixture on the part of nature. If our faces were not similar, we could not distinguish man from beast; if they were not dissimilar, we could not distinguish man from man. All things hold together by some similarity; every example is lame, and the comparison that is drawn from experience is always faulty and imperfect; however, we fasten together our comparisons by some corner. Thus the laws serve, and thus adapt themselves to each of our affairs, by some roundabout, forced, and biased interpretation.[66]

As such views gained ground, even the defenders of the ancients lost faith in the absolute value of their texts, and it became inevitable that the even more useful texts of the New Philosophy and the New Science would drive the ancient ones off the intellectual market. True, even in the seventeenth century the modes of reading that the humanists had practiced continued to exist in altered shape, the one in the form of a newly rigorous scholarship and the other in that of a newly austere classicism. But survival required transformations too sweeping to be discussed here, and in any event reading had ceased by then—if only for a time—to be the central problem of Western culture.

Recent research has not offered any powerful explanations for the tensions and contradictions we have explored. But it has provided some clues that may help us to recover the lost intellectual rules that governed and validated moves in this strange game. One crucial theme in recent research has been the central role of rhetoric in all humanist thought.[67] If we treat humanist readers as practitioners of a form of rhetoric, some difficulties vanish. If we assume that most humanists interpreted—like good rhetoricians—for a specific purpose and a particular audience, we can infer that many of them read historically or philologically while composing technical monographs and allegorically while composing commentaries or teaching students. To that extent some apparent contradictions can be explained away—though not a debate like that between Massari and Guidetti.

If we examine the humanists' hermeneutical methods, moreover, we will see that historical as well as imitative modes of reading depended on rhetoric for their tools. In telling his students to ask, as they read Scripture, "how [Christ] was born, how he was raised, how he grew up, how he acted toward his parents and relatives, how he set about the task of preaching the Gospel," Erasmus turned the rhetorician's precepts for composing a work into the questions to raise while reading it.[68] Valla did the same, even more dramatically, in his *Declamation on the Donation of Constantine*.[69] And given this identical substrate of method, historical and rhetorical readings may have seemed less dissimilar to their Renaissance practitioners than they do to us.

FINALLY, we should bear in mind that historical readings had at least one central aim in common with unhistorical ones—and that too was rhetorical. Even the most technically proficient humanist scholars expected that they would employ their discoveries about ancient literature in composing their own literary works. Poliziano took great pleasure in incorporating the new myths, facts, and variant readings he uncovered as a scholar in his own Latin poems and letters—and in pointing out that he had done so in his most rebarbative technical monographs. As late as the 1560s and 1570s, Joseph Scaliger reconstructed archaic Latin not merely because it was there but because he hoped to use it as the medium for new translations from Greek tragedy.[70] None of these facts can explain what Erasmus thought he was doing when he interpreted *Eclogues* 2.1; all of them make his tactics seem less strange, if not less reprehensible. For all of them reveal that the need to use his texts, or to make them useful to others, often gave the humanist not only a motive for studying the classics but an occasion for slipping from one interpretative framework into another. Pedagogy, prosody, and polemic, those great stimulators of historical research, often stimulated the most creative of allegories as well.

Mysteries remain. Garin rightly instructs us that humanism meant "the formation of a truly human consciousness, open in every direction, through historico-critical understanding of the cultural tradition." Murrin rightly enjoins us to recapture "the excitement which allegorical rhetoric could give to the ancient world and to the Renaissance."[71] In Book 2 of the *Secretum* Petrarch has Franciscus offer Augustine two Virgilian allegories. The first Augustine recognizes as simply true: "You've done a splendid job of finding the truth underneath the clouds." The second Augustine receives in a more complex and ambivalent way, as true philosophy but possibly false exegesis:

> I admire the secrets of poetic discourse of which, I see, you have a rich supply. Perhaps Virgil had this in mind when he wrote; perhaps he could not have been more distant from any thoughts of this kind and simply wanted to describe a storm on the ocean, and nothing else, with these verses. Nevertheless I think that what you've said about the power of the passions and the rule of the intellect is both acute and accurate.[72]

In Book 3 of the *Colloquium Heptaplomeres,* written more than two hundred years later, Jean Bodin has his spokesmen for the various religions offer a whole palette of different shades of interpretation:

> SENAMUS: Nothing has worried me longer than the allegory of the two trees and the serpent.
>
> SALOMO: It is unknown to the Greek and Latin commentators alike. And though some of the Hebrews have opened up the hidden senses of the allegory, all their efforts will be in vain if God does not illuminate our minds so that they may grasp these things.
>
> FRIDERICUS: It seems very dangerous to me to reduce the literal sense of holy Scripture to allegories; the entire history may vanish into myths.
>
> OCTAVIUS: Now, Fridericus, do you think the serpent spoke with the woman? They hate each other so direly, that the mere sight of a snake may cause a woman to suffer a miscarriage, and in a whole crowd of men a snake will seek out the single woman for vengeance. Nothing truer could be said, then, than "The letter kills, the spirit gives life."
>
> SALOMO: In scripture a straight historical account is often given, as when the people undergoes a census and each tribe receives its leader. There are also cases when a historical account is given, but in addition to the history there is a hidden allegory . . .[73]

Any full history of reading in the Renaissance will have to encompass Garin and Murrin, to make room for the full range of readings of Petrarch's Book 2 and Bodin's Book 3. And that history will be a tale of mysteries undreamed of in existing accounts of classical scholarship, hermeneutics, and literature in the Renaissance.

I conclude by passing briefly from description to analysis—or at least to interrogation. How, in the end, can we account for the humanists' ability to carry on, boldly reading and rereading, despite their divided loyalties and contradictory assumptions? I have no answer to propose, but I can raise a large question that may be suggestive. Is the split I have described perhaps not a peculiar attribute of Renaissance humanism but a normal feature of any humanist movement—any effort to renovate society and culture by returning to a distant, golden past, supposedly incarnate in a canon of classic texts? After all, recent German and American work has shown that the creators of modern professional philology, the German scholars of the period from 1750 to 1850, made similar efforts at once to read their texts historically and to treat them as ahistorical classics. Friedrich August Wolf, whose *Prolegomena ad Homerum* of 1795 was the first full-scale product of the German school, pointed out flaws in the Homeric poems that could enable them to be chopped up into the earlier strata from which they grew. Yet Wolf the destroyer could not contain his admiration for the poems in their final state, and fervently praised the very artistic unity and cohesion that his research had exposed as an illusion: "But the bard himself seems to contradict history, and the sense of the reader bears witness against it ... Almost everything in [the poems] seems to affirm the same mind, the same customs, the same manner of thinking and speaking. Everyone who reads carefully and sensitively feels this sharply ..."[74] And at least until the middle of the nineteenth century, his successors continued both to praise the canonical perfection of the Greek and Roman classics and to dissect them into their underlying sources by stressing the incompetence with which these had been fused together.[75]

To take an oddly similar case from a far greater distance, in culture if not in time or space—if we examine the effort to renew the study of the Confucian classics that grew up in the lower Yangtze region of China during the seventeenth and eighteenth centuries, we encounter many of the same phenomena that characterized the Renaissance.[76] The Ch'ing scholars set out, like the Western humanists, to return to a set of classic texts that were still being studied, but in what they considered a trivial and unrewarding way. For the "eight-legged" form in which civil service examinees cast their artificially pointed essays on the Confucian classics, the Ch'ing philologists substituted new and serious ways of encountering the sources.[77] They collated texts and studied archeology and epigraphy. Some of them made the correcting of textual errors a high intellectual calling in its own right:

> You tell me that collating texts is the great passion of your life—
> To sort them out with as fine precision as a sieve sifts rice;
> That to get a single right meaning is better than a ship-load of pearls,
> To resolve a single doubt is like the bottom falling off the bucket.[78]

They encountered opposition: "as your scholarship broadens," the pioneer Ku Yen-wu was warned, "your eccentricities will deepen."[79] But they persisted even when their research cast doubt on the authenticity of classic texts accepted as genuine for millennia. After all, many of them felt, their pursuit of technical details and their savage hunt for error were motivated by the highest of ethical ideals. Only the most rigorous scholarship could give access to the classics—and only the classics, they continued to believe, offered a message that transcended time and space.[80]

The progress of scholarship in the lower Yangtze followed many paths, as Benjamin Elman has shown in an impressive book—as many as it had followed in Renaissance Italy. What is most striking from our point of view is that some Ch'ing scholars came to feel that their moral ends could not be attained by scholarly means. And they expressed the pathos of this intellectual plight as movingly as Montaigne. The eighteenth-century bibliographer Chang Hsüeh-ch'eng, for example, admitted without reservation that he had no cogent solution for the epistemological difficulties posed by any effort to cross historical distances. In essence, he argued, no two individuals were really alike: "Each man of course lives his own life, and each man's life is unlike that of any other. And even the experiences of one life differ one from another."[81] Accordingly, he held that even the most skilled literary scholar could not fully master texts written by someone else. Citing the story of the eighth-century scholar Hsiao Ying-shih, who had read a T'ang writer's work with no name attached and gradually identified the author from its style, Chang commented: "People have always said that Hsiao . . . had a genuine appreciation for fine writing. But words are rooted in the mind, and minds are as unlike as faces. Hsiao was not able to conclude as soon as he had seen the essay that Li definitely had written it. . . So we cannot call this real understanding."[82] And he insisted, as forcefully as Lipsius or Muret, that the scholar could operate usefully only by pulling from the classics those ideas and facts which fitted current practical needs.

Are the resemblances that seem to link early modern and modern, Eastern and Western movements the result of chance or the evidence of a common cultural mechanism? I cannot say—nor can anyone else, until we enjoy much fuller knowledge of both medieval and Renaissance readings of the ancients, and much deeper studies of the ways in which political and institutional structures shaped so private an experience as reading.[83] But I can end with a further question. We too seek both to interpret texts historically and to make them accessible as classics in the present. What felt and unfelt contradictions will our readers—if we find any—perceive in our humanism?

2

The Scholarship of Poliziano and Its Context

FOR AN indeterminate period between 1513 and 1521 Claude Bellièvre, that amiable traveler and amateur of papyri, stayed in Rome. Among the many sights he saw there was the Vatican Library, where he examined the Codex Romanus of Virgil with considerable care:

> In the inner library of the Vatican we saw a very old codex of Maro, which Angelo Poliziano boasts of having seen in chapter 77 of his century of *Miscellanea*. There he uses many pieces of evidence to prove that one should say "Vergilius" with an *e* and not "Virgilius," though the latter is most commonly used. And amid the other evidence he particularly cites that of that old codex of Maro, in which "Vergilius" is written throughout with an *e*. I too, while I was carefully studying this manuscript—which by no mean means everyone is allowed to handle—noticed something else worth remembering. Now there are some recent writers, who try to write all too eloquently, who maintain that "Explicit liber primus" is improper Latin. We, in order that we may agree with Poliziano and many other good men, who think that this very old codex should be considered very trustworthy, shall also think that "Explicit liber primus" is proper Latin. For it is written in this same manuscript, at the end of each book, in clear letters, to serve as the numbering of the books. This manuscript is written in capital letters, and the story is illustrated throughout. And if

by chance you wish to see the true form of the script, it is this, which I have drawn as accurately as I could.[1]

Bellièvre's interest in palaeographical and orthographical details, his careful description of the manuscript's location and appearance, his precise citation of Poliziano's work—all these interests and practices are new. And Bellièvre derived them from the one modern source he mentions: Poliziano's *Miscellanea,* his carefully annotated copy of which is still extant.[2]

I do not know if C. Bellièvre met T. Diplovataciüs. They would probably have enjoyed one another's company. Each of them was both lawyer and antiquarian; each of them had intellectual roots in both humanistic and juristic traditions. And they had another link as well. Diplovataciüs's *De claris juris consultis* derived just as directly as did Bellièvre's note from the work of Poliziano. Diplovataciüs could never have reconstructed the lives and writings of the ancient jurists had he not been able to draw on Poliziano's letters—especially the famous letter to Jacopo Modesti da Prato, which summarized much of what could be learned from the *Index Florentinus.*[3] Nor could he have reconstructed the *fortuna* of the *Digest* without the information given in the *Miscellanea.*[4] To be sure, he had his troubles with Poliziano's Latin; but he nevertheless ransacked Poliziano's works to the best of his ability.[5]

These cases suggest—and others confirm—that Poliziano brought about a revolution in philological method. For before Poliziano we find no such work being done in the criticism of texts and sources. In this chapter I shall try to sketch the course of this revolution, drawing chiefly, but not exclusively, on the *Miscellanea.* I shall also try to put the revolution in context by comparing Poliziano's work in some detail with those of his immediate predecessors.

The humanists of the period 1460–1480 tended to employ one standard medium for recording and communicating the results of their researches. They commented on classical texts, line by line and often word by word. At first, they gave their commentaries merely as university lectures, and many of them have come down to us only in the form of *recollectae*—students' notes. From the 1470s on, however, it became more and more common to revise such commentaries after delivery and to publish them either independently or along with the texts they concerned.[6]

This style of commentary had a number of advantages. In the first place, it was traditional. Many ancient or old line-by-line commentaries adorned standard authors: Servius on Virgil, Donatus on Terence, Porphyrio and Pseudo-Acro on Horace, the scholia on Juvenal and Persius. These works provided the starting point for the earliest humanist commentaries on the works they dealt with; indeed, very often the humanists did little more than repeat what their ancient predecessors had said, merely taking care to conceal the extent of their indebtedness. Pomponio Leto, the first humanist commentator on Virgil, relied very heavily on an interpolated text of Servius; Gaspare de' Tirimbocchi, the first humanist commentator on the *Ibis,* relied equally heavily on the extant scholia.[7] Where there was no ancient or pseudo-ancient commentary, there was sometimes a later one of similar form and method—for example, the twelfth-century commentary by "Alanus" on the *ad Herennium,* which Guarino sedulously pillaged for his lectures.[8] Moreover, even for the humanist commenting on texts that were not adorned with ancient scholia, the ancient commentaries provided an obvious model for style and method, one both readily accessible and at the same time satisfactorily different from the style of the late medieval classicizing friars.[9]

This style had other advantages as well. Line-by-line commentaries inevitably bulk as large as or larger than the texts they deal with. The commentator, in other words, set out to fill a large amount of space. His audience expected him to turn any suitable word or phrase into the occasion for an extended digression: into the etymology of a word, into the formation of compounds from it, into its shades of meaning; most often, perhaps, into the justification in terms of formal rhetoric for its appearance in the passage in question. Many digressions departed even further from the text, into mythological, geographical, antiquarian, and even scientific matters. A commentary on almost any ancient author could thus become an introduction to ancient language, literature, and culture. In short, the commentary made a highly flexible instrument of instruction. Here, too, the humanists followed their ancient models. Servius, in particular, used the medium of a commentary on Virgil to impart quantities of information on almost every conceivable subject.[10]

Finally, this style appealed to students. Since the commentator felt obliged to gloss every word that might present a difficulty, he generally made his text accessible even to students of mean intelligence or poor preparation. At the same time, the student who could write quickly enough to keep up with his teacher ended up with an invaluable possession.[11] When he himself went out to teach, he could simply base his lectures on those of his teacher and so avoid the trouble of independent preparation. It is hardly surprising that students came to demand lectures of this kind: what student would not have his teacher do all the work? As M. Filetico, an unwilling practitioner of this style of commentary, wrote: "At the time [ca. 1468–1473] certain very learned men had made the young accustomed not to want to listen to anything unless they added a definition on almost every word . . . I therefore had to follow their customs."[12] The style endured. Poliziano himself employed various forms of it in his lectures, adapting the content of the excursuses to the needs of his hearers.[13] Its most preposterous result did not appear until 1489—namely, Perotti's *Cornucopiae*, in which a thousand folio columns served to elucidate one book of Martial.

But the word-for-word approach had disadvantages as well as advantages. It forced the commentator to deal with every problem, the boringly simple as well as the interestingly complex. It also forced him to waste time and pages on the donkey-work of listing synonyms—which is all that thousands of the humanists' short glosses amount to.[14] Worst of all, in a period of intense literary competition the commentary made it impossible for its author to shine. For the most noticeable aspect of all the humanists' commentaries is their similarity to one another. Especially in their printed form, the so-called modus modernus, the commentaries are nearly indistinguishable. Waves of notes printed in minute type break on all sides of a small island of text set in large Roman.[15] Even numerous digressions into one's field of expertise could not make one commentary distinctively superior to its fellows, for they were hidden by the mass of trivial glosses.

Characteristically, Domizio Calderini, one of the first to have one of his commentaries printed, was also the first to break away from the commentary tradition. In 1475 he decided that he had found a new medium. "Hereafter," he wrote, "I shall not be much concerned with commentaries." Rather, he said, he would concentrate on translation from Greek and on another work

> which we have entitled "Observations," in three books, of which the first contains explications of three hundred passages from Pliny, the second whatever we have noticed that others have omitted in [explicating] the poets; the third what we have gathered and observed in Cicero, Quintilian, Livy, and all other prose writers.[16]

This, then, was to be the new style: books written by and for scholars, books that dealt selectively with difficult and interesting problems.

Calderini published only a few notes "ex tertio libro Observationum." But they were soon followed by similar works larger in scale and more finished in form: M. A. Sabellico's *Annotationes in Plinium* (finished 1488, published 1503); Filippo Beroaldo the Elder's *Annotationes centum* (1488); Poliziano's *Miscellanea* (1489). Poliziano's work, then, belonged to a solidly established, if modern, genre, and to that extent the *Miscellanea* did not represent a break with the immediate past.

As soon as we look beneath the surface, however, the differences become far more striking than the similarities. Calderini did not claim explicitly that his "Observationes" were inherently more interesting or useful than commentaries. In fact, the work is for the most part an advertisement for himself and for his previous writings, and Calderini quoted passages from his own commentaries at length and with relish.[17] Sabellico said that he was publishing a selection from his notes on Pliny; he did not say that a selection was preferable to a full commentary.[18] Even Beroaldo was defensive about the selective and structureless character of his *Annotationes:* "Indeed, in making these annotations we did not keep any order in the contents. For we were dictating them extemporaneously ... This offspring was clearly premature, as it was both conceived and brought forth in less than a month's time."[19]

Poliziano, in contrast, claimed that the very disorder of his work gave it charm. In phrases carefully adapted from Gellius, Clement of Alexandria, and Aelian, he boasted of the variety of subjects treated in the *Miscellanea,* for variety was "fastidii expultrix."[20] Unlike his predecessors, Poliziano explicitly proclaimed the merits of the new genre, which he regarded as a revival of the ancient antiquarian miscellanies. He thus made the break with the commentary tradition sharp and open.

Poliziano also claimed that the method he employed in the *Miscellanea* was even more novel than its form. First of all, he placed a new emphasis on the quality and quantity of his sources. As early as the preface he attacked the methods of his predecessors and set up his own method as both novel and exemplary. He claimed that he had cited only genuine works by genuine ancient authors:

> But lest those men who are ill employed with leisure think that we have drawn [our conclusions] . . . from the dregs, and that we have not leapt across the boundaries of the grammarians, we have at the outset followed Pliny's example. We have put at the beginning the names of the authors—but only ancient and honorable ones—by whom these [conclusions] are justified, and from whom we have borrowed. But [we have not put down] the names of those whom others have only cited, while their works have disappeared, but those whose treasures we ourselves have handled, through whose writings we have wandered.[21]

At the same time, Poliziano argued that his predecessors had consistently misused their sources. He used Calderini's commentary on Ovid's *Ibis* as one example of incompetent work. Calderini's claim to have drawn on numerous arcane sources he refuted with contempt:

> Domizio expounded Ovid's *Ibis.* He began by saying that he wrote matter drawn from Apollodorus, Lycophron, Pausanias, Strabo, Apollonius, and other Greeks, and Latins as well. In that commentary he invents many vain and ridiculous things, and makes them up extemporaneously and at his own convenience. By doing so he proves either that he has completely lost his mind or that, as someone says, there was so great a distance between his mind and his tongue that his mind could not restrain his tongue.[22]

He backed up this tirade by dissecting Calderini's work on one line. Calderini had taken *Ibis* 569 as reading

> Utque loquax in equo est elisus guttur Agenor ("And as the talkative Agenor was strangled in the horse").[23]

"As the result of a fall from a horse," he explained, "Agenor's hand became stuck in his mouth and he perished."[24] This explanation, Poliziano insisted, was Calderini's own invention. In fact, the line must be emended to read

> Utque loquax in equo est elisus guttur acerno ("And as the talkative one was strangled in the maple-wood horse").

Citing passages from Homer and Tryphiodorus, Poliziano explained that the line alluded to the death of Anticlus, one of the Greeks who had entered Troy in the Trojan horse. Odysseus strangled him to prevent him from revealing their presence prematurely.[25] The moral of the episode was clear: from Calderini's method of inventing explanations could come nothing but confusion and error.

There was nothing new in accusing one's predecessors of misusing or inventing sources. Calderini had done just the same. In his commentary on Quintilian Valla had cited an oration of Cicero *pro Scauro*. "Indeed," wrote Calderini,

> I have read in Valerius Maximus and Pedianus that the case of Scaurus was tried in Cicero's presence. But I have never read that Cicero delivered the oration on his behalf from which Lorenzo claims that he drew these words. Nor do I believe that it exists. And I am afraid that Lorenzo, following some ignoble grammarian, recited these words rather than reading them anywhere in Cicero.[26]

So much Poliziano might have said. The difference lies in two things: in the truthfulness of the attacks, and even more in their consistency with the actual practices of the attacker. Calderini's attack was not in fact justified, for Valla had taken his quotation from a fairly reliable source: Isidore's *Etymologiae* (19.23.5). Moreover, the commentary of Asconius Pedianus—which Calderini himself cited in his attack—clearly indicates that Cicero delivered a speech *pro Scauro*. More important, Calderini's attack on Valla was inconsistent with his own practice. For in the very next section of the *Observationes* he enthusiastically retailed what he had read about Simonides "in a Greek writer"—"apud Graecum scriptorem."[27] One whose own references were so slipshod had no business correcting other people's footnotes—and can hardly have been upholding a personal ideal of full, clear, and accurate citation.

Poliziano's new standards for the use of sources represented a clear break with the methods of the last generation. Their methods were formed not in the study but in the lecture hall. They never acquired the habit of full or precise quotation from their sources, for they could not attain such precision if they were to lecture comprehensively on the wide variety of topics that their texts suggested. In particular, they seldom used extensive quotations in Greek, which would have been unintelligible to most of their students and unmanageable for most printers. Instead, they usually provided vague paraphrases, together with imprecise indications of the sources from which they were drawing. Worse still, like their ancient exemplars, they often invented explanations by back-formation from the texts they claimed to be elucidating: "misinformation is often elicited from the text by aid of unjustified inferences."[28]

When the members of this generation turned from the compiling of vast commentaries to the collection of precise *Annotationes,* they did abandon one bad habit that had characterized their lectures. They no longer set out two alternate solutions of a given problem without choosing between them—a maddening habit which had characterized the classroom lecture since the Hellenistic period.[29] Indeed, the whole point of their new genre was to show off their ability to solve problems once and for all. Unfortunately, however, they did not abandon their other habits of sloppy, inaccurate citation and unjustified back-formation. Even Perotti, who tried to be honest, abbreviated the names of the authors he cited when the resulting forms were ambiguous—for example, "Lu." Moreover, he usually failed to inform his readers whether the verses he cited came to him at first or second hand, even though on many occasions he cited not a line from an extant work but a fragment preserved by Festus or Nonius Marcellus, and on others he cited even verses from standard works at second hand.[30] Calderini, as Dunston and Timpanaro have shown, knew Plautus only at second hand, though his own words suggest first-hand knowledge.[31] Most of Perotti's contemporaries were dishonest as well as sloppy. Pomponio Leto gave it out in his lectures that he had a complete text of Ennius.[32] Calderini falsified his notes on Martial in order to refute a justified attack by Perotti. Worse still, he invented a Roman writer, Marius Rusticus, from whom he claimed to derive disquieting information about the youth of Suetonius.[33]

Poliziano's practices could not have been more different. Not only did he list his sources; even when he quoted a fragment at second hand, he generally pointed out that he was doing so and identified the intermediary source. In 1.91, for example, he quoted some verses of the ancient comic poet Eupolis. He then wrote: "Now we did not draw these verses of the poet Eupolis from the original source, since his works have been lost. But we derived them partly from a certain very accurate commentator on the rhetor Aristides, partly from a letter of the younger Pliny."[34] Poliziano's attacks on the practices of his predecessors were traditional only in form. Unlike the earlier ones, they stemmed not only from a desire to gain a reputation and to destroy those of others, but also from a genuine desire to reform the current method of citation.

Poliziano's use of his sources was also different in kind from that of his predecessors. He was the first to compare and evaluate sources in a historical way—that is, in the way still employed today. Poliziano's sources presented him with various kinds of problem, some of them fairly trivial. For example, he not uncommonly encountered ancient sources that contradicted one another about historical or mythological details. The solution in such cases was usually obvious. It was only natural to follow the most authoritative source, which in most cases simply meant the oldest one. And that was just what Poliziano did when, for example, he preferred Homer's testimony about the ages of Achilles and Patroclus to those of Aeschylus and Statius.[35]

So far there is nothing new here. Petrarch had encountered contradictions in his ancient sources while compiling the *De viris illustribus*.[36] Salutati and Bruni had uncovered discrepancies in the ancient histories of republican Rome.³⁷ Flavio Biondo had found ancient authors contradicting one another about the functions of certain ancient buildings.³⁸ And all of them had found it possible to resolve such contradictions. They assumed that the more authoritative source, or the one that rested on more authoritative sources, was correct. The divergent accounts in other texts must have resulted either from scribal errors, in which case they could be emended, or from simple slips on the part of the less authoritative writer due to bias or bad memory.³⁹

Poliziano, however, arrived at a novel insight. He saw that even a group of sources that agreed still posed a problem. Given three sources A, B, and C, all of which agreed on a given point; if B and C depended entirely on A for their information, should they be considered to add any weight to A's testimony? Poliziano insisted that they should not. In other words, even a group of concordant sources must be investigated, and those which were entirely derived from others must be identified and eliminated from consideration. The way to perform such an investigation was to arrange the sources genealogically, and then to pay attention only to the source from which the others were derived.

Poliziano stated this principle in *Miscellanea* 1.39, while explaining a riddle in a poem of Ausonius. Ausonius had employed the expression "Cadmi nigellas filias"—"little black daughters of Cadmus" (*Epist.* 14.74). Poliziano explained that it referred to the letters of the alphabet: "For Cadmus was the first to bring letters into Greece from Phoenicia."⁴⁰ Since the Latin letters were directly derived from the Greek, Ausonius could refer to them too as "daughters of Cadmus." Poliziano cited Herodotus as his authority for stating that Cadmus had imported the alphabet. He admitted that other ancient writers had said the same. But he argued that all of them were simply repeating what they had read in Herodotus. Since their testimony was entirely derivative, it must be ignored:

> I omit Pliny [*NH* 7.56] and very many others [*Suda*, s.v. *grammata;* Tacitus *An.* 11.14], who say that Cadmus brought them into Greece. For since these different men recalled indiscriminately what they had read in Herodotus [5.57–61], I think it

enough to have restored these matters to his authority. For in my opinion the testimonies of the ancients should not so much be counted up, as weighed.[41]

This elimination of derivative witnesses does not seem striking now. In context, however, it was remarkably original. Beroaldo discussed the same riddle from Ausonius in cap. 99 of the *Annotationes*. He solved it in the same way. And he too cited the sources that had informed him about Cadmus and the alphabet:

> He [Ausonius] calls the letters "daughters of Cadmus" because Cadmus is said to have been the inventor of letters. In Book 7 of the *Natural History* Pliny says that he brought sixteen letters from Phoenicia into Greece. Therefore the ancient Greeks called the letters Phoenician, according to Herodotus in Book 5. The same writer says that he saw "Cadmean letters," very similar to Ionian letters, incised on certain tripods in the temple of Apollo. Furthermore, Cornelius Tacitus avers that Cadmus was the author of letters, while the Greek peoples were still uncultured.[42]

Beroaldo shows no concern with the dependence or independence of his sources. Pliny is evidently as reliable as Herodotus, and Herodotus no more reliable than Tacitus. Since they all agree on the main point in question, he cites all of them. And he omits other accounts not because they are derivative but because they seem irrelevant.

Poliziano views the whole problem of the reliability of sources from a different vantage point. He seeks no longer, as Beroaldo had, to amass evidence indiscriminately, but to discriminate, to reduce the number of witnesses that the scholar need take into account. This new approach to source criticism made possible Poliziano's revolutionary transformation of philology.[43]

His most sweeping innovation in philological method was, as Kenney has shown, to treat textual criticism as a historical study.[44] When he found what seemed to be corrupt passages in recent manuscripts or printed texts of classical writings, he did not try to emend them by conjecture. He went back to the oldest sources—that is, to the oldest manuscripts. He recognized that they were not free from errors; but he insisted that they were the closest extant approximations to what the ancient authors had really written. The newer texts were removed by more stages of copying from antiquity, and any apparent correct readings they contained were merely the results of attempts at conjectural emendation. Such alluring but historically unjustifiable readings were less valuable to the textual critic than the errors of the old manuscripts, for at least their errors "preserve some fairly clear traces of the true reading which we must restore. Dishonest scribes have expunged these completely from the new texts."[45]

Poliziano employed this method throughout the *Miscellanea*. In the vulgate text of Virgil, for example, he found *Aeneid* 8.402 in a metrically impossible form: "Quod fieri ferro, liquidove potestur electro." He consulted the Codex Romanus, and, as he wrote: "in that volume, which is in the inner library of the Vatican, which is remarkably old, and written in capitals, you will find not 'potestur' but 'potest,' a more commonly used word."[46] Again, in the vulgate text of Suetonius he found a meaningless clause in *Claudius* 34; "si aut ornatum, aut pegma, vel quid tale aliud parum cessisset." In what he called the "veri integrique codices," however, he found not "aut ornatum" but "automaton," a reading that made perfect sense: Claudius made his stage carpenters fight in the arena "if a stage machine, or trap, or something else of the sort failed to work." Poliziano took care to identify the codices on whose testimony he relied: "Look at the Bologna manuscript from the library of St. Dominic, or another one at Florence from the library of St. Mark . . . ; both are old. But there is another one, older than either, which we ourselves now have at home . . . you will find this latter reading in all of them."[47] Poliziano's explicit citations and evaluations of manuscripts have been meticulously collected by Silvia Rizzo. Rich though these materials—the only ones available to most of Poliziano's Renaissance readers—are, they form only the long-visible tip of the vast iceberg of his notes and collations, which show almost as much rigor and attention to detail as he claimed they did.[48]

This new method could not have been more in contrast with the practices of Poliziano's predecessors. Beroaldo, for example, relied almost exclusively on conjecture. And even when he cited manuscripts he identified them only in the vaguest terms. Here, for example, is how he gives the manuscript readings of a line in Juvenal:

> The verse is to be read as follows: "Turgida nec prodest condita pyxide Lyde" [*Sat* 2.141]. There "condita" is in the ablative and is connected with "pyxide." Quite recently I found that verse written thus in a very old manuscript. And some time ago Angelo Poliziano . . . told me that he had noted the passage written that way in a manuscript of unimpeachable fidelity.[49]

Here is how Poliziano described the latter manuscript: "We found the same reading [cacoethes (*Sat* 7.52)] in an old manuscript written in Lombardic script, which Francesco Gaddi . . . made available to me for study. But that [other] verse is also as follows in this codex: 'Turgida nec prodest condita pyxide Lyde.' "[50] Poliziano's description includes the name of the manuscript's owner and a classification of its script. Beroaldo's citation—which is, if anything, unusually precise for him—gives neither.

In most cases the old manuscripts deserved more trust than the new merely because they were older, and therefore fewer stages of transmission intervened between them and the author. But Poliziano analyzed some textual traditions in a more complex and more decisive way. He applied his genealogical method of source criticism to the manuscripts of certain texts and proved that one extant manuscript was the parent of all the others. In such cases, he showed, the extant archetype must be the sole source used in establishing the text.

One case, as Timpanaro has shown, was that of Cicero's *Familiares*.[51] Poliziano had at his disposal the ninth-century Vercelli manuscript (Laur. 49,9 = M) and a fourteenth-century manuscript which he wrongly believed might have been written by Petrarch (Laur. 49,7 = P). He also consulted an unspecified number of more recent textual witnesses. In *Miscellanea* 1.25 he proved that the fourteenth-century manuscript, in which a gathering had been transposed because of an error in binding, was the parent of all the more recent manuscripts, for the same transposition occurred in all of them, without any evidence of physical damage to account for it. He also asserted, without giving the evidence, that the fourteenth-century manuscript was itself a copy of the ninth-century one. And he concluded that since the ninth-century manuscript was the source of all the others, it alone should form the basis of the text. As he wrote:

> I have obtained a very old volume of Cicero's *Epistolae Familiares* . . . and another one copied from it, as some think, by the hand of Francesco Petrarca. There is much evidence, which I shall now omit, that the one is copied from the other. But the latter manuscript . . . was bound in such a way by a careless bookbinder that we can see from the numbers [of the gatherings] that one gathering has clearly been transposed . . . Now the book is in the public library of the Medici family. From this one, then, so far as I can tell, are derived all the extant manuscripts of these letters, as if from a spring and fountainhead. And all of them have the text in that ridiculous and confused order which I must now put into proper form and, as it were, restore.[52]

A second case was that of Justinian's *Digest* or *Pandects*. Poliziano used a different method to identify the extant archetype. He received permission through his patron, Lorenzo, to collate the famous Florentine manuscript of the *Digest*. He noticed certain erasures and additions in the preface, which he thought must have been made "by an author, and one thinking, and composing, rather than by a scribe and copyist."[53] And he had read in Suetonius's life of Nero (52) a description of Nero's drafts of his own works, with their corrections entered in the emperor's own hand. Romantically he inferred that the famous and splendid Florentine manuscript must be the very manuscript that Justinian and his commissioners had first written.[54] The authors' copy must obviously be the archetype. Consequently, all texts of the *Digest* ought to be emended in accordance with the text of the Florentine manuscript. In *Miscellanea* 1.41, for example, he replaced the vulgate reading "diffusum" with "diffissum," the reading of the Florentine manuscript, in D.2.11.1(3): "Et ideo etiam lex xii. tabularum, si iudex vel alteruter ex litigatoribus morbo sontico impediatur, iubet diem iudicii esse diffissum." The passage thus made perfect sense: the Law of the Twelve Tables orders that, if the judge or either of the litigants is prevented by illness from attending court, the day of the trial "is to be postponed" (esse diffissum).[55] Again, in 1.78

Poliziano examined the reading of the Florentine manuscript at D.1.16.12: "Legatus mandata sibi iurisdictione iudicis dandi ius habet" ("A deputy on whom jurisdiction has been conferred has the right to appoint judges") (tr. S. P. Scott). Here some of the vulgate manuscripts read "ius non habet." Both readings could be supported by parallels from other parts of the *Corpus iuris,* and Accursius had discussed both readings and the juristic problems each of them posed.[56] Poliziano, however, saw no problem at all here; there could not be one. If the reading of the archetype made grammatical and juristic sense, then it must be right. Divergent readings in later codices could by definition be nothing but alterations introduced by scribes and jurists. As Poliziano put it, "in those Florentine Pandects, which, indeed, we believe to be the original ones, there is no negative at all. Therefore the Florentine jurisconsult Accursius, who also had a faulty codex, torments himself—I might almost say wretchedly."[57] In this case elimination of *codices descripti* seems to lead to the elimination of medieval legal science. There was nothing new in attacking medieval jurists—Valla and Beroaldo had done the same.[58] But the method underlying the attack was unprecedented.

Had Poliziano completed and published the second century of the *Miscellanea,* the novelty of his genealogical method would have stood out even more clearly. In 2.1 he suggested that even in the normal case, where the archetype was lost, something could still be learned about it from examination of the extant manuscripts. In this chapter he identified and corrected the transposition of Cicero *De natura deorum* 2.16–86 and 2.86–156. The transposed passages were of virtually identical length. Poliziano therefore conjectured that the same thing had happened here as in the *Familiares:* gatherings in the archetype had been transposed because of an error in binding:

> The fact that we have never had to turn over more or less than 11 [pages; that is, eleven pages in the text that Poliziano had at his disposal when writing] unquestionably shows us that an error has taken place like the one that we previously revealed in the letters: the gatherings were transposed by a bookbinder.[59]

Again, in 2.2 Poliziano proved that one manuscript of the *Argonauticon* of Valerius Flaccus was the parent of the rest. All the manuscripts contained transposed passages; these passages, in turn, were uniformly either fifty lines or multiples of fifty lines in length. The pages of the manuscript that he believed to be the parent were fifty lines long, twenty-five to a side. And in this manuscript the pages corresponding to the passages in question themselves had been transposed. Later scribes—including, to Poliziano's evident surprise, Niccolò Niccoli—had mistakenly copied the codex in the order in which they found it.[60] Such mathematical precision in the recension of manuscripts was not seen again until the nineteenth century. These chapters, with their startling anticipations of later discoveries, remained unpublished during the Renaissance; but Poliziano's new method was still revealed for those with eyes to see by the first century of the *Miscellanea*.

Informed reliance on old manuscripts had precedents in the scholarly traditions Poliziano knew. Gellius frequently consulted older manuscripts in order to correct errors in newer ones. Thus, he defended a reading in Cicero's fifth oration *In Verrem* in part because he had found it so written in "a copy of unimpeachable fidelity, because it was the result of Tiro's careful scholarship" (1.7.1). Again, he argued that scribes had replaced the unfamiliar archaic genitive "facies" with the later form "faciei" in a work by Claudius Quadrigarius. In the oldest manuscripts, he said, he had found the old reading "facies"; in certain "corrupt manuscripts," on the other hand, he had found "facies" erased and "faciei" written in (9.14.1-4). Thus, Poliziano's chief literary model had clearly seen and said that old manuscripts were more reliable than new ones.

Moreover, Renaissance humanists from Petrarch on had also sought out and copied or collated old manuscripts. Some of them had even studied the genealogy of manuscripts. Giovanni Lamola was a friend of Guarino. In 1428 he set out to collate the codex of Cicero's rhetorical works which had been discovered seven years before in the cathedral archive at Lodi. This manuscript contained complete texts of Cicero's *De oratore* and *Orator*, which had previously been known only in mutilated texts, and of his *Brutus*, which had previously been unknown.[61] All of these works were fundamental for the rhetorical teaching of the humanists. Consequently, the discovery attracted attention immediately: Poggio Bracciolini, who was then in England, knew of it within a year. And many copies of the new texts were soon in circulation. But as the Lodi manuscript was written in what the humanists called "Lombardic script"—that is, an unfamiliar minuscule—they found it hard to read. Consequently, they made their copies not from the original but from other humanists' copies. As a result both of inevitable mistakes in transcription and of equally inevitable attempts at conjectural emendation, the texts in circulation soon became extremely corrupt.[62] Lamola declared that it was necessary to return to the original source. He wrote to Guarino that he had "restored the whole work according to the earlier text." He knew that the Lodi manuscript was very ancient from its unusual script: he described it as "summae quidem venerationis et antiquitatis non vulgaris effigies." More important, its discovery had created a sensation only a few years before, and no other complete manuscript of the works it contained had been discovered. Therefore, Lamola, like every other humanist of his time, knew that the Lodi manuscript must be the parent or ancestor of all the other complete manuscripts: "from that accurate exemplar," he said, "they copied the text that is now commonly accepted." He decided that any attempt to emend the text must be based on a collation of this manuscript. He even maintained that the errors in the Lodi manuscript were worthy of preservation and study. For even the errors of so old a manuscript were preferable to the conjectures of later scribes: "I also took care to represent everything in accord with the old [manuscript] down to the smallest dot, even where it contained certain old absurdities. For I'd rather be absurd with that old manuscript than be wise with these diligent fellows."[63]

Giorgio Merula was a prolific editor and commentator of the generation just before Poliziano's. In his edition of Plautus, which appeared in 1472, he pointed out that all the extant manuscripts of twelve of Plautus's comedies were descended from one manuscript: "There was only one manuscript, from which, as if from an archetype, all the extant manuscripts are derived."[64] The archetype to which he referred was the eleventh-century Orsini manuscript of Plautus (Vat. lat. 3870). This manuscript had been brought to

Rome in 1429 by Nicholas of Cusa. Like the Lodi manuscript, it had created a sensation among the humanists, for twelve of the sixteen plays it contained had previously been unknown. Hence, it too was widely known to be the source of the other manuscripts. Merula could not collate the Orsini manuscript himself and had to content himself with reconstructing its readings by collating copies of it.[65] But even though he did not act on his knowledge of the textual tradition, Merula too understood that the manuscripts of one text could be arranged genealogically, and that the text should be based on the extant ancestor of the others.

Poliziano, however, showed how to examine a group of manuscripts and discover how they were related to one another. He showed that in some cases such an examination could identify an archetype that was not commonly recognized as such. He insisted that such an examination must be performed in all cases. He maintained that even where no archetype could be identified, "conjectural emendation must start from the earliest recoverable stage of the tradition." [66] And he backed up his statements about the history of texts with precise identifications and evaluations of the manuscripts he used. Poliziano's genealogical method went far beyond the isolated insights of Lamola and Merula. In fact, he arrived at the same set of principles with which modern scholars still set about editing classical texts.

Poliziano's passionate need for rigor and completeness enabled him to surpass his predecessors. For it led him to study many textual traditions; hence he encountered many instances where later manuscripts were derived from extant earlier ones. And the genealogical criticism of sources, which he had established for the myth of Cadmus, enabled him to see the descent of manuscripts as a particular case of a general rule about sources—that they should be weighed rather than counted up and that derivative ones should be ignored. No previous scholar had even come close to formulating the set of critical principles that Poliziano considered to be generally valid.

Poliziano applied his genealogical method to literary works themselves as well as to manuscripts. He realized that the Latin poets—whose works were his primary interest—had drawn heavily on Greek sources in a variety of ways. And he came to insist that only a critic who had mastered Greek literature could hope to deal competently with Latin.

The Romans, first of all, had frequently employed Greek words—sometimes transliterated, sometimes not—when there was no appropriate Latin one. They had employed a Greek adjective or common noun when Latin had no word for the shade of feeling they desired to express—or when there was a Latin word but it did not fit the meter. From Hellenistic pastorals they had borrowed Greek names of shepherds and nymphs. Along with Greek mythology they had taken over some Greek names of gods and heroes. And only someone who knew the original Greek could know the proper forms of these words.

Poliziano corrected many corrupt Greek words in Latin texts and defended many genuine Greek words from attempts at emendation. Thus, in *Miscellanea* 1.2 he attacked those who had emended the expression "crepidas . . . carbatinas"—"thick, rustic shoes"—in Catullus 98.4. They had replaced "carbatinas" with meaningless words of their own invention like "Cercopythas" and "Coprotinas." In fact, "carbatina" was simply a transliteration of the Greek word *karbatinē*—"rustic shoe"—used by Aristotle and Lucian. With great sense of occasion Poliziano produced the Greek authorities that supported his thesis: "We shall set out from the stockpile of Greek, as if from a storeroom, authorities whose credibility can be neither derogated nor taken away. By them the reading is preserved intact and the fog of explanation is dispelled."[67] In another chapter he again attacked unnamed opponents, this time for attempting to alter the form "Oarion"—the reading of all the manuscripts—to "Aorion" in Catullus 66.94.[68] By quoting a staggering array of Greek sources Poliziano proved that "Oarion" was a perfectly acceptable Greek spelling for Orion's name, which Catullus had simply transliterated:

> In a certain elegy [translated] by Catullus from Callimachus, Oarion is read instead of Orion. Since some are now beginning rashly to attack this word, which is still whole and intact, I must make every effort . . . against this perverse audacity of ignorant men, even by the authority of . . . Callimachus, who speaks as follows in his hymn to Diana, which is still extant:

oude gar ōtos
oude men ōariōn agathon gamon emnēsteusan [3.264–265]

But also Nicander, to a like effect, in his *Theriaca:*

boiōtōi teuchousa kakon moron ōariōni [15]

And Pindar in the *Isthmians:*

ou gar phusin ōariōneian elachen [3–4.67 Snell]

And also elsewhere:

oreian peleiadōn mē tēlothi ōariōna neisthai [*Nem.* 2.11–12 Snell].

From this Eustathius argues in his commentary on Book 5 of the *Odyssey* that he is named *para to oarizein* (*ad Od.* 5.274). Therefore the correct reading is not Aorion but Oarion.[69]

The restoration of corrupt Greek words to Latin texts was not an innovation in itself. Guarino had made something of a specialty of filling in the spaces in Latin texts where Greek words or passages had dropped out, and of correcting the nonsense medieval scribes had made of the Greek words they had tried to copy.[70] Calderini had stated just as clearly as Poliziano that "some Greek words have been accepted for our use."[71] He had explained the Greek word "nomon" ("tune"), which Suetonius used in his life of Nero (20.2), and he had restored the word "Melicus" in Pliny (*Nat. Hist.* 7.24), though he was far from clear about its meaning.[72] Beroaldo restored the Greek words "cysthon" and "coston" to a curious epigram by Ausonius (82[123].5 Peiper) and the Greek name Hecale(s) to Apuleius (*Met.* 1.23).[73]

What was new in Poliziano's work was the completeness of his mastery of Greek. When Beroaldo or Calderini explained a Greek word in a Latin text, they tended to explain fairly obvious words and to draw their explanations from obvious sources. Even when their results matched Poliziano's, the contrast between their methods was clear. Beroaldo knew the story of Hecale from Pliny and from Plutarch's life of Theseus. Poliziano knew the passages from Pliny and Plutarch. But he also cited relevant passages from Statius, the *Priapea,* the *Suda,* an anonymous Greek epigram that accompanies certain codices of Callimachus's hymns, and the old scholia on Callimachus's *Hymn to Apollo*. And he used them not only to tell the story of Theseus and Hecale but also to establish just what Callimachus's *Hecale* was and why he had written it (Callimachus's detractors had claimed he could not write a long poem; the poem was his answer to their charges).[74] For Poliziano, then, the comparative study of Greek and Latin could shed light on both languages; it could even lead to the partial reconstruction of lost works.

The Romans, however, had not confined themselves to borrowing words and names; they had also translated long passages—sometimes even whole poems. And Poliziano took advantage of this fact too in his text-critical method. Thus, he knew that Catullus 66 was a translation of Callimachus's Lock of Berenice, for Catullus said so in poem 65. The available texts of Catullus gave 66.48 in a variety of meaningless or unmetrical forms, including "Iupiter ut Telorum omne genus pereat" and "Iupiter ut coelitum omne genus pereat." Callimachus's poem was lost. But Poliziano found a quotation from it in the scholia on Apollonius Rhodius, a quotation that corresponded to line 48 in the Latin: *Chalubōn hōs apoloito genos*—"May the race of Chalybes [a race renowned as workers of iron, and thus as makers of scissors] perish." He realized that "Telorum" and "coelitum" must be scribal corruptions of the unfamiliar word "Chalybon," which Catullus had simply transliterated. The original Latin must have read "Iupiter ut Chalybon omne genus pereat."[75]

In 1.27 Poliziano explained some puzzling lines in one of Cicero's letters (*Fam.* 7.6.1). He recognized that they were lines of verse, and, further, that they rendered a passage from the *Medea* of Euripides (214ff.). He knew that Ennius had written a Latin adaption from the *Medea*, and he rightly argued that the lines must be a fragment of Ennius's work. This identification marks the beginning of the reconstruction of early Latin poetry.[76]

Yet the most striking effects of Poliziano's comparative method were not in the field of textual criticism but in that of exegesis. Poliziano was the first humanist to insist that most Latin poets had drawn very heavily on Greek models for both content and style. Proper exegesis must therefore begin from identification of the philosophical theories and myths that Latin poets had borrowed and the metaphors and grammatical constructions they had imported.

In 1.26, for example, Poliziano pointed out that Ovid had adapted *Fasti* 1.357–358 from an epigram in the Greek Anthology (*AP* 9.75). Ovid had translated the Greek "as literally as possible"—"quam potuit ad unguem"—but he had still failed to capture all the nuances of the original: "The Latin poet did not even touch that—if I may so call it—transmarine charm."[77] Since Ovid himself was "poeta ingeniosissimus," this episode led Poliziano to formulate a playful judgment about the Latin language. Quintilian had characterized Ovid's style as "lascivus" ("abundant," "Asiatic") (*Inst. or.* 4.1.77). Poliziano argued, on the basis of his comparison, that Ovid failed to equal his Greek model not from lack of ability but because the Latin language itself ran counter to his special stylistic gifts: "This is the fault of the [Latin] language, not so much because it is lacking in words as because it allows less freedom for verbal play."[78]

In other cases Poliziano combined the study of verbal borrowings with that of intellectual ones. He discovered—or learned from Landino—that Persius had modeled his fourth satire on the Platonic dialogue *Alcibiades* 1. What Landino had merely remarked upon became for Poliziano the central interest of the poem. Persius had drawn the entire philosophical message of his poem from the dialogue: "it is clear that Persius ... drew from it the discussions that Socrates there holds with Alcibiades about the just and unjust, and about self-knowledge." Indeed, Poliziano said, the words "Tecum habita," with which Persius's poem ended, summarized the dialogue as well: "When he says 'tecum habita,' doesn't he seem to have understood the meaning of that dialogue clearly—if indeed, as the commentator Proclus affirms, Plato here had in mind precisely that Delphic writing which admonishes every man to know himself?"[79] At the same time, Poliziano pointed out that Persius had made some direct allusions to individual lines in the dialogue.[80] Persius is an extremely obscure and allusive poet. Medieval readers, completely at a loss to understand the satires, had invented wild explanations for them. Persius was said to be attacking "leccatores"—gossips, gluttons, bishops and abbots who failed to live up to their vows.[81] Only after Poliziano's explanation could Persius's poem be read as its author had intended.

Poliziano, of course, was hardly the first to point out that Latin poetry was heavily dependent upon Greek models. Latin poets themselves generally claimed to be not the first to write a particular kind of poetry, but the first to introduce a particular Greek style or verse-form into Latin.[82] Moreover, Gellius had compared passages from Latin poetry with the Greek originals from which they had been adapted. For example, he compared Virgil's adaptations of certain passages from Theocritus and Homer with the originals. He pointed out that Virgil had omitted a good deal, and he argued at some length that such omissions were necessary. Mere literal translations, he argued, would be clumsy, and would therefore lack poetic effectiveness. Virgil's practice of selective translation should therefore serve as a model for future Latin poets:

> Whenever striking expressions are to be translated and imitated from Greek poems, it is said that we should not always strive to render every single word with literal exactness. For many things lose their charm if they are translated too forcibly—as it were, unwillingly and reluctantly. Virgil therefore showed skill and good judgment in omitting some things and rendering others when he was dealing with passages of Homer or Hesiod or Apollonius or Parthenius or Callimachus or Theocritus or some other poet. [9.9.1–3, tr. J. C. Rolfe.]

Similar information and arguments were available in other works as well—for example, the *Saturnalia* of Macrobius, where Virgil and Homer are compared at length, and the commentaries of Servius on Virgil.[83]

These texts, in turn, attracted the attention of various medieval and Renaissance scholars before Poliziano. Richard de Bury, Bishop of Durham in the fourteenth century, had read his Gellius and Macrobius.[84] In consequence, it is not surprising to find him asking, in his *Philobiblon,* "What would Virgil, the chief poet among the Latins, have achieved, if he had not despoiled Theocritus, Lucretius, and Homer, and had not ploughed with their heifer? What, unless again and again he had read somewhat of Parthenius and Pindar, whose eloquence he could by no means imitate?"[85] And it was a commonplace among humanist teachers in the fifteenth century that only those who had sound knowledge of Greek literature could properly understand Latin literature.[86]

Calderini, moreover, had given detailed attention and much effort to the comparative study of Greek and Latin. In his brief commentary on Propertius, for example, he arrived several times at novel and interesting results. He pointed out that Propertius 1.20 is to some extent modeled on Theocritus 8: "In this passage his particular aim is to imitate and adapt Theocritus, on the story of Hylas."[87] And his wide reading in Greek poets, scholiasts, and historians enabled him to unravel a number of Propertius's mythological and geographical allusions. For example, he rightly interpreted the phrase "Theseae bracchia longa viae" in Propertius 3.21.24: "He means the long walls, which were called [*makra teichē* or *makra skelē*, literally 'long legs'] in Greek. They ran from the city [of Athens] up to the Piraeus; a careful account is to be found in Thucydides."[88] He also noticed that Propertius often made use of variant forms of well-known myths. Thus, in 1.13.21 Propertius described the river god Enipeus as Thessalian. Calderini rightly pointed out that other ancient writers located Enipeus not in Thessaly but in Elis.[89]

Yet for all its new material, this commentary sharply reveals how different Calderini was from Poliziano, for it shows how completely he lacked Poliziano's special qualities of judgment, reflection, and system. In 4.1.64 Propertius calls himself "the Roman Callimachus." The phrase is not so simple as it appears; when a Roman poet claims a particular Greek poet as his model, he may mean that he has derived his subject or his meter from the Greek, but he may also mean simply that he is as innovative, learned, and subtle as the Greek had been.[90] Calderini, however, took Propertius's phrase in the most literal way possible. He assumed that Propertius must have written direct adaptations from Callimachus: "[Propertius] calls himself the Roman Callimachus. For he sets out Callimachus, a Greek poet, in Latin verse."[91] And he applied his interpretation at least once in a most unfortunate way. In 1.2.1

Propertius calls his mistress Cynthia "vita"—"my life." "The word," wrote Calderini, "is drawn from Callimachus, who is Propertius's chief model. For he too, while flattering his mistress, calls her by the Greek word [for life (i.e. zōē)]."[92] Calderini's arguments clearly found some acceptance, for Beroaldo repeated them—without acknowledgment—in his own commentary on Propertius, which first appeared in 1487. He too argued that Propertius had taken Callimachus as his model, his "archetypon." And he too described Propertius's use of "vita" in 1.2.1 as done in imitation of Callimachus.[93]

The notes of Calderini and Beroaldo stimulated Poliziano to produce a splendid rebuttal in *Miscellanea* 1.80, where he raised a crushing objection to both Calderini's general interpretation and his reading of 1.2.1—namely, that not one shred of evidence proved that Callimachus had written any love poetry at all, much less a love poem in which zōē was used as an endearment: "I find it astonishing that Domizio and some others after him . . . dare to write that Propertius says this or that in imitation of Callimachus. For beyond a few hymns nothing at all remains to us of that poet, and certainly there is nothing at all that treats of love."[94] Poliziano certainly had the best in the exchange. Calderini had cited no evidence to back up his theory. As a commentator on Juvenal, he knew from 6.195 that zōē was sometimes used as an endearment. The notion that Callimachus had used the word in that sense was no doubt Calderini's own invention. But Poliziano's rebuttal is even more significant when seen in the context of the chapter in which it occurred. For in addition to attacking Calderini, Poliziano there gave a demonstration of the proper way to use Greek poetry to elucidate Propertius. In 4.9.57–58 Propertius alludes to the myth of Tiresias's encounter with Pallas while she was bathing. To illustrate the myth Poliziano published, along with much other material, the first edition of Callimachus's entire poem on the Bath of Pallas, which he also translated word for word.[95] The Alexandrian poets, in other words, could be used to explicate Propertius's mythological allusions—but not, except where the parallel was certain, to explain verbal details. In 1.80, then, Poliziano showed that he understood more than just the virtues of the comparative method in exegesis—he understood its limitations as well.

He understood that different Latin poets had used their Greek sources in different ways, and that the exegete must take those differences in poetic method into account when making comparisons. For all Calderini's wide reading in Greek sources and genuine sensitivity to the nuances of Latin poetry, he could not rival Poliziano in the methodical study of the poetics of allusion and adaptation.

Modern scholars still begin the study of Latin poems by looking for possible Greek sources and models. To be sure, when they find a model they compare it with the Latin in a far more detailed manner than Poliziano did. We want to know not only that Ennius translated Euripides, and that in doing so he made alterations, but also what specific changes he made, and what purpose they were meant to serve. But it would be unhistorical to reproach Poliziano for not carrying his comparisons far enough. After all, he took them as far as his ancient models and sources had. Rather, his work on the comparative method should be seen in the same perspective as his work on manuscripts: as the lineal ancestor of the methods that are still employed.

Poliziano chose his two chief areas of study, manuscripts and Greek sources, in response to a specific historical and biographical situation. In order to maintain his position with the Medici, a position that became less intimate after 1480, he had to prove that his scholarship was something new, something distinctly better than that of the previous generation. His conscious adoption of a new standard of accuracy and precision enabled him to prove just that. At the same time, his constant references to rare materials enabled him to give public thanks to the friends who made them available to him. Above all, these references allowed him to heap praises on his patron Lorenzo, who, he said, "in his service to scholars lowered himself" even to arranging access to manuscripts, coins, and inscriptions.[96]

Finally, Poliziano's situation determined his decision to study obscure and difficult subjects. By treating the study of antiquity as completely irrelevant to civic life and by suggesting that in any case only a tiny elite could study the ancient world with adequate rigor, Poliziano departed from the tradition of classical studies in Florence. Earlier Florentine humanists had studied the ancient world in order to become better men and citizens. They sought to recover the experience of classical republicanism in order to build a sound republic in their own time.[97] Poliziano by contrast insisted above all on the need to understand the past in the light of every possibly relevant bit of evidence—and to scrap any belief about the past that did not rest on firm documentary foundations. "Custom unsupported by the truth is long-lived error," he remarked, quoting Saint Cyprian, who had applied the same words to the need to purify the traditions and doctrines of the Church.[98] This starkly dramatic comparison of the need for pure classical texts and the need for an incorrupt Christianity gave Poliziano's method a moral as well as a philological edge; it even transferred the moral charge of classical studies from the actions they had been meant to stimulate to the rigor with which they should be pursued. But it also left the further ends of such study unspecified. And the exhaustive preparation he demanded in any case left little time for good citizenship. Moreover, when he set ancient works back into their historical context Poliziano eliminated whatever contemporary relevance they might have had. Medieval scholars had made the satires of Persius relevant to their own time by misreading them. Poliziano's historical reading of the satires showed clearly that they were not concerned with the misdeeds of monks or other contemporary problems. By so removing any moral function from classical studies, Poliziano made them as appropriate an object for Medici patronage as the elegant nonsense of the neo-Platonists.[99]

Poliziano's principles did not triumph at once. Indeed, it would have been amusing to watch his reaction from a safe distance when his friend Baptista Mantuanus's tract on the spelling of Virgil's name arrived. "Vergilius," Mantuanus claimed, could not be right. To be sure, it was the spelling found in the Codex Romanus and in certain inscriptions. But that was hardly conclusive evidence. Why should the testimony of three old manuscripts outweigh that of thousands of more recent ones?[100] Beroaldo was Poliziano's friend and correspondent, and he outlived Poliziano by eleven years. Yet he failed to grasp Poliziano's insights. He was not scrupulous about reproducing sources accurately, and so he perverted the sense of more than one passage from Plato.[101] Moreover, he continued to consult manuscripts only intermittently and to report their readings with only vague indications of their origins.[102] Even Poliziano's disciple Pietro Crinito seems to have learned relatively little from his master. His own method, at least as embodied in the *De honesta disciplina,* rested on the study of the ancient Latin grammarians rather than on that of manuscripts and Greek sources.[103]

But Poliziano's new methods survived his contemporaries' lack of insight. By the middle of the sixteenth century, scholars employed them throughout Europe. The comparative study of Latin and Greek flourished especially in France; systematic recension of manuscripts preoccupied scholars in Florence and Rome.[104] And isolated insights from his work appeared in the writings of many others besides Bellièvre and Diplovatacius.

We are not yet sure of the channels by which Poliziano's methods were passed on; it is not clear whether anyone taught them in the generation that immediately followed his. One point, however, may be made. Poliziano stated and exemplified his principles in many other works besides the *Miscellanea.* In his more advanced lecture courses he made frequent references both to manuscripts and to Greek sources of Latin literature.[105] In the books in which he had entered his collations of manuscripts, he also entered the famous subscriptions in which he repeatedly stressed the need for exact collation and for systematic recension.[106] Some of his books, moreover, contained Greek parallels which he had assembled.[107] The memory of his teaching and the physical presence of his *Nachlass* must have had some effect, at least in Florence.[108] Above all, there were his letters, which portrayed the wonderful Indian summer of Italian humanism more clearly than did any other work. These too were shot through with Poliziano's innovative methods. In them we seee Poliziano boasting to Piero de'Medici of the soundness of the sources on which his new theory about the origins of Florence was based.[109] We see him questioning the spelling of a proper name in Pomponio Leto's copy of an inscription.[110] We see him studying the Florentine Pandects and beginning the study of Roman law before Justinian.[111] And we see him comparing his own introduction of Greek metrical devices into Latin with Virgil's.[112] The letters were often reprinted, both in editions of Poliziano's works and independently. And they were far more accessible than the more technical *Miscellanea*. Only when the fortunes of all these works are known will we be able to see clearly how Poliziano's revolution gave rise to a new philology.[113]

3

Traditions of Invention and Inventions of Tradition in Renaissance Italy: Annius of Viterbo

JOSEPH SCALIGER encountered two supernatural beings in the course of his long and well-spent life. He saw one of them, a black man on a horse, as he rode by a marsh with some friends. He only read about the other, a monster named Oannes with the body of a fish and the voice of a man. Yet as so often happened in the Renaissance, the encounter with Art had far more lasting consequences than that with Life. The black man tried to lure Scaliger into the marsh, failed, and disappeared, leaving him confirmed in his contempt for the devil and all his works: "My father didn't fear the Devil, neither do I. I'm worse than the devil."[1] Oannes, in the book that Scaliger read, climbed out of the ocean and taught humanity the arts and sciences. Devil Tempts Man was no headline to excite the Renaissance public; but Amphibian Creates Culture was out of the ordinary even in the sixteenth century.

The fish who gave us civilization appeared at the beginning of the account of Babylonian mythology and history written by Berosus, a priest of Bel, early in the third century B.C. Berosus drew on genuine Babylonian records but wrote in Greek, for the benefit of the Seleucid king Antiochus I Soter. He and other Near Eastern writers, like the Egyptian Manetho and many Jews, tried to avenge in the realm of the archive their defeat on the battlefield, using documents and inscriptions to show that Babylon, Israel, and Egypt were older and wiser than Greece. Jewish and Christian writers preserved Berosus's Babyloniaca. It was in the unpublished world chronicle of one of them, George Syncellus (ca. A.D. 800), that Scaliger met Berosus and his fishy pet, in 1602–3.[2]

The most remarkable thing about the encounter was Scaliger's reaction to it. As a good Calvinist he considered Babylonian gods to be abhorrent and Babylonian boasts of the great antiquity of their state to be fanciful. As a good scholar he knew that Berosus was not a name to inspire trust. In fact, as his disciple John Selden pointed out, a century before Scaliger wrote "THERE came forth, and in Buskins too (I mean with Pomp and State) . . . an Author, called *Berosus* a *Chaldee* Priest"—a forged text that had become a sixteenth-century best-seller and perverted the early histories of every country in Europe.[3] Scaliger had been one of ps.-Berosus's sharpest critics. Yet in this case he showed respectful interest in what he had every reason to dismiss as mad forgeries. Taking his first notes on the story of Oannes, he remarked only that in another account the same fish was called Oes, and added a note on Berosus himself from the early Christian writer Tatian.[4] Compiling his last large work on world history, the *Thesaurus temporum* of 1606, he included all the Berosus he could find, dated the material as precisely as he could, and boasted of the service he had performed by collecting these previously unknown texts.[5] He did not even remark—as his close friend Isaac Casaubon mildly did, when taking his own notes on the same manuscript chronicle—that "the nature of a certain animal, *Oannes*, is particularly curious."[6] Instead, he defended the work of Berosus—like that of Manetho, which he also recovered and published—as genuine near eastern historiography. True, the early sections of these texts seemed fabulous, but they still deserved the reverence that goes with genuine antiquity, and they also linked up neatly with the ancients' true accounts of later history. Challenged by the Heidelberg theologian David Pareus, Scaliger developed the latter argument into the more polemical thesis, hardly a new one in this context, that the apparently fabulous histories of the pagans clothed real events in mythical form.[7] He thus preserved and defended what we now know to have been the first genuine large-scale products of the ancient Near East to reach the modern West—works so alien to the Western tradition that they could hardly be interpreted at all until the discovery and decipherment of parallel records in cuneiform, more than two hundred years later.

How then do we account for Scaliger's divinatory prowess—his ability to shake off the prejudices normal to his period and place and see that his Near Eastern fragments, if unintelligible, were also unimpeachable? The rich historiographical scholarship of the 1960s offers us an answer. The humanists of the early Renaissance—notably Petrarch and Valla—tested and rejected many medieval forgeries. They showed that texts like the charter of Julius Caesar that exempted Austria from imperial jurisdiction or the Donation of Constantine that gave the Pope control of the Western Empire used a language different from the attested language of their supposed authors, ignored facts recounted by reputable historians, and contradicted the beliefs and aims of well-known historical actors associated with them.[8] The theologians and jurists of the mid-sixteenth century, men like Melchior Cano and Jean Bodin, were confronted by a much wider range of supposedly authoritative texts and an even more pressing set of religious and political problems. Accordingly they went much further than their predecessors. They had not only to purify the canon of its fakes but to weigh the authority of its genuine components. Accepting the humanists' isolated but valid insights, Cano and Bodin tried to fuse them into a general art of choosing and reading authorities about the past. They provided not empirical case studies but universally applicable rules for evaluating sources. And by applying these consistently to a wide range of texts, slightly later scholars like Henri Estienne, Joseph Scaliger, and Isaac Casaubon purged the classical corpus of its fakes and pseudepigrapha. They made clear the priority of Homer to his spurious Latin rivals, Dictys and Dares; the priority of Plato to his supposed Egyptian source, Hermes Trismegistus; the priority of Hesiod, who really composed for an audience of shepherds, to Musaeus, who wrote for an audience of Alexandrian grammarians. The image conjured up is of a train in which Greeks and Latins, spurious and genuine authorities, sit side by side until they reach a stop marked "Renaissance." Then grim-faced humanists climb aboard, check tickets, and expel fakes in hordes through doors and windows alike. Their destination, of course, is Oblivion—the wrecking-yard to which History and Humanism conduct all canons—and certainly consign all fakes.[9]

This vision suggests that humanist critical method was both new and modern. Two centuries and more after the Renaissance, when Karl Otfried Müller confronted the Greek account of Phoenician antiquities forged by Wagenfeld, attributed to a mysterious disappearing manuscript from Portugal (and accepted by the epigrapher and orientalist Grotefend), he needed only to apply the humanists' touchstones to make the odor of authenticity vanish. Ps.-Philo of Byblos, as presented by Wagenfeld, misunderstood and contradicted the fragments of his own work preserved by Eusebius (though he faithfully retained typographical errors from printed texts). He made many unlikely grammatical and syntactical errors, large and small ("auch im Gebrauche der Partikeln ist manche Unrichtigkeit zu bemerken"). And he believed in the gods (though the real Philo had been an atheist). Müller transcended the humanists only in his sympathy for forgery as a work of art. He praised Wagenfeld's *Geist* and *Phantasie*, the splendid aptness with which he had caught "the spirit of ancient, Greek-Oriental historiography." In other respects, however, Müller was merely doing what came humanistically.[10]

Happily, a still more recent generation of scholarship has introduced some attractive loops and swerves into this rectilinear and teleological account. Joseph Levine and others have shown that the simple accumulation of data, generation after generation, did more than method could to catalyze the fixing of a modern canon of classic texts and objects.[11] Others have emphasized the distance that lies between the humanists' exposures of fakes and those of the modern philologist. We now know, for example, that Isaac Casaubon obtained one of the great "modern" results of humanist philology—the inauthenticity of Hermes Trismegistus—precisely because he believed firmly in a traditional dogma, that no pagan could have written a book as pure, clear, and theologically correct as the *Hermetic Corpus* at the early date traditionally assigned to it.[12] History often offered not the subversion but the confirmation of dogma.

But by far the most fetching of these new directions of research is that opened up in two elegant articles by Werner Goez. Goez argues that previous historians have omitted not just an important way station but the crucial one from their account of the ride of the ancients. The Dominican Annius of Viterbo who forged the fake Berosus at the end of the fifteenth century created not only texts but rules for the choice of texts as well, general and plausible ones. These rules in turn formed the basis of all later systematic reflection on the choice and evaluation of sources. Some of the mid-sixteenth-century theorists, like Melchior Cano, rejected Annius and all his works; others, like Jean Bodin, accepted them. But all of them developed their theories of reading in direct response to the challenge he presented. Thus, a forger emerges as the first really modern theorist of critical reading of historians—a paradox that only a reader with a heart of stone could reject.[13] If Scaliger could tell that his Berosus was real, he owed his perceptiveness in large part to the creator of the false Berosus he despised.

In this chapter I propose to examine Annius's role in the development of historical and philological method, both by analyzing his own work and by inspecting the mid-sixteenth-century reactions to it. I hope by doing so to reveal my own ignorance *de omni re scibili et quibusdam aliis*—and to suggest that the development of modern responses to classical histories, real and fake, is even more crooked and complex than the tale told by Goez.

Mystery, Ancient and Modern, with Seaography

In 1498 Eucharius Silber published Annius's *Commentaries on Various Authors Discussing Antiquities*. This elegant volume contained original sources by real Greek authors like Archilochus, Berosus, and Manetho; by imaginary Greek authors like Metasthenes (a perversion of Megasthenes, the name of a Greek who wrote ca. 300 B.C. about India); and by noble Romans like Cato, Fabius Pictor, and Propertius (exceptionally, this last text was genuine). These texts, cut up into neat gobbets set in a large and impressive Gothic type, swam on a rich foam of commentary by Annius himself, giving a nice impression of classical or biblical status.[14] They looked—and read—like a comprehensive and powerful history of the world. They wove biblical history, ancient myths, and medieval

Trojan legends together into a single story. Noah—the only pious member of the race of giants that inhabited the prelapsarian world—became the father not just of Shem, Ham, and Japhet but of a pride of other giants. His sons in turn insinuated themselves into national mythologies of the most diverse kinds, which Berosus neatly laid out in genealogical diagrams ("the lawyers," Annius helpfully remarks, "imitated this example in their use of the form of trees to set out degrees of consanguinity").[15] The Tuyscon whom Tacitus made the ancestor of the Germans, the Hercules whom Spanish legend made the founder of Iberia, the Etruscans whom Tuscan scholars from Leonardo Bruni on had seen as the true, non-Roman progenitors of the modern Italian city-state, and the Dryius who founded the learned Anglo-Gallic order of Druids all found places in this rich if chaotic history of the world. And spicy details—like the story, perhaps derived from the rabbinical tradition in which it also occurs, of how Ham touched Noah's genitals, murmured an incantation and made him sterile—gave entertainment value to what might otherwise have seemed an austere and tedious narrative.[16]

Annius made his story vivid with devices too many to list. A neat illustration gave material form to ps.-Fabius's account of early Rome. An introductory appeal to the Catholic kings of Spain—those possessors of the "bravery, victory, chastity, courtesy, prudence, modesty, piety, solicitude" of Moses and David—gave the works powerful patronage.[17] A reference to two Armenian friars as the source of Annius's new texts—and the Armenian oral tradition that Noah was there called Sale "Aramea lingua"—gave it an exotic flavor highly appropriate to that age of lovers of hieroglyphs and readers of the *Hypnerotomachia Poliphili*.[18]

Above all, the content of the work made a powerful appeal to at least three period forms of historical imagination. Like one of his principal sources, Josephus, whose *Antiquitates* he plundered, Annius offered new details about Noah and the early kingdom.[19] These provided precisely the rich context for the biblical history of man that the Bible itself lacks. Martin Luther, for example, urges readers of his *Supputatio annorum mundi* to realize that events now forgotten had filled the "vacua spatia" of his chronicle. Many kings and institutions had flourished in biblical times besides those mentioned in the Bible, and most of the patriarchs—as their long life spans showed—had lived simultaneously, not successively.[20] The Annian texts offered precisely the richly detailed backdrop in front of which Luther would have preferred to see the family dramas of ancient Israel performed. No wonder that he found Annius his richest nonbiblical source—despite his discovery of occasional errors and inconsistencies in the texts. No wonder either that the chronologist Ioannes Lucidus Samotheus explicitly defended Berosus's enlarged version of the Mosaic account of the origins of the nations: "Berosus described more [founders of nations than Moses did in Genesis 10] because Berosus the Chaldean described as many rulers as there were founders of kingdoms and peoples; Moses, however, described those who had different languages..." Differences in forms, he explained, stemmed from the two authors' use of different languages, Hebrew and Chaldean—the latter in any case changed by the Latin translator.[21] On the whole, after all, Berosus stood out among ancient historians for the large amount of material he had in common with Moses; Guillaume Postel insisted that Berosus had a bad reputation precisely because "he passed down to posterity an account similar to that in the sacred [books], and thus is despised and ridiculed by men poorly disposed toward divine things, because of the very quality for which he ought to be praised and preferred to all other authors."[22]

Annius's constant emphasis on Egypt as the source of civilization, on the great journeys and achievements of Osiris, appealed as strongly to that widespread Egyptomania which inspired collectors of hieroglyphs, composers of emblem books, and painters of historical scenes from the late-fifteenth-century Vatican to the other extremes of Renaissance Europe, Jacobean London and Rudolphine Prague.[23] And Annius's heavy emphasis on the connections between the biblical Orient, Troy, and the states of fifteenth-century Europe opened up an even more fruitful field for wild speculations. Every nation and city from Novgorod to Naples felt the need for an early history that rivaled or surpassed the ancient histories of Greece and Rome, to which the humanists had given such prominence. No wonder that scholars from Toledo to Trier leaped with alacrity down the rabbit hole where Annius preceded them. True, he offered his richest rewards to his native Viterbo, that cradle of world civilization where Osiris had taught the arts and left inscriptions; to Spain, where his patrons the Borgia came from; and to the reputation of Egypt. But he also had much to offer Frenchmen, Englishmen, Germans, and other Italians. Hence, he became omnipresent in the historical fantasies and historical frescoes of the early sixteenth century—though, to be sure, the particular interests of those who vulgarized his work had the normal transformative effect.[24] Geoffroy Tory complained that greedy readers coerced him to print an edition of the Annian texts without commentary—already a notable change—for his Parisian public in 1510. Something of the nature of that demand can be glimpsed in the Princeton copy, bought hot from the press in 1510 by Robert Nicolson of London, who carefully underlined every passage dealing with the early history of the British (while either he or another reader also marked the sterilization of Noah with a lurid marginal sign).[25]

Until recently, serious scholars have occasionally tried to clear Annius of the charge of forgery—to make him a genuine transmitter of early texts preserved in Armenia or, more plausibly, the gullible victim of a forger. But in the last twenty years research on him has intensified. Italian scholars have dug out and published some of his *inedita*. And by doing so they have demonstrated that he not only commented on but wrote his *Antiquities*. Eduardo Fumagalli has exposed him citing in early letters a recognizable (but not finished) version of the ps.-Cato that he printed.[26] A work of 1491–2, the *Epitome of Viterbese History*, shows him already at work on Fabius Pictor. Describing the rape of the Sabine Women, he writes: "according to Fabius, an audacious crime in the form of a rape of women took place four months after the founding of the city, eleven days before the Kalends of September [22 August]."[27] Here Annius conflates two passages from one ancient source, Plutarch's *Life of Romulus*, where he read both that the founding took place eleven days before the Kalends of May (21 April) (12.1) and that Fabius Pictor had dated the rape to the fourth month after the foundation (14.1). By combining these two bits of Plutarch, no connection between which is evident in context, he gave an event more lurid than the foundation a nice—if spurious—precision. Later, however, he thought better of this tactic, perhaps because he noticed that his Plutarchan source gave a precise date—quite possibly Fabius Pictor's own—of 18 August for the same event. Accordingly, he made his published Fabius Pictor say only that the rape fell in the fourth month after the foundation, and transferred precise discussion of the founding date itself to another of his *auctores*, ps.-Sempronius.[28] Tactics like these reveal a Chatterton, not a Walpole, hard at work. Annius was no innocent agent or victim but a conscious artist creating a coherent piece of work.

Annius, of course, was no fly-by-night forger but a very serious man—the possessor, indeed of two of the ultimate accolades for a Renaissance Dominican, a miraculous cure that won mention in the *Acta Sanctorum* and a death by poisoning at the hands of Cesare Borgia. His fakes apparently won him papal attention, Spanish financial and personal support, and even his high office as *magister sacri palatii*.[29] The question, then, is how to appreciate and assess a fine piece of high Renaissance scholarship and art.

Here too debate has recently flared up. Goez stressed the modernity of Annius's approach, the theoretical sophistication of the arguments by which he tried to validate his fakes. But Bernard Guenée has tried to find in Annius's critical technique not the triumph of modernity but the culmination of medieval historical method. Annius merely stated in a general form rules long applied in practice by medieval scholars—especially when they needed historical evidence to adjudicate the rival claims of quarreling religious institutions or orders to a privilege or a relic. And other evidence—like Beryl Smalley's discovery that a medieval forger had been inspired long before Annius to invent a ps.-Berosus, who described the eclipse that accompanied the crucifixion—seems to confirm the traditional flavor of Annius's heady historical concoction.[30]

AMBITION

This more than any other term conveys the basic flavor of Annius's stew. He set out, as he says, on a Herculean journey: to do the "duty of a theologian: to seek, discover, confirm, reveal, and so far as possible to explicate, teach, and pass on the truth." He claims to offer *nuda veritas*, unadorned by rhetoric and uncontaminated by falsehood. And he claims that this truth embraces all events in human history, the creation of all significant arts and sciences, and the origins of all peoples. This effort to enfold in a single encyclopedic history the origins of society and culture harks back to the world of the fathers of the church, when Julius Africanus, Eusebius, and Isidore of Seville set out to provide similarly comprehensive records. Like them, Annius, is a compiler with an ideology. He wishes to reveal the truth; but the truth he reveals is as polemical as it is comprehensive. It displaces Greek culture from its central place in human history and connects the modern West directly to the biblical Near East.[31]

Annius's ambition, moreover, makes itself felt in many details as well as in the larger framework of his enterprise: chiefly when he forges ancient precedent for practices and institutions of his own time. A theologian of distinction, he had opinions on a good many controverted problems of his day: the licitness of the *monte di pietà*, the freedom of the will, and the influence of the stars on history.[32] And in the last case at least Annius built into his ancient texts support for a highly up-to-date practice. In 1480, as Cesare Vasoli has shown, Annius reacted to the Turkish landing at Otranto by issuing a back-dated prediction, based on both the *Revelation* of John and the revelations of the stars, that the time of Muslim dominance had reached its end and a Christian triumph was approaching. He supported this view with a general argument about the role of the stars as intermediaries between God and this world and a specific argument about the role of the zodiacal sign of Leo in the victories of the Turks. These prophecies—like the many others that circulated through Italy in the last two decades of the fifteenth century, heralded by the preaching of Savonarola and the strange streetcorner agitation of men like Giovanni Mercurio— were entirely serious both in detail and in their underlying assumptions. And in his *Antiquities*, almost twenty years later, Annius defended his early position retroactively. He made Berosus describe Noah as foreseeing the flood *ex astris*, seventy-eight years in advance; his commentary made other giants predict the Flood from the stars; he even quoted the opinion of "more competent Talmudists" that Noah was an expert astronomer as well as a giant. He thus made the kind of astrology he practiced not merely old but antediluvian—an enviable disciplinary history, since it chartered astrological prediction as a part of the primeval knowledge of the patriarchs.³³

DERISION

If Annius hoped to replace the texts of the ancients with his own new canon, he had to prove the fallibility of his opponents as well as the antiquity and purity of his own authors. And that he did, *ad nauseam*, at every opportunity conceivable and some that were not. Sometimes he gives the Greeks the lie direct, as in this comment on his forged Xenophon *De equivocis*:

> Berosus writes in Antiquities 5: "In the fourth year of Ninus [king of Babylon] Tuyscon the giant trained the Germans in letters and laws, Samotes trained the Celts, and Tubal trained the Celtiberians" . . . Therefore the Iberians, Samotheans, and Tuyscons were clearly the fathers of letters and philosophy, more than a thousand years before the Greeks, as Aristotle attests rightly in his Magic, and Senon, and not the Greeks, as lying Ephorus and the dreamer Diogenes Laertius imagine but do not prove.³⁴

Annius often gave free rein to his enviable talent as a composer of invective. Graecia for him was preeminently Graecia mendax; the Greeks stand guilty of inventing new doctrines instead of sticking to the old ones of the giants, of vanity, levity, and virtually every crime but mopery: "Greece and the Peloponnesus were called in antiquity Aegialea, that is, Goaty (*Hircina*), because they produce many goats and because the race of men there is dirty, fetid, and goatish (*hircinum*)."³⁵ Elsewhere, though, he took a more measured tone, using the Greeks' inability to agree as the outward and visible sign of their interior corruption: "The Greeks fight and disagree with one another, as is not surprising, and they have entirely ruined history as well as philosophy with their civil war."³⁶ This combination of invective and argument gives texts and commentary alike a hectoring, even menacing tone, that helped to ensure Annius and his boys a friendly reception not only in the Catholic Italy he knew but in the Protestant and puritanical north a few decades later. No wonder that Luther, who believed that all of Aristotle compared to the Bible as darkness compares to light, preferred Annius's Berosus to Herodotus and his ilk.

UGLIFICATION

For our purposes, however, Annius's methods matter more than his attitudes. Some of his tools were simple enough. Composing the history of "Myrsilus of Lesbos," he derived his matter from the extraordinarily rich first book of Dionysius of Halicarnassus's Roman history, recently translated into Latin. He then admitted the similarity of their accounts—"though one who reads book 1 of Dionysius seems to read Myrsilus as well"—and explained them handily: "Dionysius follows Myrsilus consistently in the first book, save in the time of Enotrius's arrival in Italy." "In any event," he wound up for the benefit of the unconvinced, "both of them will be clarified by my commentary."[37] Composing Berosus on Germany, he borrowed Tuyscon from Tacitus and went on his way rejoicing.[38]

Elsewhere he transplanted what he read in one source to alien ground. If Diodorus Siculus described the pillar on which Osiris recorded his expeditions, Annius both forged an inscription left by Osiris in Viterbo and made his Xenophon—not the proper Xenophon, of course, but another writer of the same name—describe the similar inscription of Ninus of Babylon.[39] Much of the grist for his mill came from the great Jewish and Christian writers who had assembled evidence to prove the antiquity and priority of the Judeo-Christian tradition—above all Josephus and Eusebius, from whom he took the very names of Berosus and Manetho and much else. And one of his handiest tools, the euhemerist interpretation of classical myths as reworkings of genuine events, had already proved an interpretative Swiss Army Knife to legions of interpreters pagan and Christian, from whom Annius could learn how great men did great things and then became gods.[40]

Sometimes Annius moved from the low plane of craft to the higher one of art. When it came to tying the early histories of modern nations to the Bible, the ancient sources left him cast up on a dry beach, clueless and uninformed. Accordingly, he invented, assuming that etymology provided the key that could decipher any tribe's or city's name, making it reveal the lost name of its founder. Text and commentator work together closely in these cases.

> In the time of Mancaleus [says Berosus] . . . Lugdus ruled the Celts; province and people took their names from him. Lugdus [says Annius] was the one who settled the province of Lugdunum, as his name proves.[41]

"He invents kings like this whenever he has to," so Annius's most brilliant critic of the next generation, Beatus Rhenanus, cried in disgust. " . . . But Lugdunum isn't derived from Lugdus. Dunum is a suffix, like German *berg* or *burg*."[42] But such reservations were rare. On the whole, Annius's method—itself no doubt derived from a careful reading of another ancient source, the *Etymologies* of Isidore—carried conviction, even when he went to the imaginative extreme of deriving Hercules from Egyptian *her*, "covered with skin," and Hebrew *col*, "all," "since he used wild beasts' skins to cover his whole body in place of arms, since arms were not yet invented."[43] Etymology ensured authority.

It would be wrong to assume—as some interpreters have—that Annius's novel information all came from his mother wit. Sometimes he made creative use of others' mistakes. In Book 5 of his histories Diodorus Siculus mentions those wise men "whom the Gauls call Druids." Poggio Bracciolini rendered the clause in question as "quos vocant Saronidas"—"whom they call Saronidae."[44] And it was from this very up-to-date textual foundation, built of only the best sand, that Annius derived his wise King Sarron, who taught the barbarous Celts their letters.

Sometimes, too, his information comes straight from sources that were to him classic—but that we have ceased to read. One problem modern scholarship has not solved is this: Why does Annius make Archilochus—a poet rather than a scholar, and one about whose character the most alarming information circulated—tell the story of the eight different Homers who inhabited the ancient world, the last of whom, Archilochus's contemporary, won at the Olympics and became the official reformer of the Greek alphabet? Who ever heard of a scholarly Archilochus? The answer here is simple. At some point in the Hellenistic age, a scholar compiled several divergent opinions about the date of Homer. He attributed these, using a normal late Greek phrase, to *hoi peri* Aristarchus, *hoi peri* Eratosthenes, and others. Literally *hoi peri* means "those about" or "the school of," but in late Greek it is often merely an elegant periphrasis for the proper name that follows the preposition *peri*. This must be the sense required here. The anonymous scholar also noted that others date Homer to the time of Archilochus. His compilation, in turn, found a place in Tatian's *Oration against the Greeks* (ca. A.D. 150) and in Eusebius's *Chronicle* two centuries later, in condensed form: "*heteroi kata Archilochon*" ("others date him to the time of Archilochus").[45] The problem arose when St. Jerome translated Eusebius into Latin. Quite reasonably he turned "*hoi peri Aristarchon*" into "Aristarchus," "*hoi peri Eratosthenēn*" into "Eratosthenes"; but when he came to the end of the list, he misread "*heteroi kata Archilochon*" as an elegant variation on the earlier construction, and made Archilochus the author of the last opinion rather than its chronological benchmark.[46] In this case Annius's mistake was by no means his own. It came from that dramatic era, the late fourth and fifth centuries A.D., when Christian scholars like Jerome treated the classical heritage much as their unlearned followers were treating the pagan philosopher Hypatia.

In a larger sense, too, Annius's effort to fill in and add color to the biblical narrative has deep roots in scholarly tradition. Jewish scholars had spent centuries devising *aggadot*, supplementary tales about the characters in the Old Testament that rounded out their lives and clarified their motives—like the tale of Ham's enchantment of Noah. And late medieval Mendicants, of course, had done the same for the characters in the New Testament and their favorite saints. Late medieval Dominicans enriched the biography of St. Jerome with the story that he had pushed an insufficiently respectful abbot to the edge of a cliff and allowed him to live only when he promised to dedicate a church to St. Jerome. Swiss Dominicans in Bern just after Annius's time adorned a statue of the Virgin Mary with drops of varnish, to show that it wept (and thus possessed holy powers); they even put a speaking tube between its lips and made it issue prophecies and commands.[47] The production of serious fiction designed to fill out the holiest of records was no novelty in 1498. Annius's methods of uglification, then, do not qualify him either as medieval or as modern. Classical, patristic and medieval, popular and learned, Christian and non-Christian ingredients are blended here into a stew so rich and complex that the original ingredients and spices are often beyond retrieval.

DISTRACTION

Like any forger, Annius had to keep his readers' confidence—to distract them from the holes in his material, the contradictions in his texts, and the obvious anachronisms of his style ("Alexander," writes Metasthenes, "... transtulit imperium in Graecos"—a splendid piece of Greek historical writing).[48] He used a variety of means. Some were the normal resources of the ancient and medieval forger—like the claim to have derived his information from a source so distant as to discourage verification and so exotic as to compel belief. If Geoffrey of Monmouth could attribute his account of Brutus et al. to "a very ancient book in the British tongue" offered him by Walter, Archdeacon of Oxford, Annius could certainly borrow some texts from his Armenian confreres and ask advice on Hebrew and Aramaic from his Jewish friend, the still unidentified "Samuel the Talmudist," who told him that "Alemannus" comes from Scythian Ale (river) and the name Mannus.[49]

But Annius was both more polemical and more imperious than most medieval forgers. Like other defenders of oral tradition in a scholarly age, he had to provide his own textual warrants of authority. Moreover, he was a scholar in his own right, one who wanted not only to complement but to replace the Greek historians. And to bring this about he insinuated, into both his forged texts and his commentaries, a set of rules for the choice of reliable sources. Metasthenes states these clearly:

> Those who write on chronology must not do so on the basis of hearsay and opinion. For if they write by opinion, like the Greeks, they will deceive themselves and others and waste their lives in error. But error will be avoided if we follow only the annals of the two Kingdoms and reject the rest as fabulous. For these contain the dates, kings, and names, set out as clearly and truly as their kings ruled splendidly. But we must not accept everyone who writes about these kings, but only the priests of the kingdom, whose annals have public and incontrovertible authority, like Be-

rosus. For that Chaldean set out the entire Assyrian history on the basis of the ancients' annals, and we Persians now follow him alone, or above all.⁵⁰

Annius's comment described the ancient priests as "publici notarii rerum gestarum et temporum," whose records deserved as ready belief as the notarial records in a modern archive. And his other authors referred to, repeated, and expanded on these injunctions, creating a sticky, cohesive web of mutually supportive fictions about authority. After working his way through Myrsilus, Berosus, and Philo, the reader knew that each of the Four Monarchies, Assyrian, Persian, Greek, and Roman, had had its own priestly caste and produced its own sacred annals. Only histories based on these deserved credit; and any given historian deserved credit only for those sections where he drew on an authoritative set of records.⁵¹ For example, Ctesias the Greek (a real ancient, but also a forger) "is accepted for Persian history and rejected from Assyrian history," since he drew his account of the former from the Persian archives (in fact, of course, he invented it . . .) and made the latter up.⁵²

These principles have attracted more attention than any other segment of the corpus Annianum. Rightly, too; for they do seem a prescient effort to separate history, the record of events, *res gestae*, from history, the literary work of an individual, *historia*. And they certainly mark an effort to replace the empirical, case-by-case practices of the early humanists with a general theory. Yet at the same time they have an eerily traditional quality. Guenée points out that when the monks of Saint-Denis and the canons of Notre-Dame disputed around 1400 on the burning question of which of them had which bits of Saint Denis, both sides cited and assessed historical evidence. One of the advocates of Saint-Denis argued that the *Grandes chroniques de France*, which supported his case, should prevail. After all, it was an "approved and authorized" history, preserved in a "public archive." Was new humanist old canonist rewritten in fetching macaronic Latin?⁵³

In fact, here too Annius drew on a one-time classic now read only by specialists, as Walter Stephens has shown in detail.⁵⁴ In the last years of his life, the Jewish historian and honest traitor

Josephus wrote a polemical work in two books against the grammarian Apion, who had defamed the Jews. In the course of this he repeatedly emphasized the novelty of Greek and the antiquity of Jewish civilization. And to nail this point home he emphasized that the Jewish and Near Eastern texts he quoted rested not on individual opinion but on archival documents recorded by a caste of priests:

> The Egyptians, the Chaldeans, and the Phoenicians (to say nothing for the moment of ourselves) have by their own account a historical record rooted in tradition of extreme antiquity and stability. For all these peoples live in places where the climate causes little decay, and they take care not to let any of their historical experiences pass out of their memory. On the contrary, they religiously preserve it in their public records, written by their most able scholars. In the Greek world, however, the memory of past events has been blotted out. (1.8–10)

Josephus elsewhere praises Berosus for "following the most ancient records" (1.130), the people of Tyre for keeping careful "public records" (1.107), and the Egyptians for entrusting the care of their records to their priests (1.28). And if the *contra Apionem*, available in Latin since the time of Cassiodorus, was little read in the Middle Ages, Annius certainly used it heavily. In fact, in his comment on Metasthenes Annius made clear—*more suo*—what his source was. He explains that "Josephus used Metasthenes' rules to make a most valid argument" against the Greek views on the origin of the Greek alphabet.[55]

This reuse of ancient scholarship, though unusually extended, is far from unique in the corpus. In ps.-Sempronius, Annius lists a series of opinions, including that of the astrologer L. Tarrutius Firmanus, about the year of Rome's foundation. He draws his datings, as O. A. Danielsson showed long ago in what remains a very useful article, from the Roman compiler Solinus (third century A.D.).[56] He then declares his—or Sempronius's—preference for *Eratosthenis invicta regula*—"the unvanquished rule of Eratosthenes." Danielsson read with amusement what he took to be this "echt annianische Phrase." In fact, Annius took it directly from Lampugnino Birago's Latin translation of Dionysius of Halicarnassus's *Roman Antiquities*. At 1.63 Dionysius explains that he has elsewhere shown the *canones* (chronological canons or tables) of Eratosthenes to be sound; Birago rendered Greek *canones* as latin *regulae*, rather than *tabulae* or *laterculi*, and thus misled at least one reader. Annius's rules, then, were neither medieval nor modern. They were instead a classical revival, for the most part a restatement of that partly justified Near Eastern pride in great longevity and accurate records that animated so much of the resistance to Hellenization and to Rome—and gave rise in its own right to so many forgeries. We will not find in Annius the culmination of medieval historical scholarship or the origins of modern historical hermeneutics.

Nothing ages so quickly as one period's convincing version of a still earlier period. Annius's antiquity looks intensely quaint and entirely Renaissance now. Indeed, it looked quite modern to some of his early readers. Beatus Rhenanus, for example, recognized at once the single authorship of texts and commentary. "While the one milks the he-goat," he drily commented, "the other holds out the sieve." Pietro Crinito, who preferred to take his fragments of Cato from genuine Roman sources like Macrobius, had no trouble condemning Annius. Nor did Juan Luis Vives, who inserted a powerful attack into his commentary on Augustine's *City of God*, where it found a surprising number of readers. Nor did the anonymous skeptics of whom Postel bitterly complained.[57] But these exceptions do not disturb the general rule. For every Rhenanus there was at least one Trithemius, eager and willing to embroider on Annius in the most fanciful ways (Trithemius's version of *broderie allemande* took the form of his own invented text, that of the Scythian historian Hunibald, who recounted the deeds of the Germans from Marcomir on)—even if he did point out, in an uncharacteristically critical moment, that it was absurd for everyone in Europe to boast of Trojan ancestry, as if there weren't a good many older families in Europe and as if the Trojans hadn't included some rascals.[58] In history as in the economy, bad currency drives out the good.

Seven Types of Assiduity:
Readers, Rules, and Annius in the Mid-Century

More than seventy years ago, Friedrich von Bezold called attention in a brilliant essay to the great vitality and interest of mid-sixteenth-century historical thought. As he saw, intellectuals of very different origins and types—from the Spanish Dominican Melchior Cano to the irenic lawyer François Baudouin—all confronted the same set of theoretical and practical problems. All had to find guidance for churches split on points of dogma, kingdoms split along multiple social and religious fault lines, and families divided by both religious and political questions. And all agreed that the authoritative canon of ancient texts, biblical and classical, should provide the remedies needed to heal the fissures in church and state and quell the European trend toward religious and civil war. Reading was urgent; but reading unguided by rules led only to chaos, as the Reformation clearly showed. Accordingly, the mid-century saw a massive effort to rethink and regulate the reading of the ancients—particularly the historians, those preeminent guides for practical action in the present. Which sources are which? This simple question burned and stimulated for two decades.[59]

Some modern scholars have made even larger claims. They have taken the mid-century theoretical writers—especially Cano, Baudouin, and Bodin—as doing more than raising questions—as formulating a modern set of rules for weighing sources.[60] But they have not in general examined these in detail in the light of the Annian rules that their authors knew, and they have abstracted the texts on historical method from the wider body of sixteenth-century scholarly literature on related points, as if visionaries like Guillaume Postel and chronologers like Johann Funck did not attack the same problems, respond to the same Annian stimuli—and profoundly influence the theorists. A broad look at mid-century scholars' use of Annius will enable us, I think, to refine and moderate some of the claims that have been made on behalf of individuals or about the modernity of the movement they supposedly made up.

We can begin with Postel, that strange man, half visionary and half philologist, who started out in religious life in the early Jesuit order and wound up honorably confined as a madman in a French convent. A real scholar, a man who knew Greek well enough to compile a pioneering study of Athenian institutions and knew Hebrew and other Eastern languages better than almost any other European of his time, Postel had prejudices even more overpowering than his erudition.[61] He saw classical Greek and Roman culture as a perversion of an earlier, Near Eastern revelation, best entrusted in his own day to the virtuous Gauls; he condemned Romulus as a descendant of Ham who had tried to extirpate the virtuous laws and customs established in Italy by Noah, a.k.a. Janus. He knew that some doubted the authenticity of Berosus and the rest, but he maintained the positive stoutly, accepting the texts and Metasthenic rules as givens: "Though Berosus the Chaldean is preserved in fragments, and is disliked by atheists or enemies of Moses, he is approved of by innumerable men and authors expert in every language and field of learning. Hence I grant him the faith deserved by any accurate author."[62] At the other end of the spectrum we find Baudouin, writing in 1560, expressing his surprise that so many of his contemporaries had accepted as genuine the "farrago" of Berosus, with its many obvious falsehoods.[63] On the one hand unquestioning faith and reverence, on the other the disgust of a gardener confronted by a poisonous spider; neither position, as one would expect, rests on elaborate argument.

Neither man, of course, can be simply taken as "modern." Postel, for all his literalist insistence that Berosus's closeness to Genesis was his great virtue, also made an elegant historiographical point: Berosus sometimes told stories that redounded to the discredit of the Chaldeans, and a witness testifying against his own interest deserves belief.[64] Baudouin, by contrast, enjoys great credit now as a theorist of source-criticism. Yet his modern-sounding argument that while all historians tell lies and make mistakes, all histories are not therefore fabulous, was in fact a quotation from an ancient forgery—the *Scriptores historiae Augustae*.[65] Baudouin took it from the bravura dialogue at the beginning of the life of Aurelian by "Flavius Vopiscus" (*Divus Aurelianus* 2)—one of the several reflections on *fides historica*, good sources and archival documents that adorn the *Scriptores*. In fact, the "rogue scholar" who forged these texts, with their alluring references to what could be found in "Bookcase 6 of the Ulpian Library," may well have taught Baudouin the principle that a good historian relies on original documents—something that "Vopiscus" claimed he had systematically done (*Scriptores Historiae Augustae, Divus Aurelianus* 1; *Tacitus* 8.1; *Probus* 2). Still, on Annius at least Postel and Baudouin took uncompromisingly opposed positions.

Between the extremes, positions grow even more complex, and the supporting arguments—or at least the supporting attitudes—more subtle. On the side of credulity we find a writer like John Caius of Cambridge—a skilled Hellenist, like many sixteenth-century medical writers, and one with a sharp interest in questions about lost and inauthentic medical works from the ancient world. In the 1560s he became embroiled in a dispute with Thomas Caius of Oxford about the antiquity of the two universities.[66] Trying to prove the antiquity of learning in England, he cited Berosus copiously about the giants Sarron, Druys, et al., who founded public institutions of learning in England and Gaul around the year 1829 after the Creation, a bit more than 150 years after the Flood. Yet for all his apparent belief in the learned Sarronidae and Berosus "antiquae memoriae scriptor," he took care to indicate that the giants had not founded Cambridge—that came later—and, more important, that the giants had been so called not because they were huge but because they were aborigines, *gēgeneis*.[67] True, one or two of them, like Polyphemus and Gogmagog, had reached great heights, but on the whole "giants, like modern men, came in a variety of sizes," even if nature brought forth stronger and bigger offspring in those purer days. By confining his use of Berosus to this very early period, by rationalizing away some of his more bizarre ideas, and by faith Caius could avoid applying to the myths that supported his own position the cutting-edge philological *Kritik* he applied to Oxford myths about the academic beneficence of Good King Alfred. And a similar attitude—of distrust mingled with unwillingness to give up such rich material—can be found in others, like the historian Sleidanus, the historical theorist Chytraeus, and—perhaps—Caius's younger Oxford contemporary Henry Savile.[68]

On the side of criticism we find a number of writers—like the theologian Cano, the Portugese scholar Gasper Barreiros, and the

Florentine antiquary Vincenzo Borghini—piling up evidence to prove the falsity of the Annian texts. They rapidly found in his richest ancient sources ample evidence of his mistakes. Berosus, in Josephus, explicitly denied the Greek story that Semiramis had converted Babylon from a small town to a great city; the Berosus in Annius affirmed it. Josephus's Berosus wrote three books, Annius's five.[69] And in any event Josephus's Berosus knew only about events before his own time, while Annius's Berosus mentioned the founding of Lugdunum, which took place two hundred years after his death.[70] These critics, moreover, did not confine themselves to pointing out blunders of organization and detail. They also showed that Berosus wrote the wrong kind of history for his age and place. The Greeks of his time, after all, knew nothing about western lands like Spain; how could Berosus, still farther east than they, know more?[71] And as to the "annals" of the Greeks and Romans, Cano pointed out in a brilliant historiographical essay that none existed. Josephus, Annius's main source, denied that the Greeks had had designated public historians. And Livy, the main source for early Roman history, showed by his infrequent citation of public records and his many errors and hesitations that "there were no public annals in the libraries and temples of the gods." Cano's conclusion was lapidary and remorseless: "They who say that the Greek and Roman monarchies had public annals against which other histories must be checked say nothing ... For it has been shown that no Greek or Roman public annals existed. Therefore there were no authors who described deeds or times in accordance with those Greek and Roman annals."[72] Here the limits of Annius's own historical imagination told against him. A more refined notion of the practice of classical historians revealed that they were rarely if ever "public recorders of events."

Still more complex were the reactions of the Wittenberg chronologer Johann Funck. A student of Philipp Melanchthon and a friend of Andreas Osiander, Funck attacked the records of the ancient world with both philological and scientific tools. These soon enabled him to chip away the authority of one of the deadliest Annian writers, Metasthenes, who covered the centuries after the Babylonian exile of the Jews for which neither the Bible nor any pagan author offered a full, coherent, and acceptable narrative.

Like Copernicus—and some earlier Byzantine writers—Funck set out to use the data preserved by Ptolemy, the great ancient astronomer. Like them, he wrongly identified Salmanassar, a king of Assyria mentioned in the Bible, with Nabonassar, the king of Babylon from whose accession on 26 February 747 B.C. the Babylonian astronomical records used by Ptolemy began. Unlike them, he systematically teased out the implications of astronomy for history. He identified the biblical Nabuchodonosor (incorrectly by modern standards) with the king Nabopolassar mentioned by Ptolemy. He pointed out that Ptolemy fixed the beginning of Nabopolassar's reign absolutely, since he dated a lunar eclipse to "the fifth year of Nabopolassar, which is the 127th year from Nabonassar (= 21/2 April 621 B.C.)" (*Almagest* 5.14, tr. Toomer). He found a different epoch date for Nabuchodonosor in Metasthenes. And he concluded that Metasthenes—or the archives he had used—must be rejected: "Do not let his authority stand in your way. Rather examine how far he stands in agreement with Holy Scripture and Ptolemy's absolutely certain observations of times. That way, even if you do not manage to reach the absolute truth you may approach it as closely as is possible."[73] Having examined a full range of texts, he also decided that ancient historians could lead where astronomical records gave out, so long as they were critically chosen: Herodotus and Eusebius, not Ctesias and Metasthenes, should be preferred.[74] Funck pioneered the way along what remains the only path to absolute dates in ancient history. Though he, like the reader he addresses, did not reach the absolute truth, his footing was remarkably sure. Yet Funck found no stimulus in his examination of Metasthenes to raise wider questions about Annius's writers or their achives. Where the early pages of Luther's *Supputatio* offered white spaces, Funck's swarmed with the deeds of the giants and the first seven Homers, all derived from Annian sources. He considered Berosus "the most approved history of the Babylonians" and copied him out joyfully, invention by invention.[75] Thus technical methods of a strikingly modern kind could coexist with an equally striking credulity.

Bodin, whose *Method for the Easy Comprehension of History* of 1566 has proved a textual Greenland that has killed off interpreters for centuries, struggled mightily with Annius's texts and Funck's ideas. He knew enough to add guarded references to the possible falsity of Berosus's and Manetho's fragments in his bibliography of historians—but not enough to do the same for Metasthenes or ps.-Philo (or, indeed, for Dictys and Dares).[76] He quoted Metasthenes' advice about choosing historians without a word of caution, and praised Metasthenes as a historian who used archival sources and wrote about a people not his own (about which he could be objective).[77] And when it came to the problems Funck raised, he showed a shattering lack of perceptiveness. Berosus and Metasthenes disagreed with "the rule of celestial motions" not because they made mistakes or used bad sources, he argued, but because they had not recorded the years and months of interregna. If only they had done so, like that "scriptor diligens" Ctesias, all discrepancies would drop away and all good sources hang together in one great Happy Historical Family.[78] If Bodin's willingness to accept pagan attacks on Christianity as the product of milieu and education rather than moral debility marks him out as an unusually perceptive reader, his use of Metasthenes sets narrow limits on his critical faculties and reveals that Annius helped to inspire—and even to shape—his notion of critical method. And even his insistence that the accuracy of historians be judged case by case, not assessed for all time by a single verdict cast in stone—his belief that Dionysius of Halicarnassus, for example, described Roman foreigners more objectively than his fellow Greeks, and therefore should be read in different ways at different points—even this is no more than a development of Annius's argument that a single historian could be accepted for one kingdom and still rejected from the reliable sources for another. Bodin's rich tapestry of methodological admonitions reveals many gaudy Annian splotches when held up to the light. Despite his comprehensive curiosity and psychological insight, Bodin's limits are more striking than his strengths—especially when he is compared to the forgotten Johann Funck, whose work he knew so well.

The most complex—and one of the most influential—of all the mid-century readers was Joannes Goropius Becanus, the Flemish doctor whose *Origines Antwerpianae* of 1569 mounted the shrewdest attack of all on Annius, and drew in doing so on much of the literature we have surveyed. To refute the forgeries he collected, in Greek, as many as possible of the fragments of the real authors Annius had travestied and as many collateral testimonia as he could find. Some of his finds were conventional—and perhaps derivative—like his use of Josephus to show that the real Berosus did not think that Semiramis made Babylon great.[79] Others, however, showed far more penetration. Attacking Archilochus—from whom his predecessors had discreetly withheld their fire—he showed that no one ascribed a work on chronology to him. Finding in Tatian the original of the passage on the eight Homers that had inspired Annius to create ps.-Archilochus, Goropius printed it, showed that Eusebius must have quoted it in the form that Tatian gave, and argued that the Latin text of Jerome's translation of Eusebius was corrupt and must be corrected or filled out "not as our antiquity-hawker wished, but by reference to what Tatian recounts."[80] The original reference had been not to Archilochus's theory that Homer lived in his own time but to someone else's theory that Homer and Archilochus were contemporaries. By diligent search in Clement of Alexandria, Goropius even managed to give that someone a name, Theopompus.[81] Goropius, in short, found inspiration in Annius not to advance theories but to collect fragments and elucidate them. The *Origines Antwerpianae* are the distant ancestor of *Die Fragmente der griechischen Historiker*.

Yet Goropius had more in mind than negative criticism and technical philology. He had his own new history of the ancient world to advance—one in which the Dutch were the remnant of the antediluvian peoples and their language, with its many monosyllables, was the primal speech of Adam. To prove this he offered evidence of many kinds—notably the famous experiment of King Psammetichus, who locked up two children, did not let them learn any words, and found that they spontaneously asked for *"Bekos,"* the Phrygian word for bread—thereby identifying the Phrygians rather than the Egyptians as the primeval race (Hdt. 2.2). This showed, Goropius reasonably argued, that the Dutch were the oldest; after all, "they call the man who makes bread a *Becker*. That king's ancient experiment shows that the language of the inhabitants of Antwerp must be considered the oldest, and therefore the noblest."[82] This revision of world history—which, as even Goropius admitted, rested on novel readings of the sources—was closely related to Goropius's attack on Annius. An essential element of his history of the migrations lay in the denial that Noah and his fellows had been giants; and thus prejudice as well as precision inspired Goropius's sedulous work as collector and exegete.

Enough has been said to make several points clear. The mid-century certainly saw a concerted effort to reshape the history of the world and to rethink the sources it should be derived from. But this effort took place as much in the tedious and technical pages of chronologies—and the terrifying and bizarre ones of historical fantasies—as it did in those of writers on the uses of human historians. No single writer, no single genre held a monopoly on the relevant forms of criticism; fantasts on some points were the grimmest and most exacting of realists on others. And twenty years of ardent speculation, most of it provoked by Annius, left his forged texts and his tarted-up ancient rules firmly in command of large parts of the historical field as most scholars viewed it. Even those who attacked him most ardently often did so in a partial and half-hearted way; even those who accepted some of his forgeries did splendidly at unmasking others.

Goez is triumphantly right to point to the pervasive stimulus Annius afforded, but wrong to overemphasize Annius's isolation and originality. And any effort to ground in the thought of the mid-sixteenth century the rise of a new and operational method of source-criticism risks committing what has well been called a "hagiographical anachronism"—the fallacy of attributing to the original and learned of the past ideas and methods consistent with what we now believe in.[83] Baudouin and Bodin, Postel and Goropius are thinkers individual and original enough to need no *ex post facto* rescue operations designed to prove that they were modern as well.

MEANWHILE, back in Leiden, how did Joseph Scaliger manage not to reject the real Berosus as he had the false one? None of the writers we have examined could have taught him to accept as somehow generally reliable a text much of whose factual content was false. Whence came enlightenment?

In part—insofar as it reflected Scaliger's unique ability to recognize that the Greek of Berosus and Manetho, while peculiar, was not that of a later forger—it derived from his unique familiarity with the other preserved fragments of their and similar texts, which he had studied in his chronological works since the 1580s. In the second edition of his *De emendatione temporum* Scaliger included an appendix of fragmentary Greek historical texts preserved by Josephus and Eusebius, and strongly defended their general veracity against both Catholic and Protestant critics. He argued that they generally matched the Old Testament and the data of astronomy very well. To that extent his whole career prepared him to confront the materials in Syncellus with expertise and sympathy.

In the *Thesaurus*, however, Scaliger offered not just ordinary sympathy but powerful support, and in this case for accounts that certainly did not match—that even contradicted—the Bible. Whence came his willingness to defend what would surely have seemed as indefensible to most scholars of his day as it did to Casaubon, who dismissed Manetho's dynasty lists as inventions? The answer is clear and definite, though unexpected: it came from nearby Friesland. There earlier sixteenth-century intellectuals had developed a model *Urgeschichte* of the province. They argued that three Indian gentlemen, Friso, Saxo, and Bruno, had left their native country in the fourth century B.C. They studied with Plato, fought for Philip and Alexander of Macedon, and then settled in Frisia, where they drove off the aboriginal giants and founded Groningen. The image is enchanting: three gentlemen in frock coats sitting around a peat fire, murmuring politely in Sanskrit.[84] But around 1600 it enflamed the temper of a critical humanist whom Scaliger esteemed, Ubbo Emmius. He denounced Friso and his friends as fables and the sources they came from as spurious.[85] And Suffridus Petri, who had given the Frisian tales currency in elegant Latin, mounted a brilliant defense. He claimed that ancient texts now lost and popular songs like the *carmina* of the early Romans and Germans, long familiar from Livy and Tacitus, could have preserved the origins of Frisia even if formal historians did not. And he insisted that even if such popular sources contained fables, they should be analyzed not scarified: "A good historian should not simply abandon the antiquities because of the fables, but should cleanse the fables for the sake of the antiquities."[86] Oral tradition, in short, needed critical reworking, not contempt.[87]

Scaliger knew these debates because Leiden friends of his like Janus Dousa plunged into them, trying to purify Holland of its origin myths as Scaliger tried to purify Egypt and Babylon. What is remarkable, again, is his reaction. He praised Emmius but imitated Petri. The tolerant and eclectic attitude Petri recommended for Friso informed Scaliger's approach to Berosus and Manetho. When Scaliger published the Babylonian *Urgeschichte* and defended it, urging that it deserved at least the *reverentia* that Livy had shown for ancient stories and arguing that it was a mythical transfiguration of real events, he used a forger's and a fantast's tools to integrate the real ancient Near East into the Western tradition. Even if the forger was Petri rather than Annius, he too was a forger who gave philology new intellectual worlds to conquer.

Forgery and philology fell and rose together, in the Renaissance as in Hellenistic Alexandria; sometimes the forgers were the first to create or restate elegant critical methods, sometimes the philologists beat them to it. But in either event one conclusion emerges. The rediscovery of the classical tradition in the Renaissance was as much an act of imagination as of criticism, as much an invention as a rediscovery; yet many of the instruments by which it was carried out were themselves classical products rediscovered by the humanists. Paradox, contradiction, and confusion hold illimitable dominion over all; we wanted the humanists to give us a ticket for Birmingham but they have sent us on to Crewe. The only consolation is to sit back, relax, and enjoy the leather upholstery and gaslights that made old-fashioned journeys so much more pleasant than modern ones.[88]

4

Scaliger's Chronology: Philology, Astronomy, World History

THE SUBJECT of time could make tempers flare in the sixteenth century. Philipp Melanchthon, for example, was a man of equable disposition, but even he could produce some memorable invective when time was discussed in a sufficiently provocative way. "Once," he recalled,

> I had a debate in hall at dinner time with a doctor who began to criticize the study of mathematics. Since he was sitting next to me, I enquired if it was not necessary [to know] the divisions of the year. He replied that it was not all that necessary, since his peasants knew perfectly well when it was day, when it was night, when it was winter, when it was summer, and when it was noon without any knowledge of that sort. I said in reply: "That answer is clearly unworthy of a doctor." What a fine doctor he is, that uneducated fool; one should shit a turd into his doctor's hat and put it back on his head. What madness! It is one of God's great gifts that everyone can have the weekday letters on his wall.[1]

True, hardly anyone would have agreed with Melanchton's opponent. As the astronomer Erasmus Reinhold asked, "What obscurity would there be in the past had there been no distinction of times? What chaos would there be in our present life if the sequence of years were unknown?"[2] Clearly such questions were purely rhetorical. Chronology was essential to civilized life. As one of the eyes of history (geography being the other) it gave order and coherence to man's past. It offered essential help to the theologian reading his Bible, the doctor reading his Galen, and the naturalist reading his Pliny. "Once you have grasped what needs to be known about the year, the months, and the days," Pierre Haguelon boasted to the readers of his *Trilingual Calendar* of 1557,

> you will find it far easier to understand the points on which lawyers consult Hippocrates. You will find it far easier to understand Aristotle's view on when the salpa, the sar, the ray, and the angelfish give birth. You will find it far easier to understand the passage [*Historia animalium* 8.12, 597a]—not to mention others—where Aristotle says that quail migrate in [the month of] Boedromion, but cranes in [the month of] Maemacterion.[3]

Who could ask for anything more? And anyone who did could be silenced by the even more authoritative testimony of the early church. "Those whose chronology is confused," so ran Tatian's memorable phrase, "cannot give a true account of history."[4] No wonder then that the field seemed important and proved attractive throughout the Renaissance. "No Frankfurt fair goes by," Joseph Scaliger justly complained, "without its crop of chronologers."[5]

But the widespread agreement about the ends and merits of chronology was not accompanied by a similar agreement about its methods and results. Every aspect of the field excited controversy. Should the chronologer frame careful definitions of time and eternity or simply apply the normal units of measurement to specific problems? Should he cast his results in eloquent prose or austere tables? Should he draw on the Bible alone or on the classics as well? Every imaginable position on these questions had adherents and opponents. And even sharper arguments attended the problems of historical detail that the chronologer was expected to solve—for example, the determination of the number, names, and order of the kings of ancient Israel, a problem that, as Scaliger admitted, no one of sound mind could hope to solve in a fully satisfactory way.[6]

It was only natural that chronology created more problems than it solved. Time itself is a notoriously elusive entity, one more likely to grasp us than we are to grasp it. Censorinus, writing in the third century A.D., made some of the basic problems vividly clear:

> *Aevum* [time as eternity] . . . is without limits, without beginning and without end. It has always been the same and always will be, and belongs no more to one man than to another. It is divided into three periods, the past, the present, and the future. The past has no beginning and the future has no end. As to the present, which is situated between the other two, it is so narrow and evanescent that one cannot attribute any duration to it, and seems to be merely the juncture between the past and the future. It is so unstable that it never stays the same. And it plucks everything that it runs across from the future and attaches it to the past.[7]

The rich Tarot deck of images with which humanists and artists evoked the qualities of time gave vivid expression to Censorinus's paradoxes. Time was at once fleeting—"post occasio calva"—and destructive—"tempus edax rerum"; at once the unique, passing instant—"hora ruit, tempus fluit"—and the ever-recurring cycle—"Atque in se sua per vestigia volvitur annus."[8] And fixing the connections between time's regular cycles and the human past—the earliest stages of which were obscured by layers of myth—seemed as hard in the Renaissance as it had to Varro himself.[9] Whatever else they might disagree on, most chronologers found common ground in complaining about the difficulty of their enterprise: "You will find it easier to make the wolf agree with the lamb," Iacobus Curio lamented in 1557, "than to make all chronologers agree about the age of the world."[10] Yet time's difficulties only served to elicit greater and greater struggles for mastery from chronologers—just as they evoked greater and greater efforts at transcendence from Renaissance poets.

No intellectual tried harder to understand and correct time's behavior than Joseph Scaliger. His *Opus novum de emendatione temporum* of 1583 transformed the study of time almost as sharply as its title claimed. So much is well known; but much less is known about the larger background and immediate context from which Scaliger's work sprang. And it is these, rather than Scaliger's work *per se*, which I shall discuss here.

Reconstructing a discipline as it was just before some original system of science or scholarship transformed it is a demanding task. It forces one to read bizarre and obsolete books. It requires one to see the sense in what now seem misguided answers to outlandish questions. And it leads one into opposite but equally dangerous temptations—on the one hand, to make one's hero seem more creative than he really was; on the other, to make him seem no greater than his predecessors and contemporaries.

In Scaliger's case the former temptation is the more obvious and attractive. Most scholars have treated chronology as Scaliger's own invention. Before he wrote, so Mark Pattison declared in characteristically memorable phrases, "the utmost extent of chronological skill which historians had possessed or dreamed of had been to arrange past facts in a tabular series as an aid to memory." Scaliger came to the field fresh from his brilliant 1579 edition of the Roman astrological poet Manilius. Preparing this had made Scaliger an expert on ancient astronomy and chronology. He was thus able to establish single-handed the nature of all ancient calendars and dates of all celebrated events. The *De emendatione temporum* led its readers along what Pattison's master Jacob Bernays called a "neu gebrochenen Bahn für die Wissenschaft."[11]

This lucid account can be disposed of without difficulty. Scaliger could not come to chronology through Manilius, for two good reasons at least. In the first place, Manilius says nothing about the subject and explicating him requires no knowledge of it. In the second place, and far more important, Scaliger had already begun to study the rich body of existing chronological literature long before 1579.

Two strands of work on chronological problems had been spun out by scholars long before the *De emendatione temporum* was born or thought of. On the one hand, editors of classical texts and antiquarians tried to explicate the ancients' references to dates and calendars. On the other hand, more specialized students of chronology—including astronomers, geographers, calendar reformers, and orientalists—compiled systematic treatises on ancient and modern calendars and epochs. The two traditions overlapped and sometimes became tangled together. When systematists treated historical or linguistic points they drew on the antiquarians' work; and the antiquarians found astronomical data in the systematists. But on the whole the differences are more striking than the similarities. I shall try to keep the two traditions distinct by referring to the former as "humanistic" and the latter as "systematic" chronology.

Scaliger's earliest notes on chronology, previously unknown, predate his work on Manilius by several years. They clearly belong to the humanist style in chronology. The *De emendatione temporum*, published some years after the notes were compiled, belongs as clearly to the systematic style. I shall dip a spoon into both dishes, the small hors-d'oeuvre of the notes and the vast mannerist wedding cake of the *De emendatione*. By analyzing the materials I remove, I shall try to show that Scaliger's concoctions included more traditional ingredients and followed more traditional recipes than previous students have realized. And by making that point I shall hope to use Scaliger's work as a means of entry into a curious and largely forgotten scene: the world of chronological work in the century or so before 1583. At the same time, however, I shall try not to let the newly discovered merits of the forgotten masters Scaliger learned from obscure the novelties of his thought. I hope to revise Bernays and Pattison, not to make an equal and opposite mistake.

Scaliger came to chronology as he would later come to astronomy, as a humanist who wanted to edit a classical text that dealt with the topic. In August 1573 he set out to produce a collection of Roman grammatical works: the *Noctes Atticae* of Aulus Gellius, the *Saturnalia* of Macrobius, and the *De die natali* of Censorinus.[12] To this end he read Censorinus in the 1568 edition of his friend Elie Vinet, the great student of Ausonius from whom Scaliger stole some good conjectural emendations in that author.[13] Scaliger found Censorinus puzzling, for good reason. The text is so concise as to be cryptic even in its full form. And Vinet had chosen as the base text for his edition Badius Ascensius's text of 1524, which descends from Filippo Beroaldo's inept and lacunose *editio princeps* of 1497, rather than the fuller text offered by the Aldine of 1528, which derives from Calco's edition of 1503. In Vinet's lucid but elementary commentary Scaliger found few solutions to the problems the text posed. So he borrowed a manuscript of Censorinus from François Pithou. He kept this for several years, despite what seem to have been continuing pleas for its return.[14] He collated it with his printed text, filling the worst of the lacunae (in some cases he could have done as well by collating the Aldine edition). He solved one textual problem by a brilliant conjecture.[15] And he jotted down, on a loose sheet, notes on a scattering of the most pressing problems that caught his eye in Vinet's text and annotations.[16] His work on these shows him at the very beginning of his interest in chronology.

The first of Scaliger's notes seems unpromising. Censorinus remarks (22.11) that Varro had proposed to derive the name *Aprilis* from the verb *aperio* ("open") rather than from the name Aphrodite. After all, April falls in spring, when nature and living things open up. Scaliger considered Varro's etymologies wrong-headed and chauvinistic—too concerned with finding Roman origins for Roman words.[17] Of this one he remarks:

> I do not see how the month April can be derived from *aperio*. First of all, since the year had only 10 months at first, the months must always have wandered [through the seasons] and had no fixed positions in the year. Thus April would fall only once in the spring, and would come around to it again after 12 Romulean years—that is, 10 lunar ones. Then *Aprilis* cannot be derived from a verb. For when words that terminate in this way are derived from a verb they shorten their penultimate syllable—for example *probabilis*, *utibilis* or *utilis*, *facibilis* or *facilis*. But if they come from a noun they lengthen it: *Equus*, *equilis*; *ovis*, *ovilis*; *caper*, *caprilis*; *aper*, *aprilis*. There are a few exceptions: *Humilis*, *similis*, etc. Therefore *Aprilis* comes from *aper*, that is, pig. As the Athenians had *elaphēboliōn*, that is, the month of the deer, so April is *kaprēboliōn* or *kaproboliōn* [the month of the boar]).[18]

This note smacks more of lexicography than of chronology, and it is typical in precisely that respect. For a great deal of humanist chronology concentrated on providing information about terminology. Vinet's commentary on Censorinus, for example, shows as much concern for spelling as for science. Where Censorinus mentions the *annus Metonticus* (the "great year of Meton," or 19-year luni-solar cycle), Vinet remarks: "This *Metōn*, an astronomer and contemporary of Socrates, takes the genitive *Metonos* in Suidas, Plutarch, and Ptolemy, *Almagest* 3, and elsewhere. Hence the word should be *Metonicus*, not *Metonticus*."[19] Haguelon's *Trilingual Cal-*

endar promised to reveal the secrets of all the ancient systems in one lively dialogue. But here is what its protagonists, Alphesta ("the entrepreneur") and Cyanophrys ("the beetle-browed"), found to say about an essential feature of the Roman month:

> ALPH: According to Macrobius and Polydore Virgil Calends is derived from *kalō* [to call]. But Theodore Gaza wrote that according to Plutarch the Calends are so called from the preposition *clam*.
>
> CYAN: Ha ha he.
>
> ALPH: What are you laughing at? Do you take me for a laughingstock?
>
> CYAN: Perish the thought. But when you tell me that a Latin wanted to derive this word Calends from Greek, while a Greek wanted to derive an etymology for it from Latin, you make the stupidity of certain men so very vivid that even Crassus, who never laughed, or Heraclitus, who was overcome with melancholy, would die laughing.[20]

The grisly scholastic humor aside, Haguelon's approach shows a clear family resemblance to Scaliger's. Both apply the method for dealing with technical subjects that had satisfied grammarians since ancient times: define the terms, argue about etymologies, and shut off discussion before some pertinacious student or reader asks how to apply a given tool or concept in practice. No wonder, then, that humanist chronologers from Cyriac of Ancona on had found the conflicting ancient etymologies of *Aprilis* that month's most interesting attributes.[21]

Yet the analogy needs qualification. Scaliger's argument does have a quantitative and analytical component. He points out that Romulus's year had only ten months (it started, as we will see, in March). Even if April fell in spring at the beginning of one year, it must occur next not in spring but in the winter with which the next Romulean year will begin, ten months later.[22] Why then call it the month when everything opens? In a second and related note Scaliger goes more deeply into the secrets of the first Roman calendar:

The sixth year of Romulus is intercalary, so that it will match the lunar year. That was why the Roman *lustrum* [5-year purification cycle] was founded. And 6 Romulean years take up 5 lunar years, just as an intercalation took place in Greece during the Olympic contest. For that was why it had been founded.[23]

In both passages Scaliger tries to deal with the most vexed and intractable of all problems of early Roman chronology. Censorinus tells us that as early as the second century B.C. the Romans could not agree about the nature of Romulus's year:

> Licinius Macer and Fenestella after him wrote that the original Roman year was [a natural one] of 12 months; but it is better to trust Junius Gracch[an]us, Fulvius, Varro, Suetonius, and others who have worked out that it was 10 months long, as was the year of the Albans from whom the Romans sprang. They had 10 months, containing 304 days. (20.2-3)

This calendar supposedly lasted until Numa replaced it with a modified lunar year. Scaliger, like Censorinus, accepts the majority opinion. But he also tries, as Censorinus did not, to work out its technical implications. It would obviously be difficult to keep a year of 304 days synchronized with the seasons. Scaliger suggests that Servius Tullius had introduced the 5-year lustrum cycle in order to keep the Romulean civil year aligned with the lunar year, and thus with the seasons as well.

What Scaliger conceals here is even more important than what he tells us. Modern scholars had been trying for a century to explicate the puzzling 10-month year, and Scaliger's effort to do so drew on their competing arguments. Those who accepted Livy's vivid portrait of early Rome as a virtuous but primitive society also tended to accept Romulus's 10-month year. It was the sort of mistake the early Romans could legitimately have made, especially since, as Ovid said, they were better at fighting than at science (*Fasti* 1.29). Juan Gines de Sepulveda wrote that

> the Romans set out from the very foundation of the city to follow the natural and convenient method [of making the year's length equal to the period of the sun's course around the earth]. But in their ignorance of the heavenly motions they could not imme-

diately attain what they sought. For Romulus set the length of the year at 10 months only . . .²⁴

Similar views appear in L. G. Giraldi's standard survey of ancient calendars, in Onofrio Panvinio's great commentary on the Roman *Fasti*, and in Vinet's commentary on Censorinus.²⁵ But many saw ancient culture in devolutionary rather than evolutionary terms. They held that astronomy and the calendar had been revealed in all their complexities to the virtuous Jewish patriarchs, Egyptian sages, and Gallic Druids. They found the notion of gross error in so early a calendar as Romulus's unacceptable. And in trying to leap the gap that separated their convictions from the apparent testimony of the sources they performed some remarkable mental acrobatics. The debate was long and complex, and evoked interest from systematic as well as from humanistic chronologers.

At one extreme Guillaume Postel, whom Scaliger knew, insisted that Romulus had deliberately invented a faulty calendar. Doing so formed an integral part of his campaign to extirpate the memory of Noah, known as Janus in Italy, where he had established a godly and virtuous society.

> Pretending that he wished to establish a beginning for the year in Mars's honor, he stole January from the head of the year. He thus destroyed the memory of Janus, which was connected with the first degree of the sign of the sun returning to us. And thus that great rascal and tyrant completely overthrew the laws of time, by creating his new and impossible 10-month year. Hence it is his fault that the beginnings of the signs and the beginnings of the months, which by Janus's system always went together, are now 21 days apart.²⁶

By contrast Ioannes Goropius Becanus, whom Scaliger read, held that astronomy had passed without loss or interruption from Saturn to the early inhabitants of Italy. This was clear from the famous statue of Janus dedicated by Numa, which represented the god as counting to 365 (or 355) on his fingers (Pliny *NH* 34.33; Macrobius *Sat.* 1.9.10; *Suda* s.v. *Ianouarios*). Hence Romulus could not possibly have done what Postel claimed. His own followers would have prevented it:

It is extraordinary that some have made the grave error of writing that there were only 10 months in Romulus's day. Perhaps, as they say, Romulus was a man of barbarous and soldierly cast of mind. He was after all the offspring of Mars, that enemy of all right reason and culture. Yet it cannot be rightly inferred that there was no one else at the time who knew better and could reckon the year's length more accurately.[27]

Clearly almost any inference could be drawn from the murky and indecisive testimony of the ancients. The year of Romulus was less a question to be settled than a Rorschach blot which called forth a scholar's preconceived vision of the larger shape of Roman—and human—history. Postel believed in the 10-month year because he considered Romulus a brute who had set out to destroy religion and civilization. Goropius Becanus did not believe in the 10-month year because he thought more highly than Postel had of the attainments of the early Romans.

Scaliger reacted with characteristic perversity to the spectrum of opinions he confronted. He tried to make opposites coincide—to show that the early Romans could have framed a 10-month year without being ignorant of astronomy. He sought a regular astronomical cycle under the apparent chaos of the early calendar. And this suggests two conclusions at least. The first is that his grasp of the issues was shaky. No cycle of 5 or 6 years could have brought the 304-day Romulean year into line with the 354-day lunar year. Scaliger's own suggestion—that additional days or months be inserted into every 6th Romulean year—is ill-advised. If he has true lunar years in mind, it will not work:

6×304 days $= 1,824$ days

5×354 days $= 1,770$ days.

Six Romulean years, then, is a *longer* period than 5 lunar ones. Lengthening each sixth Romulean year, as Scaliger proposed, would only have made the interval between the beginnings of the Romulean and lunar years longer—and thus made matters worse. And even if he is thinking of a lunar year adapted to the solar year by the periodic intercalation of months, his solution has little to commend it. If, as would be likeliest, he assumed that the earliest Romans used the 8-year luni-solar cycle that he ascribed to the

Athenians later in this page of notes, then in the first 6-year cycle the same problem would recur, since once again

$$6 \times 304 \text{ days} = 1{,}824 \text{ days}$$
$$5 \times 354 \text{ days} = 1{,}770 \text{ days.}$$

True, in the second cycle the Romulean years would come out shorter than their lunar counterparts, since three intercalary months will be added to the eighth lunar year, and

$$5 \times 354 \text{ days} + (3 \times 30 \text{ days}) = 1{,}860 \text{ days.}$$

But even then the cycles would not produce their intended effects if added together, since

$$2 \text{ Romulean 6-year cycles} = 2 \times 1{,}824 \text{ days} = 3{,}648 \text{ days}$$
$$2 \text{ lunar 5-year cycles} = 1{,}770 + 1{,}860 \text{ days} = 3{,}630 \text{ days}$$

and no extension of the Romulean years would be needed. Had Scaliger's arithmetic been better he might have anticipated the theory later supported by B. G. Niebuhr, that great spinner of imaginative hypotheses. Niebuhr took the *lustrum* as a cycle of 5 *solar* years designed to correspond to 6 Romulean years.[28] This is elegant:

$$5 \times 365 \text{ days} = 1{,}825 \text{ days,}$$

only one day off the mark. But it fails as badly as Scaliger's theory to account for two points more significant than questions of arithmetic. No ancient source connects Romulus's year with a regular astronomical cycle; in fact, the sources say that it lacked one. Moreover, Censorinus makes clear that the *lustrum* itself was at first irregular in length, and cannot have provided a regular basis for anything (18.13–15). Whatever their other follies, Scaliger's predecessors bore these facts in mind. Those who believed in the 10-month year either admitted or asserted that it had been astronomically unsound; those who believed the Romulean year had possessed a reasonable length did not believe it had only contained 10 months. Both groups were more faithful than Scaliger to the sources.

The second conclusion is that the creation of a truly technical chronology required not the solution but the abandonment of problems like this one, as mathematically imprecise as they were historically enticing. Scaliger admitted as much in the *De emendatione*. For there he dismissed all previous treatments of the Romulean year, and the problems that had inspired them, as unworthy of the serious chronologer's attention: "Let us reject that 10-month year which is of no use for any purpose whatsoever, and let us hiss off the stage the thesis that transformed Romulus the shepherd and peasant into Meton and Callippus. And let us then take refuge in valid argumentation."[29] "Valid argumentation" turns out to be a detailed attack on the system of intercalation that regulated the Roman lunar calendar until Caesar reformed it. Scaliger quite unjustly assumed that this would provide the key to the earliest Roman calendar as well, and attacked the rich and detailed documentation concerning it with a dexterity and attention to detail that his first efforts lacked. For our purposes, it matters most that in redefining the questions in this way he rejected a whole tradition in chronological studies—and, indeed, his earlier self. We shall see soon that he lost in some ways as he gained in others by this act of renunciation.

A third note seems merely to underline the distance that separates the chronology of the humanists from the rigor and assurance of the *De emendatione temporum*:

> If the beginning of [the Athenian first month] Hecatombaeon began from the new moon immediately after the [summer] solstice, I do not see how intercalation could take place every 8th year. For this was impossible, since for that they used a 19-year cycle which leaves no room for intercalation, since that is taken care of by years with additional months. But they had perceived that in every 8th year there were 30 additional months.[30]

Here we see Scaliger enmeshed like Laocoon in the grip of authoritative but contradictory witnesses. Plato says in the *Laws*, a text Scaliger knew well, that all magistrates must assemble in one temple "on the day before the day on which the new year is about to begin with the new moon that follows the summer solstice."[31] This suggests that each new year in Athens began on a date set by calculation or observation—that of the first new moon after the summer solstice. But Macrobius says that the Greeks synchronized their lunar year of 354 days with the solar year of 365¼ days by adding three months to every 8th lunar year (*Sat.* 1.13.9–10). If he is right (he is not) Plato must be wrong. For since the lunar year is 11¼ days shorter than the solar year, if no intercalation takes place in the first 7 years of each cycle, then, as Jean Lalamant explained in 1570,

> the beginning of the Attic year—or, if you prefer, the first day of their first month, Hecatombaeon—will not always fall on a given day, or even in a given month [of the solar year]. But it will move forward from one year of the 8-year cycle to another and will occur 11 days earlier each year.[32]

And since the first new moon of the first year cannot fall more than 30 days after the summer solstice, in most years of the cycle Hecatombaeon will begin before, not after, the solstice. Only the 3 intercalated months of the 8th year will restore the beginning of the year to the place Plato calls for, and then it will move forward again in the same way. Meanwhile many other sources mentioned the 19-year cycle of Meton, which provided more accurately than the 8-year cycle could for the prediction of new moons within solar years, but which could not coexist in one working calendar with Macrobius's system.

Scaliger has at least realized that his sources contradict one another. That is an achievement in itself. Lalamant simply pretended that Plato had been referring only to the first years of each cycle; Matthaeus Beroaldus, whose *Chronicum* appeared in 1575, cited Plato and Macrobius on the Athenian year as if they complemented each other nicely.[33] Yet Scaliger can do nothing more than call attention to the difficulty. It defies solution. And this helplessness before a not very difficult historical and technical problem could not contrast more starkly with the resourcefulness Scaliger shows in the *De emendatione temporum*. The last sentence is especially revealing. Scaliger says that the Greeks had found 30 additional months in each 8th year. He is trying to summarize Macrobius, but pen or memory has failed him: the correct figure would be 3 months of 30 days.

Scaliger's difficulties with arithmetic are not just evidence that he had not yet worked through Ptolemy and Euclid with care. They are typical of the humanist chronologer. Scaliger's friend Aldo Manuzio, Jr., provides a neat parallel. In 1576 he published a detailed reconstruction of Meton's 19-year cycle. No astronomer would have found it hard to describe the cycle and its purpose briefly and crisply. Henry Savile, for example, gave his students this account in his 1570 Oxford lectures on Ptolemy:

> Meton saw that the equinoxes and solstices could be restored to their original positions more or less accurately by assuming a period of 19 lunar years, in which 235 synodic months pass. The method used was that in addition to the day added by old custom in each Olympiad, 7 months would be added to 19 lunar years, 6 of 30 days and the 7th of 29. This period the Greeks called the 19-year cycle, and the great year of Meton.[34]

In modern terms: to bring the beginnings of the lunar and solar years together periodically and to predict when in each solar year new moons will fall, we may use the following relationship:

19 solar years = 19 × 365¼ days = 6,939¾ days

19 lunar years = 235 lunations = 6,939⅔ days (approximately).

This is slightly inaccurate as an account of Meton's cycle, as Scaliger himself was to show.[35] But it was close enough for most purposes, and could be found as standard doctrine in many books.[36] Yet it was too much for Aldo. He knew that "all the Greeks, except the Arcadians and Acarnanians, reckoned the year at 354 days." But the ancient sources all agree that Meton was a Greek, and not an Arcadian or an Acarnanian. Hence, he concluded, Meton must have used the standard Greek lunar year in his cycle; and that, in turn, must have consisted of

19 × 354 days = 6,726 days.[37]

Aldo cheerfully proposed to emend on these grounds Censorinus and other ancients who mentioned the cycle—even though a cycle of 6,726 days would have left the solar and lunar years in complete confusion, the 19th lunar year ending while the 19th solar one had some 7 months to run. Yet Aldo's version of the cycle was no more confused than Vinet's. His commentary on Censorinus described the Metonic cycle as identical to the 19-year cycle of the modern computus; but his text stated that it had consisted of only 6,440 days.[38] Evidently large numbers baffled humanists.

If a balance were to be drawn now, the value of humanist chronology and its importance for Scaliger's development would seem slight. But a fourth note shows that the humanist style had virtues to offset some of its faults.

> Dio's remark that Julius Caesar learned his method of intercalation from the Egyptians is . . . quite absurd. For he took the idea from Lucan. And Plutarch and Dio took as statements of historical fact every utterance that Lucan made about the civil war, invoking poetic licence. But in fact the Egyptians did not intercalate except when a year of the dog-star [a period of 1,460 Julian years] was over, which happened only once in human memory. The Greeks intercalated every 4th year. Why could Caesar not have learned from them?[39]

The immediate stimulus for this remark lay in one of Vinet's annotations: "[Caesar] adjusted the year to the course of the sun, as he learned to do from the Egyptians in Alexandria; so Dio, 43."[40] But as Scaliger's note suggests, the Egyptian origins of Caesar's calendar reform were attested to by a host of respectable ancients. Dio says flatly that Caesar took his innovation "from his stay in Alexandria" (43.26.2). Lucan shows him discussing astronomy with the Egyptian sage Acoreus. "Even Eudoxus's calendar won't outdo my year," Caesar boasts. Good, replies Acoreus; in that case Caesar might find it helpful to know that the sun divides time up into years (10.187, 201). Appian, even more explicitly, compares Caesar to Alexander as a lover of Eastern wisdom and says that Caesar "changed the year, which was still irregular . . . to follow the course of the sun, as the Egyptians reckoned" (*BC* 2.21.154). It was only natural, then, that most Renaissance scholars agreed with Vinet and Manuzio: Caesar had imitated "the Egyptians' year."[41]

In this case Scaliger's note stands out for its independence from traditional stories and opinions. He rightly points out that the Egyptians did not intercalate fractions of days as the Julian calendar did. Rather they allowed their fixed year of 365 days to drift slowly through the seasons until its beginning came around again to a given point in the solar year—a process which took 1,460 Julian years.[42] The Egyptians knew that the solar year was about one-fourth of a day longer than their conventional one, but made no effort to take account of that fact in their working calendar. And since all calendars that do take account of the true length of the solar year are imprecise in effect and inelegant in length, it was the drifting uniform Egyptian year that proved, in Reinhold's words, "most appropriate for observations of the heavenly motions"—and that could not have provided the model for the fixed Julian year with its added intercalary day in every fourth February.[43] In arguing that "the Greeks" could have provided Caesar's model, Scaliger points to a truth that his predecessors (and some of his successors) missed. The astronomical basis of Caesar's calendar reform was banal. It could have come from any serious Greek or Babylonian astronomer, and romantic stories about Caesar's dalliance with Egyptian sages are quite irrelevant. What matters is not where he found his simple idea but how he put it into practice. In this case Scaliger's well-developed historical sense enabled him to reduce the fabled wisdom of the East to its true, modest proportions—and so to give a sound and original answer to a genuine chronological problem.

In fact, Scaliger's early answer was more judicious than the one that replaced it in his mature work. By the time he wrote the *De emendatione temporum* he had read in Diodorus Siculus (1.1.2) that the Thebans used a year of 365¼ days. He had read in Strabo (17.1.29, C806) that Eudoxus and Plato had learned this secret during their long stay in Heliopolis. And he had decided that the chronologer must find precise origins for every calendar in earlier systems of similar constitutions.[44] So he gave way, as he had not when younger, to his natural tendency to undersimplify. Following the late and footling reports of Diodorus and Strabo with touching faith, he reversed his earlier views and announced that Caesar had in fact devised his reform "after working for some days with the sages of Egypt"—even though he continued to insist that the Greeks had known the length of the solar year before Caesar's or Eudoxus's times.[45]

The balance we might strike now would not be clear or simple. Chronology of a sort evidently formed part of what one might call the normal science of humanism. It was monographic, not systematic, in character; it dealt with calendrical and chronological problems touched on in well-known texts or connected with famous men rather than with the structures and details of the main ancient calendars. It rested on wide knowledge of classical sources but often suffered from arbitrary assumptions. The humanists had little mastery of ancient astronomy and lapsed into wooliness when the arithmetic that faced them grew complex. Yet they sometimes did yeoman work at what was, after all, their profession: offering judicious interpretations of the sources. Scaliger's technical chronology is very far from his earlier humanist chronology; yet without his humanistic training some of his technical triumphs—like his arrangement of the months of the Attic year in their proper order—would have been impossible. And on some points the young humanist saw farther than the old systematist. This should remind us that a revolution in any discipline produces broken eggs as well as omelettes—and involves forgetting things of value as well as making discoveries.

I TURN to systematic chronology—to the *De emendatione temporum* and the earlier treatises that Scaliger plagiarized and pilloried with equal zest. Like many epoch-making books, the *De emendatione* does not on first encounter reveal much about its sources and formation. Above all, it gives an impression of meticulous preparation and startling originality. The first four books deal with the principal calendars, ancient and modern, solar and lunar. The fifth and sixth establish the most important dates from the Creation to rather more recent times. The seventh presents texts and translations of medieval Jewish, Ethiopian, and Byzantine treatises on the calendar, known as computuses. The eighth, which Scaliger introduces as the natural culmination of the rest, explains the bearing of Scaliger's researches on calendar reform in his own time. Throughout Scaliger takes great pains to emphasize that he has explored and mapped an unknown territory.

Enough evidence survives, both inside and outside the *De emendatione temporum*, to show that this apparently neat picture only serves to obscure a more complex reality. By ignoring the impression Scaliger's own words create, by watching him learn to do chronology, and by identifying the derivative elements in the finished compound of his book, we can establish that even Scaliger's masterpiece drew on and responded to current developments in a well-established discipline.

Scaliger's informal letters to his friends Claude Dupuy and Florent Chrestien show him working on chronology as early as 1579–80. The *De emendatione* makes its first appearance in the modest form of a "little computus" which Scaliger's usual publisher, Mamert Patisson, hesitated to print. By July 1580 Scaliger was sketching out some "petites *diatribai*" to accompany it.[46] An unpublished letter to Chrestien brings the picture into better focus. Scaliger, depressed and isolated in "Arabia Deserta," his term for the Limousin, thanks Chrestien for writing liminary verses for what he calls "computus Aethiopicus."[47] Evidently the "little computus" that Scaliger first set out to publish—and that provoked his first extended writing on chronology—was the Ethiopian text that appears with Scaliger's translation and commentary in book 7 of the *De emendatione*.[48] Scaliger's most comprehensive and systematic treatise began, paradoxically, as an edition of and commentary on a single short text.

The letter reveals still more. It actually offers a glimpse of the *De emendatione* no more than half conceived:

> You will have to make a good many more [verses]. For I have more computuses on paper than in my purse. I have all the computuses of the Christian and Muslim peoples, and what's more, I've dug deep into the earth and found the computuses of the ancient Greeks. Please remember what I'm telling you: I have changed the times, as Daniel says of Antiochus Epiphanes.[49]

Clearly Scaliger already saw himself as creating something staggeringly new and learned; hence his comparison, comic in its enormity, between his own reconstruction of lost calendars and Antiochus's deformation of the religious calendar of Israel. A coy quotation from Horace about the need to conceal "arcana" bears out this impression. At the same time, however, it is also clear that as yet Scaliger only intended to produce something on the order of books 1-4 and 7 of the *De emendatione*: a systematic treatment of past calendars. He does not mention epochs or calendar reform. Evidently the clear and symmetrical structure of the finished book, with its cross-references and logical progression, was not the one that Scaliger first set out to build. That too casts doubt on the traditional historiography.

A second document, now in Leiden, proves the slow and indirect nature of the path that Scaliger followed. In 1580 he entered a long note in the endpapers of his twelfth-century Arabic manuscript of the four gospels.[50] Here he used the colophon to the gospel of John as the springboard for a leap into chronological waters that almost proved too deep for him. The colophon reads in part:

> This holy gospel was completed thanks to our Lord and our Savior Jesus Christ on the 31st day of the month Tamuz, on Tuesday, the twelfth indiction, the year 6687 from the Creation.

Scaliger analyzed this with characteristic energy and resourcefulness:

> The scribe says that the book was written on the last day of July (for that is what the Christian Syroarabs call Tamuz) on the 3rd day of the week, in the 12th indiction. Now the Greeks and Syrians, following Eusebius and the Septuagint, set the epoch of Christ at 5500 [years] after the Creation. Our friend says that the book was written in A.M. 6687.
>
> $$6687 - 5500 = 1187.$$
>
> But in that era, the last day of July will not be the 3rd day of the week, nor will the indiction be the 12th. It therefore seems that they make the epoch of the world 8 years older than the Syrians, and hold that 5508 years had passed after the Creation when Christ was born. Thus the book in question was written in
>
> $$[6687 - 5508] = \text{A.D. } 1179.$$
>
> This era has indiction 12 and the number 12 in the Solar cycle Its Dominical letter was G. Thus B, the last [ferial] letter of July, was the third day of the week.[51]

To paraphrase the argument in more modern (and explicit) terms: the colophon date initially suggests that the scribe finished work on 31 July 1187. But if we apply the standard formula or consult a table of indictions, we find that 1187 was indiction 5, not 12.[52] Moreover, if we apply the standard formula or consult a table of Dominical letters, we find that 1187 had Dominical letter D.[53] But 31 July in the Julian calendar has ferial letter B; and in a common year with Dominical letter D, all days with letter B fall on the 6th day of the week, Friday, not on the 3rd, Tuesday.[54] The date 31 July can fall on Tuesday only in years with Dominical Letter G. Scaliger's solution is simple and elegant. If we count backward in the standard tables, we find that A.D. 1179 had indiction 12 and Dominical letter G, both of which were sought. The scribe must have finished then. Accordingly, he must have dated his book on the assumption that Christ was born

$$6,687 - 1,179 = 5,508$$

years after the Creation.

Scaliger goes on to list other years that share this combination of indiction and Dominical letter. He remarks that the combination will recur after

$$28 \times 15 = 420$$

years, and that in the year he is writing, 1580, the indiction is 8 and the solar cycle 21. Hence both cycles will simultaneously reach year 1 in 1588, after 7 more years (as both had reached 12 in 1179). He concludes by dithering a bit about when during the year his scribe takes the indiction to begin and by collecting some material about indictions from a variety of sources. Bits of this note reappear in the *De emendatione*; it may well be one of the *diatribai* that Scaliger mentioned to Dupuy in 1580, part of his original commentary on the *computus Aethiopicus*.[55]

The note seems most impressive. It shows Scaliger accurately translating from Arabic, dexterously solving a technical problem, and beginning to investigate recurrent patterns in calendrical cycles—the very line of inquiry that would soon lead him to devise the Julian Period, his most enduring contribution to technical chronology. Yet these appearances deceive. On closer examination, the note reveals how little Scaliger yet knew about basic calendrical rules and facts. To begin with, the note reveals a curious—and consistent—error in computation. The years that Scaliger lists as sharing the characteristics of 1179 in fact do not have them, as a table will show:

Year	Indiction	Dominical letter
820	13	AG
865	13	G
910	13	G
955	13	G
1000	13	GF
1179	12	G

All the indictions are wrong by 1. For 1000 the Dominical letter is wrong as well. In a leap year with letters GF, 31 July will behave like a day in a common year with letter F; hence it will fall not on Tuesday but on the 4th day of the week, Wednesday. More curiously still, the note ascribes to Eusebius the belief that Christ was born 5,500 years after the Creation. In fact Julius Africanus held this view, apparently for eschatological reasons, and Eusebius rejected it.[56] He built a different—and undramatic—sum, 5,198 years, into his *Chronicle*, as Renaissance editions of it made clear. True, Scaliger had an authority for his view. A fragment attributed to Eusebius in Gentian Hervet's Latin translation of the *Quaestiones et responsiones* of Anastasius of Sinai does ascribe the figure 5,500 to Eusebius.[57] But no one well acquainted with the subject would have preferred this witness to Jerome's translation of the *Chronicle*.[58] In any event, Scaliger's trial-and-error method of determining the era shows that he did not yet know that the most common era for all Eastern Christian writers was the Byzantine, which set Creation at 1 September 5509 B.C.[59] And the later section of the note, in which he labors to identify the starting-point of the indiction, is addressed to what he should have seen was a nonproblem. In any standard indiction system 31 July 1179 would fall in the 12th year of the cycle.[60]

Ironically, these criticisms of Scaliger's note redound to his credit. The errors show inexperience, not incompetence; anyone who works in chronology makes mistakes in computation and learns the basic epochs only gradually. In any event, he did solve the colophon correctly, and within a year or two would solve far harder problems with far less effort. The importance of the errors is historical. They prove that one could undertake a serious and elaborate study of ancient astronomy—as Scaliger did while editing Manilius—without mastering the literature of the computus—which Scaliger here struggled to control, more than a year after Manilius had reached the bookshops. At the same time, the note gives a strong hint about how Scaliger mastered his new field. The wording of the last phrases in the note suggests that he used a standard textbook, the *Opusculum de emendationibus temporum* ascribed to one Ioannes Lucidus Samotheus; this would have given Scaliger all the calendrical information he deploys here. And a second fraction of the note reveals another, more unexpected modern source. Scaliger writes:

> In the time of Bede the year of the
> creation was, according to the Greeks,   6276.
> The year of Christ in the times of Bede   703.
> Difference—years from Creation to Christ   5573.
> But this is corrupt. Below, however, Bede clearly says that 778 years had passed between Christ's time and his.
>
> $6276 - 778 = 5498$.
>
> This era is 2 years lower than the common Greek one, 5500.[61]

The text Scaliger struggled to correct was not in fact a genuine work of Bede's but a calendrical work from late-eighth-century France which Ioh. Noviomagus had included in his edition of Bede's *Opuscula de temporibus*.[62] Its first canon reads:

> To know the year by the Greek system, multiply
>
> $15 \times 418 = 6270$.
>
> Add the indiction of the year in question (this year it turns out to be 2). The sum is
>
> $[6270 + 2 =]6276$.[63]

Scaliger rightly stigmatizes the passage as corrupt. But his remedy is worse than the disease. A simple change from 6276 to 6272 would make the sum correct and produce a result (A.D. 764) that yields indiction 2; Scaliger's proposed solution, A.D. 778, has indiction 1. Worse still, Scaliger's solution implies that Bede could have written something in A.D. 778; but Renaissance scholars knew perfectly well that he had died around A.D. 735.[64] Yet here too criticism of Scaliger misses the main point. The note reveals that he learned standard techniques by working, slowly and painfully, through standard books. When Scaliger needed to deal with a computus, he read similar works by Bede—and perhaps set out at first to produce something rather like the corpus of computuses and technical commentaries that Noviomagus had assembled half a century before.

Once we acknowledge that big works of scholarship necessarily embrace much tralatitious matter, Scaliger's achievement seems far less mysterious than his letter to Chrestien suggests. Western calendar reformers had produced a host of full analyses of the Julian ecclesiastical calendar. Astronomers had attacked the Egyptian, Persian, Syrian, and Islamic calendars used in the classic works of the Greek and Arabic astronomers, long available in Latin. Hebrew scholars, like the Jews they studied, had long felt the need to understand the relation between the lunar and solar years and the religious laws that sometimes interfered with the normal functioning of the Jewish luni-solar calendar. A number of writers—above all the French doctor Jean Lalamant and the Basel mathematician Erasmus Oswald Schreckenfuchs—produced synthetic works that explicated the Oriental calendars and tried to do the same for those of the Greeks and Romans, even though the classical sources for these were scanty and superficial. This body of literature gave Scaliger both a model for the structure that he first set out to build and a good many of the building blocks that went into it.

Even the finished *De emendatione* owes a considerable debt to these books. One small case study will confirm and enrich the conclusions we have drawn from Scaliger's unpublished notes. In Book 7 Scaliger discussed the postponements of the Jewish New Year sometimes required by religious law. He explained that the first day of the month Tishri could not fall on the 4th or 6th weekday. If it did, then Yom Kippur—a day on which Jews could not work or cook—would fall immediately before or after the Sabbath, on the 6th or 1st weekday.

> But two Sabbaths in a row would be inconvenient in the region of Jerusalem, because of the cooked foods, which it would be dangerous to keep for two full days in hot regions, especially at that hottest time of year. Furthermore the bodies of the dead could not be kept [for two days] without incurring a risk of decay.[65]

The argument seems sensible but is wrong in several ways. It misrepresents what the rabbis actually taught. In tractate *Rosh Hashanah* of the Babylonian Talmud, the argument about food and the argument about corpses are attributed to different sources and treated as contrasting, not as complementary (20a). Moreover, the argument about food concerns *yarəqayaʾ*, ("greens"); and Rashi's commentary makes clear that the problem was that *raw* greens would wilt and become inedible in the course of two days. Of cooked foods the Talmud and the commentators make no mention. Finally, the Talmud makes clear that decomposition of food posed a problem only in Babylon—not, as Scaliger thought, in Palestine, which had a more temperate climate.

Scaliger's juxtapostion of irreconcilable arguments, his misconstrual of one of them, and even his embroidery on them can all be explained with ease—so long as we seek out his secondary source. In 1527 Sebastian Münster had published his *Kalendarium hebraicum*, a lucid exposition of the Jewish computus based on the profound work of Moses Maimonides. Münster wrote as follows about postponements:

> If the 1st day of the month fell on the 4th or 6th weekday, they would then have to observe two Sabbaths in a row, since Yom Kippur would fall on the 6th or 1st weekday and they consider it as holy as the Sabbath. This would be hard for them, so they say, because of the pot-herbs (*olera*) and the dead. For they would have to keep cooked pot-herbs until the 3rd day, and similarly the corpses of the dead, which cannot be done safely in hot regions.[66]

Scaliger took the two arguments not from the Talmud but from Münster, not checking to see whether they in fact agreed. He also drew from Münster the mistaken notion that the food that could not be kept for two days was cooked. And he simply rephrased Münster's vague reference to hot climates as a specific, rationalizing argument about the climate of Palestine. His only real addition was his transformation of Münster's precise translation, *olera*, into the more general (and less accurate) *cibos*.

Scaliger's treatment of the Jewish calendar as a whole marks a great improvement on Münster. Where Münster, like Maimonides, went deeply into the functioning of the calendar, Scaliger reconstructed its history. He argued, apparently for the first time, that the calendar of the Jewish computus was not that of the Pentateuch, and that it must rest in large part on the arithmetic and astronomy of Babylon. And he admitted that he had learned far more from Münster than from any other Western writer.[67] What he did not admit—here or elsewhere—was the extent to which the threads that he wove together so deftly had been spun by others. Once these have been traced back to the reels from which he unwound them, the whole nature of his enterprise becomes clear. At least in its first stages, as a survey of calendars, the *De emendatione* was a masterly addition to a well-established technical genre. Scaliger's clearest allusions to this fact occur in the numerous passages where he calls attention to his debt to a living practitioner of the Oriental computus tradition, the Syrian Jacobite patriarch Ignatius Na'matallah, whose two long letters to Scaliger were his basic source of information for the Oriental 12-year animal cycle, the Syrian calendar, and much else.[68]

De emendatione 5–6: The Fixing of Epochs

What then of the scholarly tradition that makes Scaliger the first chronologer to have dated historical events by astronomical methods? His command of astronomical texts and techniques certainly informs the *De emendatione* as a whole, and does much to make Books 5 and 6 impressive. Pocked with long series of numbers, bristling with tables, these formidable books include computed Julian dates for many eclipses and conjunctions, converted Julian dates for dozens of events, and many efforts to connect datable astronomical phenomena with human history. As early as the first edition, the *De emendatione* ties the beginning of Alexander's world empire to the lunar eclipse that preceded the defeat of Darius at Gaugamela—and thus substitutes for the vague traditional date, Olympiad 112, the startlingly precise one "'after 20 September 331 B.C."—a date that remains the cornerstone of later Greek history.[69] Even Scaliger's sharpest critics admitted his originality in this respect. Tommaso Campanella, for example, held that the cycles of the planets were both too irregular and too poorly recorded to serve as the basis of chronological research. But even he admitted that most of his contemporaries disagreed: "The Germans admire Scaliger's chronology, and many of our countrymen follow it . . . for he wished to correct the count of years from the eclipses and lunar cycles mentioned in the histories of older times . . . "[70] No wonder, then, that less hostile witnesses have found Scaliger so praiseworthy.

In fact, however, chronologers had seen the need to draw on astronomy long before Scaliger. Early in the fifteenth century Pierre d'Ailly tried to correct the history of the world—and to reconcile the divergent chronologies of the Hebrew and Septuagint texts of the Bible—by tying great events on earth to the datable great conjunctions of Saturn and Jupiter.[71] Later in the century the world chronicler Werner Rolevinck and the editor of Eusebius's *Chronicle*, J. L. Santritter, both proclaimed that historians needed to know some astronomy.[72] It was easier to praise astronomy than to use it. When Santritter tried to make Eusebius's figures match "the clear truth according to Alfonso, the King of Castile," he ran into severe difficulties—not least because he missed by some 1,500 years when trying to give Alfonso's date for the Flood.[73] And at best this earlier work amounted only to the imposition on historical data of numerical patterns that astronomers and astrologers had already woven.[74] Yet it did reflect a widespread interest in the applications of astronomy.

What transformed astronomy from an object of uninformed admiration into an auxiliary tool for historical research was the appearance of Copernicus's *De revolutionibus orbium coelestium* in 1543. For Copernicus revived in passing one wrong but fertile idea: that the Babylonian king Nabonassar, from whose accession on 26 February 747 B.C. Ptolemy dated his observations, could be identified with the biblical Salmanassar, the scourge of Israel.[75] Though Copernicus did not draw them all out, this synchronism seemed to have dramatic implications that the less precise ones of Eusebius lacked. It gave a biblical figure an absolute date based on a pagan science. It connected the dated astronomical events in Ptolemy to the Bible. It seemed to give some order and coherence to one of the most confusing periods in the history of ancient Israel.

In 1551 Theodore Bibliander built his chronology around one form of "this key," as he called it, "to world history."[76] In the same year Erasmus Reinhold incorporated Copernicus's view—with refinements—into what became the standard reference tool for all astronomers, the *Prutenic Tables*. By arguing the case for Copernicus's suggestion in detail, Reinhold showed what the application of Ptolemy could do to give biblical history a sturdy quantitative backbone: "As far as history is concerned, Mardocempad, whose years as king of Babylon Ptolemy reckons in 3 lunar eclipses, and whom Ptolemy sets 26 years later than Nabonassar—he can be no one but the one called Merodach both by Metasthenes and other writers and by sacred Scripture..."[77] Johann Funck, finally, included an accessible version of this argument in a work that reached the entire scholarly public. Finding the Hebrew and Assyrian kings impossible to arrange in a sensible order, he turned for help to his father-in-law and close friend, Andreas Osiander. Osiander, who had written the notorious disclamatory preface for the first edition of Copernicus's book, knew all about Nabonassar and Salmanassar. He told Funck that ancient astronomy could provide the Ariadne thread to lead him out of the labyrinth of biblical chronology. And indeed Funck found in Ptolemy just the guidance he had previously lacked. He set out, so he told his readers, to found "chronology on the absolutely certain eras of Ptolemy the astronomer... beginning from the great era of Nabonassar, which all followers of Ptolemy take both as absolutely certain and as the oldest of all."[78] Funck's learned book converted Catholics and Protestants alike into practitioners of technical chronology of the new style. In 1557 Iacobus Curio—a firm believer in the tenet "Ptolemaeus non fallit"—believed equally firmly that "our friend Ptolemy calls Salmanassar Nabonassar, a name that sounds Egyptian."[79] In 1566

Jean Bodin made the equation of Salmanassar and Nabonassar part of the foundation for his *Method for the Easy Comprehension of History*.[80] By 1583—when Scaliger showed that Nabonassar and Salmanassar were two different kings of two different kingdoms—he was only composing a variation on a well-known theme; for in his system too Era Nabonassar provided the firm pedestal on which all lofty hypotheses rested.[81]

Chronologers before Scaliger grasped not merely the general relevance of astronomy but the crucial importance of eclipses. These, so the imperial astronomer Petrus Apianus wrote in 1540, "make it possible to fix all events to specific years, before Christ as well as after."[82] As an example he tried to redate Alexander's empire by locating the Gaugamela eclipse—only to go wrong because he took Boedromion, the Greek month in which it occurred, as equivalent to the Julian June. The eclipse he chose was the wrong one, and his date—326 B.C.—was actually farther from the truth than the Eusebian one, 328 B.C., that he hoped to correct.[83] But neither this problem nor those encountered by Gerardus Mercator, who rested his *Chronologia* of 1569 on the authority of a canon of dated eclipses, proved more than a minor obstacle. In 1578 Paulus Crusius's posthumously published *De epochis* made correct dates for Gaugamela and much else accessible between two covers.

In this short, forgotten book by a long-forgotten Jena professor of history and mathematics, Scaliger encountered the sixteenth-century tradition of serious historical chronology at its best. The result was a revolution in his thinking. As late as the fall of 1581, as the letter to Chrestien shows, Scaliger meant to write a book about calendars. But in the spring and summer of 1581 he had repeatedly asked Dupuy to find him a copy of a book by a German "en forme de correction de Cronique. C'est comme un instruction *corrigendi epochas temporum*." By 4 September he had received and read Crusius's book, "lequel est fort bon."[84] When he then turned, as he apparently did for the last months of 1581 and the first half of 1582, to reconstructing the chief epochs of world history, he did so with Crusius as his guide. Scaliger's precise dates for the epoch of the Olympiads, the Peloponnesian war, and the era of Diocletian—as well as the Gaugamela eclipse—are merely verifications of or slight refinements on Crusius's results. One specific case will show exactly what Crusius meant for Scaliger. In the commentary on Manilius, Scaliger had discussed the complete solar eclipse that, according to Varro's astrological friend L. Tarrutius of Firmum, accompanied the conception of Romulus: "The astrologer L. Tarrutius says that Romulus was conceived on 23 Choiac of Ol. 2, 1, at the third hour. That is December, when the full solar eclipse took place. But his actual birth . . . was on 21 Thoth, at sunrise. That is in September."[85] This summary of Plutarch, *Romulus* 12, is accurate enough, but it hardly penetrates the surface of the source. Scaliger simply assumes that the eclipse took place. He offers no Julian equivalent for its Olympiad date and only a rough Julian equivalent for the month in which it fell—a conversion based on the unsupported (and unjustified) assumption that Plutarch had given the date in the fixed calender of Alexandria.

In the *De epochis* Crusius had treated the same passage far more proficiently. He pointed out that Tarrutius's date could correspond either to 24 June 772 B.C. (in the Egyptian calendar used by Ptolemy and other astronomers) or to 19 December 772 B.C. (in the fixed Alexandrian calendar). He used the *Prutenic Tables* to determine that (by coincidence) conjunctions of the sun and moon had taken place on both dates, but no visible solar eclipse on either of them—or any 23 Choiac even close to 772 B.C. He concluded—rightly—that Plutarch or his source "could not reach the truth, since they lacked a detailed command of astronomy."[86] And he thus achieved as much for the elucidation of Tarrutius's eclipse as the battalion of modern chronologers who attacked it again in the late nineteenth and early twentieth centuries.[87]

In the *De emendatione*, after reading Crusius, Scaliger ignored the eclipse of Ol. 2, 1, since he knew that it had not happened. He turned his own attention to a second eclipse mentioned by Plutarch in the same passage, which supposedly accompanied the founding of Rome on 21 April, Ol. 6, 3. He now used the *Prutenic Tables* to locate the nearest relevant solar eclipse, which he dated with great precision, as falling in Ol. 7, 2—2 Julian years and 309 days before 1 Thoth Nabonassar 1 (= 19 hours after noon of 21 April 750 B.C., at Alexandria). And he argued that Tarrutius must have had this eclipse in mind as the one that dated the founding of Rome.[88] The argument is wrong; Plutarch does not suggest that Tarrutius had anything to do with the eclipse of Ol. 6, 3, and in later versions of the *De emendatione* Scaliger concentrated on the noneclipse of Ol. 2, 1, which he treated less fully and carefully than Crusius had.[89] What gives the passage importance is its effort at precision; for it shows that reading Crusius, not editing Manilius, taught Scaliger to appreciate the importance of exact eclipse dates. It seems reasonable to infer that *De epochis* inspired Scaliger to attempt an even larger collection of epochs—his own Books 5 and 6.

Contact with Crusius—and, no doubt, with Funck and Bodin as well—is not enough to explain *De emendatione* 5 and 6. For throughout them Scaliger does more than apply astronomical evidence. He also consistently confronts the biblical history of Israel and other nations and the patristic histories of early Christianity with the evidence of nonbiblical texts. His aim, made clear by bold direct statement as well as by example, was to show that the Bible was neither complete nor self-contained as a history of man. The chronologer could not date the events it mentioned—far less work out the histories of the non-Jewish nations it described—without constantly referring to nonbiblical sources. And the chronologer dared not accept many traditional authorities, like the *Ecclesiastical History* of Eusebius, which contained both gross errors and deliberate misrepresentations. This aggressively independent and critical approach to the sources has won Scaliger as much praise as his more technical achievements.

To assess Scaliger's source-criticism is in one sense harder than to assess his analyses and calendars and epochs. He had not dealt with this kind of question in his earlier work, and he did not leave rough notes to indicate the ways in which his method developed. Yet here too some comparisons with works he knew may prove enlightening. The general view that the Bible could not stand alone was well established in the 1580s. Guillaume Postel, Scaliger's first master in Oriental studies, had argued in 1551 that the full truth about antediluvian history would not be known until the Ethiopic Book of Enoch was available in Latin (the relevant bit, an allegory about sheep, might have disappointed him).[90] Wolfgang Lazius had made an even more aggressive claim in 1557. He rewrote the history of central Europe on the basis of Hebrew inscriptions found "in the district of Gumpendorf, at the first milestone from Vienna." One of the new texts, which learned Hebraists transcribed and explicated for Lazius, recorded the death of "Mordechai, a descendant of Gog [that is, a giant], the great warrior, in the year 2560 from the creation of the world." "The month and day are missing," Lazius complained, "but the monument is quite old; for only 560 years had passed since the promulgation of the law."[91] Only new evidence as striking as this, and the most up-to-date philological analysis, could have enabled Lazius to prove his thesis that the Germans were directly descended from—of all peoples— the Jews.

More striking still, in the 1560s and 1570s a number of critics anticipated specific arguments later associated with Scaliger. His savage campaign against the forged Berosus and Manetho, for example, added little new substance to the damning indictments already laid by such Catholic scholars as Melchior Cano and Gaspar Barreiros.[92] A particularly prominent example occurs in Book 5. After drawing from the Hebrew Bible the standard period from Creation to Flood, Scaliger draws a piece of corroboratory evidence from the Greeks and Babylonians:

> Moreover, according to Simplicius, Callisthenes reported that in the year when Babylon fell to Alexander, when he asked the Chaldeans about ancient times and the origins of Chaldea, they had records for no more than 1,903 years. Babylon fell in the year 4383 of the Julian period, which precedes the first of the Callippic period. Subtract from these the duration of the Chaldean past—1,903 years—and the remainder makes the year 2480 of the Julian period, which is only 60 years from the Flood. Therefore, considering the great obscurity of these events, the Chaldean computation differs little from the Mosaic. Clearly the one we set out is true.[93]

Callisthenes's report suggests that Babylon came into existence immediately after the Flood. What could confirm the very similar account in the Bible more appositely or neatly?

Scaliger's argument omits two crucial points. The first is that he has falsified the evidence. Simplicius, in the passage Scaliger refers to, attributes the flaws in the astronomy of Aristotle and Callippus partly to the fact that "there had not yet arrived in Greece the observations that Callisthenes sent from Babylon, as Aristotle had instructed him to do; by Porphyry's account these had been preserved for 1,903 years, up to the time of Alexander of Macedon."[94] Callisthenes sent back *paratēréseis*—"astronomical observations." But Scaliger had learned from Ptolemy that the Babylonians did not compile observations systematically until the reign of Nabonassar, after 747 B.C. He therefore could not accept the letter of Simplicius's account. Yet as the same time he could not bring himself to forgo such clear support for the Bible. So he transmuted Simplicius's *paratēréseis* into *archaiologiai*—astronomical observations into antiquarian information. "Who's to be master, the man or the text?"—so he might have cried as he wrestled with the Greek.

The second point Scaliger omits is even more revealing. In fact he probably never did wrestle with this bit of Simplicius's Greek;[95] for his argument was as derivative in substance as it was flawed in execution. Simplicius's data about Callisthenes had been known in the West since the time of Pietro d'Abano. Pico had used the story in his attack on the pretensions of the astrologers.[96] And from the 1560s on, Callisthenes and his observations occupied an undeservedly prominent postion in scholars' mental maps of Greek cultural history. Bodin used the story to undermine pagan claims to possess astronomical observations (and cultural traditions) that reached back for hundreds of thousands of years. Ramus used it to flesh out his account of Near Eastern astronomy and to give weight to his attacks on Eudoxus and Callippus. Ioannes Freigius incorporated Ramus's version of it in his pocket-sized world history, the *Mosaicus*, which appeared simultaneously with—and was therefore independent of—the *De emendatione*.[97] Most interestingly, Henry Savile not only plagiarized Ramus's account in his own Oxford lectures on Ptolemy but also pointed out that Ptolemy's testimony seemed to contradict it: "He seems to indicate in [*Almagest*] 3.7 that in his time observations existed only from the era of Nabonassar (that is for 400 years before Alexander)."[98] Yet even Savile did not draw the proper inference. He explained the absence of earlier observations as due to human incompetence or the burning of the library at Alexandria—not to their nonexistence at any time.

For our purposes, however, the most important early student of Callisthenes was Goropius Becanus. Scaliger referred to him with contempt, as the man who "was not ashamed to criticize Moses for drawing etymologies from Hebrew rather than Dutch."[99] Yet the evidence suggests that Scaliger met Callisthenes not in Simplicius, not in the respectable works of Bodin or Ramus, but in Goropius's fantastic *Origines Antwerpianae*:

> Since Aristotle had his favorite disciple Callisthenes with Alexander he asked him to make diligent inquiries among the experts in Babylonian antiquities for their observations of the stars and chronology. He gladly obeyed his teacher in this matter and sent all their observations back to Greece. Porphyry says that they were for just 1,903 years; see Simplicius's full and learned review of the opinions of the ancients on the number of the heavenly spheres, in his commentary on *De caelo* II. Given that Alexander took Babylon in A.M. 3636, we will have to admit that their observations went back to the year of the world 1731—when Nimrod would have been about 45 years old, if Chus, who was born immediately after the Flood, produced Nimrod when he was 30 . . . Accordingly we have remarkable evidence of agreement between the Chaldeans and those whose computations rest on the Bible . . . Hence we shall boldly reject those extra years that the 70 translators added to the Hebrew computation . . . [100]

The numbers differ slightly, and Goropius embroiders, as Scaliger did not, on the relations between teacher and pupil. On the whole, however, the connection between the two seems evident. Goropius gave Scaliger not just the evidence of Simplicius but a suggestion about how it could be distorted to include historical data; like Scaliger, and unlike Bodin, he used the story not to beat down arrogant pagans but to lend elegant technical support to the authority of the Bible. All Scaliger did was to deal with the source even less honestly than his predecessor. The episode does not lack ironies. The most up-to-date historical scholarship of the late Renaissance ended up restoring to history that characteristic figure of medieval legend, Nimrod the astronomer (not to mention his observatory, the Tower of Babel); and there he would stay until Vico dealt with him once and for all in the *New Science*.[101] But no irony is greater than the fact that Scaliger learned his method of source-criticism partly from the man whose name became the source of the standard eighteenth-century term for coining silly etymologies—"*goropizer.*"

But in another respect *De emendatione* 5 and 6 did offer substantial and disturbing novelties. In writing chronology Scaliger addressed himself to a Europe-wide, Latin-reading public; but he took a special interest in, and often wrote with an eye toward, a narrower audience of Protestant intellectuals, in Geneva and elsewhere. In the Calvinist circles to which Scaliger professed allegiance some of the questions and texts he discussed in the *De emendatione* were already contentious. His successor at the Geneva Academy, Matthaeus Beroaldus, had published in 1575 a *Chronicum* that claimed to rest entirely on the authority of Scripture and denounced any use of pagan sources as impious. Many Protestants had published chronological works that went to another extreme, drawing heavily not only on the Bible but on the forged works of Annius. Abraham Bucholzer, whose *Isagoge chronologica* of 1577 seemed learned to Scaliger's Genevan correspondents, used pseudo-Metasthenes as well as genuine Ptolemy and Censorinus. Even earlier Jacob Grynaeus had published an edition of Eusebius's *Ecclesiastical History* that treated that work as absolutely complete and accurate and denounced the efforts of the Magdeburg Centuriators to replace it with something more modern and reliable.[102]

In the *De emendatione* Scaliger demolished Beroaldus, Annius, and Eusebius with equal pleasure. In Book 6 he used the weeks of Daniel as a pretext for attacking both the Annian forgeries and Beroaldus's belief that the history of the Greeks and Persians could be established from Scripture alone, and he used the life of Christ as the occasion for a long and lively demolition of Eusebius's belief that the *Therapeutae* had been the earliest Christians. Using Philo brilliantly, Scaliger revealed the Therapeutae to have been a branch of the Essenes, and bitter enemies of Christianity. And he treated the whole episode as a case in point of the need to criticize authorities: "See what happens when authority is preferred to truth; everyone who reads this thinks it must be true, since it comes from

Eusebius."[103] When Beza and a lesser Genevan scholar, Corneille Bertram, sent him a critique of his work, he reacted merely with irritation at their efforts to dissuade him from taking controversial positions on points of importance to the church. Bucholzer, whose chronology they recommended to him, was a mere amateur, who swallowed "ce Metasthenes de foin forgé par cest imposteur dominicain Annius" and did not understand the real application of Ptolemy to historical research: "il n'ayde de rien au principal." And if Scaliger had to admit that they were right to call him an amateur in ecclesiastical history, he showed no disposition to cease writing it. In his letter to Beza, he remarked that he would show that he had studied "les histoires sacrees" to some purpose.[104] And in his working copy of the *De emendatione*, he wrote in the margin of his fiercest attack on Eusebius: "We especially wanted to deal with these matters because of the beginnings of Christianity. And they are not out of place here."[105] Within the narrow world of French Protestantism, if not the larger one of European learning, Scaliger had some justice on his side when he complained: "Voila comment tout le monde n'entend point ce mestier."

Calendar Reform

So far we have not considered the motive that led Scaliger to pick chronology for special study. The evidence available so far does not enable us to do so. But it does enable us to refute one common theory inspired by Scaliger himself, who writes in the beginning of Book 8 that the reform of the calendar could well seem to be the chief motive for his work.[106]

In fact, Scaliger's early references to the *De emendatione* do not suggest that he intended to deal with calendrical problems of a practical kind systematically. Moreover, his letters show that the manuscript he sent off to Patisson in June 1582 included "*seven* big books *de emendatione temporum.*"[107] And the internal dates in Book 8 all come from the very end of 1582 and the beginning of 1583—and thus prove beyond doubt that Book 8 was written last.

Like the other books, 8 drew heavily on the existing literature. And in the case of calendar reform, even more than in the others, a treasure trove of madness, error, and perfectly sensible suggestions stood ready to be looted. Unlike the other books, however, 8 added few novelties of any substance. Scaliger told Dupuy that Zarlino's book on calendar reform—which he actually read in the summer of 1582, after finishing Books 1-7—showed merely that "in all of Italy no one understands this subject."[108] Yet his own book shared one of its major technical suggestions with Zarlino's—and it proved to the competent that Scaliger had only a gifted amateur's knowledge of the field.[109]

A Balance

In the case of systematic chronology, it would seem, the novelty in the substance of Scaliger's work lay in large part in the synthesis he created of existing elements. For the most part, astronomers and more historical students of chronology had had little to do with one another even when they worked on the same problems. Sometimes the contacts that did take place were merely head-on collisions. A hard-nosed believer in astronomy like Panvinio deplored the use of rabbinical sources, which he took as an effort to base chronology on the Cabala.[110] A hard-nosed believer in texts like Gilbert Génébrard repaid the astronomers in full, for their attacks on the rabbis, their frequent use of Annius, and their misguided notion that chronology should be "a matter not of reading but of computing." In his case, understanding the ways in which others used the *Almagest* did not lead to emulation. In Ptolemy, he explained, "the names and times of Babylonian and Assyrian kings seem false or corrupt, so nothing certain can be founded on him; far less can he be used as the source for a chronology—save perhaps a fictional one—even if some do try to use him so."[111] And even those who tried to combine mathematical and textual skills generally failed. Even Crusius, who interpreted Ptolemy with great historical skill and insight, fell victim to the Annian forgeries.[112] Even the sharpest Catholic critics of the Annian forgeries did not know how to use Ptolemy against them.

In his willingness and ability to fuse astronomy, Oriental studies, and classical philology, Scaliger had no predecessor. In fact, by doing so he himself became the model after which the polyhistors of the next century and more tried to shape themselves. By 1620 it was clear across Europe that a serious scholar needed the humanities, Hebrew, and mathematics.[113] The uniting of these in one head and one book was Scaliger's achievement.

It would be wrong to treat Scaliger's most influential book as a philological *Biographia literaria*—a patchwork of derivative materials—without also making plain that Scaliger, like Coleridge, not only brought together what had previously been separate but also reshaped everything he borrowed to fit new purposes. The chief novelty in Scaliger's work, however, is difficult to describe with precision; for it consists less in the presence of something new than in the absence of something virtually all his predecessors had shared.

Chronologers before Scaliger tried not merely to establish the outlines of history but to reveal its meaning and predict its outcome. They found, in the eras and intervals they computed, evidence of the constant action of a guiding force in human affairs. On the one hand, many resorted to the traditional ordering schemes of the Bible, the Talmud, and the Fathers. The 4 Monarchies and 70 Weeks of Daniel, the 1,260 years of Revelation, the 6 days of 1000 years that made human history correspond to the 6 days of Creation, the 12 hours of 500 years that made it correspond to the 12 hours of Christ's suffering on the cross, the 3 ages of the Talmudic prophecy of Elias, and the 50-year Jubilees of the *Chronicle* of Eusebius and Jerome—all these and more authoritative patterns, one sure if another failed, competed for the chronologer's attention.[114] Those who wanted to discover a divine order in the facts could follow the example of Orosius, who used the common duration of Rome and Babylon—1,164 years—as proof that God had intended them to be seen as parallel empires (2.3). And those who wanted a ready-made pattern not sanctioned by ecclesiastical authority could find it in the rich and colorful stock of Joachim of Fiore.[115]

Cacoethes computandi flourished in the sixteenth century as seldom before.[116] Glimmerings of a divine plan danced behind the most apparently insignificant numerical patterns. Bucholzer—one of the most serious Protestant chronologers—tabulated intervals that cropped up repeatedly, like the 532 years that separated the Flood from the death of Eber, the birth of Eber from the death of

Jacob, and the death of Jacob from the fall of Troy. He circulated, first in manuscript and then in print, a "chronological game" which showed that years whose numbers included repeating combinations of numerals often housed dramatic events. Thus the Ten Tribes went into captivity in A.M. 3232, Babylon fell in 3434, and the Sicilian Vespers rang out in 5252.[117] Yet Bucholzer did not go so far as the Catholic Goropius, who found a message hidden in the Holy Ghost's mistakes. Christ, he pointed out, died on the sixth day. Time consists of six ages. And six is an interesting number, the sum of its factors:

$$1 \times 2 \times 3 = 1 + 2 + 3 = 6.$$

When set in ascending order, moreover, these factors symbolize the members of the Trinity. Consider in that light the difference between the Hebrew and Septuagint computations of the period between the Creation and the Flood: 1,236 years. Surely, Goropius reasoned, the Greek translators added "this mystic number" deliberately; and surely it opened, at least for learned readers, a "deep chasm of contemplation."[118]

Those not content with theological modes of interpretation could borrow a complementary one from the astrologers. Since antiquity, pagans and Christians alike had been tempted to see the stars as the agents, or at least the clearest indicators, of the divine will that gave order to history. Tarrutius not only dated the eclipse for Romulus's conception but drew up a horoscope for the city of Rome, and—according to Cicero—"did not hesitate to divine the city's future" from the position of the moon (*De divinatione* 2.98–99). Most early Christians found in the Star of the Magi confirmation of the validity of astrology—at least for the period before Christ's birth.[119] Firmicus Maternus, who had the benefits of both Christianity and astrology to rejoice in, drew up a hypothetical horoscope for the world itself, a *thema mundi* (*Mathesis* 3.1). And in the twelfth century, Latin Europe inherited from its Sassanian inventors and Arab preservers the elegant and powerful theory of the great conjunctions, which held that the conjunctions of Jupiter and Saturn marked all turning points in human history.[120] From the fourteenth century on, vernacular and learned writers alike tried to show that the history they knew matched this scheme.

Pierre d'Ailly, for example, computed his own *thema mundi* and then deduced from the planetary positions at the Creation the dates of the great conjunctions that followed—and that, in turn, determined those of great events (A.D. 1789 was marked for something dramatic in one of his schemes).[121] Though he eventually lost faith in his original *thema mundi* and drew up another, quite different, one that required a different dating of conjunctions and events, he never lost faith in the general principles he had adopted at first, and even continued to argue that his first, wrong scheme had something to contribute toward the interpretation of history.[122]

The new astronomy of the sixteenth century produced an occasional criticism of this scheme, but few efforts to abandon it as a whole or even to rid it of its most obviously arbitrary assumptions. G. J. Rheticus—so Christopher Rothmann complained to Tycho Brahe—criticized traditional methods, but replaced them with new ones just as poorly founded: "While Rheticus writes that al-Battānī too freely misuses the mysteries of astrology, he commits the same error himself. For how can changes in the eccentricity of the Sun produce changes of empires?"[123] Yet Rheticus's scheme—according to which the periods of 858½, 1717, and 3,434 Egyptian years that governed the motion of the center of the earth's orbit also determined the rise and fall of empires—had won the enthusiastic interest of Osiander and provoked serious criticism from Bodin and Campanella.[124] Even those least committed to the authoritative traditions of research in astronomy rarely denied the authority of the heavens over the earth. If Campanella insisted that the heavenly motions were far less regular than Copernicus thought—were, indeed, too irregular to be computed accurately—he found in this very irregularity firm evidence that the sun was coming closer to the earth; and in that he saw the sure sign of an impending transformation of society.[125]

Little historical astrology or numerology is to be found in Scaliger. In Book 8, written hastily and last, he does remark that a great conjunction had occurred or would occur after each reform of the Julian Calendar (the Augustan and the Gregorian).[126] In the same passage, he makes a more explicit—and more puzzling—comment about the Jewish week of years, a 7-year period in which every 7th year was a Sabbatical one: "ON THE SABBATIC YEAR If we divide a given [number of] years of the world by 7, the remainder is the [position of] the year in the current week. And it is clearly marvelous that through God's hidden plan the observance of the Sabbatic year is concealed in the years of the world."[127] But neither this numerological principle nor any other seems to play a significant role in the body of the *De emendatione*. Angels, tables of great conjunctions, and the future are not among those present. Debate about the meaning and nature of time Scaliger relegates to those whose métier requires them to discuss such matters—the philosophers. And at least once he made his opposition to the search for meanings in the past extremely clear. When he read Bucholzer's *Chronologia*—which followed Eusebius in ordering history by 50-year Jubilees, and found a "secret analogy" between these and the years of the world—Scaliger denied both Bucholzer's reconstruction of the cycle and the consequences he drew from it:

> He has made a good many mistakes, as when he makes the Jubilee consist of 50 complete years, a thing which contradicts both the authority of Scripture and reason. For how can 50 be divisible by 7? It would not be a serious matter simply to have believed this, for the old Greeks also held this view—at least, I remember coming across this fantasy in one of the old Greek ecclesiastical writers. That would be tolerable, if only he did not base allegories worthy of Origen on these 50-year Jubilees.[128]

In condemning Bucholzer's numerological speculation more harshly than his technical error, Scaliger reveals the true novelty of his attitude. The dates he establishes are dates, not lessons. And this deliberate austerity, this concentration on the technical and the soluble, is the feature of the *De emendatione* that would have seemed most unusual to a practiced reader of earlier chronologies.

This revision of Bernays and Pattison may well seem somewhat unsatisfying. For their simple, majestic portrait of a great sage contemplating the past it substitutes a complex tableau. And even that is less a *School of Athens* than a *Raft of the Medusa*, where Scaliger figures as the topmost in a heap of writhing bodies. The result is neither elegant nor simple. But it does rest on the documents now known, and it even suggests a possible explanation for his decision to explore this new and rocky field. Perhaps what mattered to him was precisely the ability to collect and arrange facts without imposing a larger meaning on them. His closest friends, the jurists de Thou and Dupuy, reacted to religious war and the dissolution of society by turning to honest scholarship and accurate history. In the compiling of data, rather than the forging of an ideology, they found an escape from chaos if not a way to reduce it to order. And in scholarship they found a forum where they could carry on reasonable discussion with men of different faiths.[129] Scaliger too found in chronology at first a field in which he could escape the bounds of his Calvinism; and many features of the first *De emendatione*, like the attack on Annius and the use of materials supplied by a Jacobite living in Rome, give the book a distinctly ecumenical feel.[130] Perhaps he too saw in the diligent and accurate collecting of information of no practical importance not only a way to build himself a unique monument but also a way to escape from the miseries of *Arabia Deserta*.

5

Protestant versus Prophet: Isaac Casaubon on Hermes Trismegistus

NO ANCIENT WRITER had an afterlife more active, more paradoxical, or more crammed with incident than that of Hermes Trismegistus.[1] And nothing in Hermes' afterlife became him more than his leaving of it. This episode—Isaac Casaubon's exposure of the *Corpus Hermeticum*—is as rich in foibles and follies as any other chapter in Hermes' long story. New documents make it possible to reconstruct Casaubon's arguments, their context in his life, and their reception, more fully than has been done before.[2]

To meet Casaubon on his own ground we must journey through some of the worst-mapped and most forbidding territory known to intellectual history: the controversial theology of the years around 1600, when Calvinists, Jesuits, and Gallicans competed to attain new levels of invective and inhumanity. It was in the course of his attempt to demolish the *Annales ecclesiastici* of Cesare Baronio that Casaubon denounced the *Corpus*.[3] In fact, the entire attack on Hermes takes up only a few of the 773 vast pages of polemics that Casaubon dedicated to his last patron, James I. Most of his bolts were directed at quite different targets—above all at Baronio's efforts to support papal claims to spiritual and temporal authority. In this setting an attack on Hermes necessarily formed an excursus, aimed at the scholars among Casaubon's readers. Casaubon would not have made exorbitant claims for its novelty or importance, or even have seen it as central to his work.

Casaubon's attack, however, was neither accidental nor insignificant. His whole life and awesome self-education had prepared him to grapple with Hermes Trismegistus. His studies with Franciscus Portus in Geneva and the preparation of his long series of commentaries on classical Greek texts—Diogenes Laertius, Strabo, Athenaeus, Aeschylus—had made him, as Joseph Scaliger said, "le plus grand homme que nous ayons en Grec."[4] His years as Garde of the Royal Library in Paris and his spells of work in Bodley had given him an unrivaled knowledge of unpublished Greek texts.

But Casaubon made an even more appropriate Lord High Executioner for Hermes than these details indicate. An extract from his monumental diary (the year is 1597) will bring his character into sharper focus:

Feb. 20. I rose at five (alas, how late!) and at once entered my study. There, having prayed, I was with Basil . . .

Feb. 21. Though I rose at five, I had to leave in order to see to some errands in the town, and only entered my study—oh sorrow!—at seven. I prayed and spent on Basil the time that remained before lunch. From then on I prepared my lesson, though to no avail, for I didn't hold my class. So the rest of my time was given to Basil, to dinner, to bed after prayers to God.

Feb. 22. I rose at five. Entering my study and praying to God, I gave myself over to Basil. After lunch friends kept me busy. After dinner prayers, bed . . .

Feb. 26. Before five I entered my post and prayed; I worked on Basil only until eight. From then until lunch I unwillingly saw to other matters. From lunch friends had me; at last Basil. Dinner, Basil, prayers, rest . . .

March 11. Early prayers, Basil, whom I finished reading before eight, with God's help.[5]

As this passage suggests, Casaubon was—in cast of mind and sensibility, if not in all points of doctrine—a Protestant so dyed-in-the-wool that it would take a Weber to do him justice. He saw God's hand in everything—not only in the sparrow's fall but in his daughter's fall into a fire, which left her uninjured thanks to direct providential intervention ("tu fecisti," Casaubon told his God, "ut non prona sed supina caderet").⁶ He writhed with guilt for time spent on secular pursuits, and found blasphemy of any sort intolerable.

The passage also reveals where Casaubon's truest interest lay: in the literature, liturgy, and antiquities of the early church. Casaubon criticized Calvin himself for his unwillingness to preserve ancient usages. He searched for a brand of Protestantism more faithful than Calvin's to tradition.⁷ He studied the Bible and the writings of the Fathers in preference to any other ancient text or texts. "What stirs his soul," Mark Pattison remarked, "is Christian Greek." One would have to be stirred in order to read, as Casaubon did between 19 February and 11 March 1597, the 698 exceptionally close-packed folio pages of Froben's 1551 edition of Basil.⁸ He had a commentator's familiarity with the New Testament.⁹ And no one before, hardly anyone since, could have rivaled his knowledge of the times, customs, and languages of the Greek Fathers. Casaubon's mastery emerges from every page of his *editio princeps* of Gregory of Nyssa's third letter (1606), which bulges with notes like this one:

> The Greek Fathers refer to the mystery of the Incarnation of Christ the Savior by various terms, including these: *hē tou Christou epiphaneia, hē sōmatikē epiphaneia, hē despotikē epidēmia, hē dia sarkos homilia, hē di'anthrōpon or di'anthrōpotētos phanerōsis, hē tou logou ensarkōsis, hē parousia* . . .¹⁰

Moreover, Casaubon was no novice at distinguishing genuine antiques from faked ones. His rich notes on the content and diction of Gregory's third letter were designed not merely to illustrate the text but to prove that it was authentic even though it did not occur in the Codex Regius of Gregory.¹¹ And he had given more than passing attention to late-antique pseudepigrapha throughout his career. Near the beginning of his notes on Diogenes Laertius, as we will see, he assigned the *Hero and Leander* to a late-antique grammarian named Musaeus rather than to the legendary contemporary of Orpheus. In 1603 he edited a whole collection of fakes— the so-called *Scriptores historiae Augustae* (to whom he gave this collective name). He revealed some of their many inconsistencies and improbable statements. He used considerations of style and content alike to argue that the works ascribed in the manuscripts to Aelius Spartianus, Aelius Lampridius, and Iulius Capitolinus could more plausibly be ascribed to a single author. He showed that the collection had been edited and revised, though the job had been done by an incompetent. He denied that the date or purpose of the revision could be precisely fixed: "Only a prophet could divine what moved the maker of this collection to arrange it in this form."[12]

Casaubon read with a critical eye, in fact, even when he was browsing more or less casually through a text that he had no intention of editing. As he read the *Batrachomyomachia* in Henri Estienne's *Poetae Graeci* of 1566, incongruities became apparent at once. The poet who said (in line 3) that he had written "in tablets on [his] knee" was, Casaubon remarked in the margin, "certainly not blind." And the poem that included *exololuxe* (in line 101), which Homer "never used thus," was "certainly not by Homer or any poet of the first rank."[13] Clearly Casaubon had internalized the rules of higher and lower criticism, long before Hermes crossed his path.

Casaubon became interested in writing a reply to Baronio well before he came to England in 1610; but the political situation of his French patron, Henri IV, did not permit a public response to be made. James, however, liked this sort of thing and lacked Henri's political obstacles. He encouraged Casaubon to set to work. In late April 1612 Casaubon finished reading and taking notes and began to write.[14] One of the very many points in the first half of Baronio's first volume that caught Casaubon's eye was a reference to the pagan prophets who had predicted the coming of Christ: Hydaspes, the Sibyls, and Hermes Trismegistus. He had already, so it seems, decided the *Hermetica* were forged.[15] But he now acquired a copy of Turnèbe's 1554 edition of the *Corpus* and worked through it as critically as he had worked through the *Batrachomyomachia*—and far more systematically.[16] He realized at once that the text could not be what it claimed to be; and in the cryptic form and awkward scrawl characteristic of his marginal annotations—Grotius called Casaubon's notes and annotated books "true Sibylline leaves"— he recorded his initial reactions.[17] By following him through the book, and comparing his jottings with his more developed final attack on Hermes, we can penetrate to the emotional wellsprings from which Casaubon's work emerged.

As Casaubon read the *Corpus hermeticum*, he was repeatedly struck by the presence of facts and doctrines familiar from the Bible. In 1.4 the Hermetic creation story begins with a "darkness that presses downward" and that is transformed into "a moist nature."[18] "Tohu et bohu," wrote Casaubon, obviously referring to the parallel Creation story in Genesis. In 1.31, where God is described as "inexpressible, unspeakable, to be named in silence," Casaubon found a reference to the Jewish doctrine that the tetragrammaton must not be pronounced. Hence his note: "ad nomen *yhwh*." In 4.4 he found a clear description of baptism.[19] These apparent correspondences filled him with discomfort. For in 1.16 Pimander says that "This is the mystery that was hidden until this day." Casaubon drew the logical implication of this in an unusually long remark: "Note that if this is true, and this fellow wrote before Moses, then God revealed his mysteries through him, not through Moses."[20]

The correspondence between the Bible and the *Corpus* had pleased such earlier scholars as Ficino and Lefèvre d'Etaples, since they supported the validity of Revelation and revealed the piety of the pagans.[21] Casaubon, however, like many French Protestants, was sensitive above all to the difference between Christian and pagan ideas and beliefs.[22] The margins of his copy of Lucretius, to take an extreme case, swarm not only with exclamations about the beauty of Lucretius's style but also with imprecations against the impiety of his beliefs: "Lucretius, you great fool; do you believe that worms and men share the same sort of soul?"[23] True, Casaubon found some pagans more acceptable than others. Aristotle's doctrine that "Nature does nothing in vain" was far superior to "the rubbish of the Epicureans, according to whom the world was created by nature, at random, to no purpose, in vain."[24] But even Aristotle and Plato had not received a separate and equal revelation; far less so the authors of the *Corpus*.[25] "I am moved above all," he wrote in the published *Exercitationes*, "by the fact that it seems contrary to God's word to think that such deep mysteries were revealed more clearly to Gentiles than to the people that God chose as peculiarly his own."[26] After all, "we know what slender rays of the truth had shone on the Jewish people in the earliest times."[27] Paul called the period before Christ *chronous agnoias* (Acts 17.30) and explicitly described the doctrine of man's salvation as a "mystery which hath been hid from ages and from generations, but now is made manifest to his saints" (Colossians 1.26).

> How can these testimonies of Scripture and others like them stand if it is true that several—and indeed the chief—mysteries of Christian doctrine were revealed to the Gentiles even before Moses? For there is no doubt, if the standard accounts of him are certain, that Mercurius Trismegistus lived before Moses.[28]

Casaubon's insight, then, grew from the rich soil of his bigotry. He wanted to show that Baronio had erred wildly on the side of generosity to the pagans.[29]

But what began as a purely emotional act of negation was soon transmuted by Casaubon's learning. He drew on his years of reading in early Christian sources to show that the language and content of the *Corpus* could not possibly fit the early Egyptian world from which it claimed to come. Throughout the margins of his copy he singled out words and phrases that could not belong to the Greek of the *prisci theologi*. 1.19, "He who loves his body, which is born of the error of love, he remains in darkness, wandering," recalled John 12.46: "I am come a light into the world, that whosoever believeth in me should not abide in darkness."[30] 4.6, "If you do not first hate your body, child, you cannot love yourself," had a much more striking parallel in John 12.25: "He that loveth his life shall lose it; and he that hateth his life in this world shall keep it into life eternal."[31] Other parallels were equally close.[32] To Casaubon, such weaving of Scriptural texts into a passage was redolent of the age of the Fathers.

So, even more, were some of the nonscriptural words that he encountered in the Corpus. Early in 2 and again in 12, the *Corpus* touches on the question whether God has an essence. Both passages are couched in an abstract and rebarbative terminology: "Whether you refer to matter or body or essence, know that these too are energies of God, and the energy of matter is materiality, and that of bodies corporeality, and that of essence essentiality."[33] Casaubon knew this diction intimately. It was that of Dionysius the Areopagite, whose *De divinis nominibus* included such "elegant" passages—the adjective is Casaubon's—as this: "The superessential infinitude transcends essences."³⁴ It continually used abstract nouns Hermes could not have known—and even echoed liturgical phrases from centuries after his time.³⁵ It could not have differed more from what Casaubon called the *simplicitas* of Herodotus and other early Greek writers, whose style seemed to him to share something of the purity of that of ancient and modern Oriental writers.³⁶ The *Corpus*, in short, condemned itself:

> The style of this book could not be farther from the language that the Greek contemporaries of Hermes used. For the old language had many words, phrases, and a general style very different from that of the later Greeks. Here is no trace of antiquity, no crust, none of that patina of age that the best ancient critics found even in Plato, and even more in Hippocrates, Herodotus, and other older writers. On the contrary, there are many words here that do not belong to any Greek earlier than that of the time of Christ's birth.³⁷

Or, as he put the point more succinctly elsewhere: "that imposter liked to steal not only the sacred doctrines but the words of Sacred Scripture as well."³⁸ And he stuffed his final attack with even more parallels than he found as he first went through the text.³⁹

Linguistic detective work was not Casaubon's only line of approach. He also went through the text looking for doctrines—not novel and profound ones but the exact reverse, doctrines that he recognized, thanks to his long experience with Diogenes Laertius *et al.*, as the ordinary stated views of Greek philosophers. Sometimes he recorded what he took to be Hermes' sources in the margin. At 4.3 the *Corpus* states that if God did not grant all men *nous*, it was not from jealousy of men, "for jealousy does not come from on high." Casaubon immediately connected this with *Timaeus* 29e (the Creator, being good, "never feels any envy for anything"), and wrote "Plato in the *Timaeus*" in his margin.⁴⁰ Elsewhere he merely entered summaries. 8.1 reads in part: "If the world is a second God, and an immortal being, nothing in the world perishes. For it is impossible for part of the immortal being to perish." Here Casaubon wrote only "The world is a second God," and "Nothing that is created perishes."⁴¹ Such cryptic phrases really are Sibylline leaves. But in the final form of the *Exercitationes* Casaubon bound them up and treated them in full. The doctrine that created things are indestructible became the central target of a solemn attack:

> I shall set out one dogma, not peculiar to Plato or the Platonic school, but common to many Greek sages; that Egyptian Mercury treats this in a manner that proves *ipso facto* that he was a pure philosopher, steeped in Greek learning, not the mysteries of ancient Egyptian wisdom. It was the opinion of the most ancient Greeks that no created things die; they merely undergo change. Death is an empty name without a thing attached to it. For what is commonly called death is really a transformation . . . Hippocrates . . . writes: "Nothing perishes, or is created that did not exist before; things are changed by being mixed together or separated . . . " [*On Regimen*, 1.4] That false Egyptian Hermes often refers to this Greek teaching . . . At the beginning of the eighth discourse he explains why nothing perishes: because the world is an eternal being, a second God. All things that are in it are members of that God, and man above all. He repeats the same argument in 12, where he adds: "Dissolution is not death but dissolution of a mixture. It is dissolved not in order to be destroyed but that it may be made new" [12.16]. But I would have to copy the whole book here if I wanted to go through one by one the heads of doctrine that the fake Mercury has turned to his own use from the Greek philosophers. For except for the points derived from Scripture, everything that he has is from them.⁴²

Finally—and here his experience with fakes came into play—Casaubon also searched the text for anachronisms, inconsistencies, and evidence of a desire to deceive. He found seams showing in the *Horoi* and elsewhere. References to the sculptor Phidias (18.4)⁴³ and the *citharoedus* Eunomus (18.6)⁴⁴—both securely dateable to a time well after Hermes should have lived—received marginal notes in Casaubon's working copy and full treatment in his text. More complex and more interesting, however, is the case of 16.1-2:

> When the Greeks will later wish to translate [my books] from our language into their own, the result will be very great confusion in the texts and unclarity. For when it is expressed in its native

language, the discourse retains the clear meaning of its words. And in fact the nature of the sound and the power of the Egyptian words contains in itself the force of what is said.[45]

The paean to the mystical splendors of the Egyptian language inspired Casaubon only to derision. He connected it with a similar (and near-contemporary) bit of puffery in the *De mysteriis Aegyptiorum* of Iamblichus, which he knew in Ficino's Latin version.[46] And he attributed its presence in the text to the forger's desire to make his offerings even more attractive, by pretending that they had originally been written in Egyptian. "Note that 'later,'" Casaubon wrote in his margin; "what a lover of drama it was who wrote this."[47] In his published attack he dwelt on this passage. The *Corpus*, he argued, was clearly written in Greek. Consider Hermes' etymology of the word *kosmos: kosmei gar ta panta* (9.8). "Are *kosmos* and *kosmei*," Casaubon asked, "words from the ancient Egyptian language?"[48] The forger's tools, in short, were not only dull but double-edged. His claim to be offering Egyptian wares was merely pretense.

Moreover, Casaubon knew that these methods were not confined to the forger of the *Corpus*:

> Since all sciences and disciplines were believed to have sprung from the invention of Hermes Trismegistus the Egyptian, almost all of those who wrote on all the sciences either celebrated his glory or sought fame and favor for their own works with his name. We should not then be surprised that in the first centuries of Christianity, when books with false titles were invented every day with complete license, someone barely acquainted with our religion thought that he should attempt the same in the science of theology and matters of faith.[49]

Here Casaubon set the text firmly into its real context, as part of the pullulating mass of pseudo-ancient, pseudo-Eastern literature that does such discredit to the minds of its Greek authors and readers. No one has since managed to remove it from that unpleasant position.

At this point two further sets of questions arise. First, were Casaubon's arguments and conclusions orginal—and if not, was he aware that they were not? Second, and more important, were they convincing in the light of the information then available and the standards of philological argument then accepted?

The first point is deceptively hard to establish. Frederick Purnell has shown that Casaubon was not the first scholar to doubt the early date and authenticity of the *Corpus*.⁵⁰ Turnèbe had already suggested that the passage of the *Horoi* that mentioned Phidias could not be by Asclepius.⁵¹ His student Génébrard had argued against the antiquity of Hermes in his *Chronographia* of 1580, using the reference to Phidias and an apparent reference to the Sibyls to redate the work.⁵²

Moreover, Casaubon knew that his arguments were not all original. Among the quotations from and references to parallel sources that he gathered on the title page of his copy of Hermes occurs the sentence "On this author Beroaldus has something, in his *Chronicle*, p. 23."⁵³ In fact, the *Chronicum, Scripturae Sacrae autoritate constitutum* (1575) of the highly orthodox Calvinist Matthaeus Beroaldus includes a sharp condemnation of Hermes, which anticipates Génébrard in objecting to the reference to the Sibyls:

> Some claim that Hermes is older than Pharaoh, as Suidas said. It is clear from the work called *Paemander* that is ascribed to him that this view is false. For it mentions the Sibyls, who were many centuries after Pharaoh. Also, the Aesculapius to whom Hermes is writing mentions Phidias, who lived in the time of Pericles. It is thus clear that that book of Mercurius Trismegistus is forged. Much that is said in it cannot be supported by any suitable argument or authority. Let it be sent to Ilerda with its author.⁵⁴

In the margin of his own copy of Beroaldus, now in the British Library, Casaubon wrote "Contra Mercuri. Trismegist." beside this passage.

Casaubon may well have known other Renaissance criticisms of the *Corpus*, such as that of Goropius Becanus, even though he did not mention them.⁵⁵ For he was always sparing in his references to modern scholars. In his discussion of the *Hero and Leander*, for example, he wrote as if he were the first to question Musaeus's authorship:

> The style impels me to believe in the accuracy of the title that Michael Sophianus found in the MSS: "By Musaeus the gram-

marian." For if anyone compares that poem with the poems of other late Greek poets—especially Nonnus, who wrote the *Dionysiaca*—as I have done very carefully, he will not doubt that this opinion is correct.[56]

Yet Casaubon knew perfectly well that Henri Estienne had put these arguments forward in 1566. He took his information about Sophianus and the MSS directly from a note by Estienne (who, incidentally, was Casaubon's father-in-law):

> I have put Musaeus after Tryphiodorus because, like Tryphiodorus, he too was a grammarian. For Michael Sophianus once told me in the presence of several others that at Genoa he had seen an old text that had, among other poems, that one, with the title "The Hero and Leander of Musaeus the grammarian." And he named the owner. But when he told me this he confirmed my own view. For when I said that the poem did not seem ancient to me, and that I found none of the true patina of ancient poetry in it, he removed my doubts and those of the others with this piece of evidence.[57]

Casaubon did not mean to pretend that he had not drawn on his father-in-law's note. Rather, he expected his readers to recognize his source. And in Beroaldus's case he had a special reason not to give credit: he considered Beroaldus exactly the sort of Calvinist he disliked most, both ignorant of and intolerant toward classical culture, and he could indicate his views by noncitation—a form of politics by footnote that scholars still play. As we are less well informed than most of Casaubon's readers about the Latin scholarly literature of the sixteenth century, it is impossible to be sure that he did not draw on others as well as Beroaldus.

Indeed, even in late antiquity some readers evidently found it hard to believe that the *Hermetica* were genuine translations from the Egyptians. Iamblichus wrote defensively in *De mysteriis* 8.4 that "The works that circulate under the name of Hermes contain Hermetic views, even if they often use the language of the philosophers; for they were translated from the Egyptian language by men who had some knowledge of philosophy." This remark, which suggests that someone had asked pointed questions, and which is repeated in the popular *Bibliotheca sancta* of Sixtus Senensis,[58] might well have set other readers to wondering.

The second set of questions is also complex. How convincing was Casaubon's demonstration? One way to answer this is to examine the testimony of his dedicated enemies—the three Catholic scholars who wrote formal replies to the *Exercitationes*, Heribert Rosweyde, Andreas Eudaemon-Joannes, and Julius Caesar Bulenger. The tone these gentlemen took is well indicated by one of Rosweyde's opening remarks: "Sophocles was absolutely right to say 'A scorpion lurks under every rock.' Clearly you are a scorpion, Casaubon, you are a scorpion."[59] All three treated Casaubon with scorn, trying as desperately to find weak spots in his work as he had tried to find them in Baronio's. Yet, as Casaubon's defender Jacques Cappell later pointed out, none of them tried to defend Hermes. Eudaemon-Joannes and Rosweyde ignored the issue. Bulenger accepted Casaubon's claim: "Many ancient and modern authors support Baronio's judgment on Hermes Trismegistus. But I do not disagree with yours ... An author who mentions Phidias and Eunomus Locrus the *citharoedus* must be late ... "[60] Instead, all of them attacked Casaubon at length for the next chapter of the *Exercitationes*, in which he had cast aspersions on the Sibylline Oracles.[61] Two conclusions may be drawn. First, the Sibylline question seemed far more vital to contemporaries than the Hermetic question—as is not surprising, given the great prominence that the Sibylline prophecies had enjoyed in patristic literature and the careful attention they had won from Renaissance scholars. As Chapter 6 will show, we need to study the history of the Sibylline tradition if we are to see the Hermetic tradition in its true proportions. Second, Casaubon's opponents evidently could not fault the evidence and arguments he brought to bear on Hermes. His defender Cappell not only accepted Casaubon's conclusions but applied a similar method to the Sibylline Oracles as well. If there had been a flaw to be found—by seventeenth-century standards—in Casaubon's work, one of these men would have turned it up.

A second test is to follow the afterlife of Casaubon's thesis in the scholarly literature of the seventeenth-century Republic of Letters, which Casaubon had helped to create. By and large, Protestant scholars seem to have accepted Casaubon's views. G. J. Vossius, who saw all of pagan mythology as a debased retelling of the Old Testament, naturally found a rejection of the *Corpus* congenial. Thomas Stanley incorporated Casaubon's arguments, briefly summarized, into his history of ancient philosophy. Even his uncle John Marsham, who believed that most of Jewish law and ritual came from Egypt, and that a Hermes really had educated the Egyptians, wrote of the *Pimander*, "Casaubon has shown that that book was assembled from Plato and Holy Scripture." Marsham's strict Calvinist opponent Witsius agreed with him on this point as strongly as he disagreed on most others.[62]

Moreover, even those who took a special interest in the problem and criticized Casaubon on points of detail did not challenge the central lines of his solution. Hermann Conring—to take a reputable example first—agreed with Francesco Patrizi that the *Corpus Hermeticum* was not the unified corpus that Ficino had considered it. He criticized Casaubon for assuming that all the texts had been assembled by one author and for one purpose. But he agreed with Casaubon that several books of the *Corpus*—notably 1 and 4—were Christian in language and content. More generally, he also agreed that no section could be used as a source for the history of genuine Egyptian philosophy:

> It is clear that the texts that Marsilio Ficino was the first to publish and translate into Latin under the single name "Pimander" are of different qualities and by different writers. Some of them seem to have been invented and falsely ascribed to Hermes by Christians, others by Platonists; very few of them give off the smell of ancient Egyptian doctrine. True, Plato may have owed his theology to the Egyptians . . . but the Egyptian religion had a good many points that he apparently did not agree with. And even those few bits which we said were more Egyptian than the rest are not from the original Hermes or any ancient writer. Casaubon's view of all sections of that work was quite correct.[63]

Conring quoted Casaubon's condemnation of the diction of the *Corpus* with full approval. And he went on to develop a general thesis with which Casaubon would have been much in sympathy: namely, that the extant evidence did not permit the reconstruction of Hermes' natural philosophy, if there had ever been a Hermes, which Conring doubted.

Even the view from the lunatic fringe was not much different. Isaac Vossius, whose interests ranged from Chinese civilization to

Greco-Roman prostitution, brought to the study of the *Corpus* and the *Sibyllina* wide reading, experience in dealing with Hellenistic and late-antique sources, and no powers of judgment whatsoever. He argued that the *Corpus*—and everything else from the Sibyllines to the pseudepigraphical works that circulated under the names of the patriarchs—was a divinely inspired prophecy of Christ's coming. Yet even he did not try to dislodge the *Corpus* from the general chronological niche that Casaubon had carved for it:

> All these books [Vossius wrote] were, in my view, written toward the end of Daniel's weeks by Jews everywhere in the world, whom God led to reveal Christ's advent to the Gentiles. Therefore they published an enormous number of works, partly under the names of their patriarchs and prophets, . . . partly under the names of those who enjoyed great reputations with the Gentiles, such as Hystaspes, Mercurius Trismegistus, Zoroaster, the Sibyls, Orpheus, Phocylides, and many others.[64]

Even Casaubon could not have composed a more appropriate list of companions for Hermes.

There were defenders, to be sure. Cudworth, though he accepted the philological arguments Casaubon had advanced, drew different inferences from them. Like Conring, he treated the *Corpus* as a congeries of disparate materials, some genuinely Egyptian. Unlike Conring, he claimed that the genuine matter could be distinguished from the spurious, and that the presence of parallels with Plato did not prove that a text was non-Egyptian (after all, Plato went to Egypt; he had presumably borrowed the doctrines in question). Moreover, even the forged parts must have some genuine elements. If the forger had invented everything, contemporary readers would have seen through his handiwork at once.[65]

These arguments have a limited measure of plausibility. Casaubon himself as a young man had speculated on the Egyptian origin of some of Solon's laws. And he had insisted upon the merits of the philosophical achievements of non-Greeks.[66] An inquiry as to how he would have met a posthumous attack is a fruitful field for wild speculation. Yet I cannot help suspecting that his reply would have moved along the lines later taken by the historian of philosophy Jacob Brucker: "though one can hardly deny that there are some remains of ancient Egyptian doctrine in the Hermetic books, we cannot tell them from the spurious bits."[67] Cudworth fails to meet the argument from diction.

Of Catholic reactions, after the immediate ones, I know less. One major opponent was that maddest of polymaths and most learned of madmen, Athanasius Kircher, who believed throughout his life that he was destined to revive a lost Egyptian wisdom. He argued that in antiquity, so far as he could tell, "The Greek and Egyptian languages differed in much the same way as Italian and Spanish or French differ from Latin. Hence it was not hard for Greeks to learn Egyptian, or Egyptians Greek."[68] Only thus could one explain the ease with which Greek philosophers had undertaken their trips to study with Egyptian priests. Only thus, too, could one explain some of the stunning etymologies that Kircher was privileged to discover for the names of Egypt's Gods: "For what is Orisis, but *hosios hiros*, sacrosanct? What is Isis but the prudent movement of nature, a name derived from *iesthai* . . . ?"[69] Even if Hermes did make Greek plays on words, he could have been a genuine Egyptian. Kircher suspected that he had originally written in Coptic.[70]

Kircher built whole edifices on these foundations of sand. He drew from the *Corpus* as well as Horapollo and Iamblichus the building blocks for those monuments of baroque wit, his readings of the hieroglyphs incised on Roman obelisks and the Tabula Bembina.[71] But Kircher's etymologies convinced few. Warburton reflected a widespread opinion when he ridiculed him for "laboring thro' half a dozen Folios with Writings of late Greek Platonists, and forged Books of Hermes which contain Philosophy not Egyptian to explain old monuments not Philosophical."[72] And Casaubon's position found a characteristically sharp exposition in the standard works of Brucker:

> Iamblichus bears witness that this was the way of things: the Egyptian priests made a habit of ascribing their own works and ideas to Mercury. This both reveals the arrogance and mendacity of these men and explains how Egypt brought forth so many Hermetic books. But this probably happened at a late period, under the Ptolemies, when the Pythagorean-Platonic philosophy had become dominant in Egypt. The original monuments of the

first Mercury sketched out theogonies. When the key to these was lost, anyone could easily make his wild ideas salable by disguising them [as Hermes'], and perverting the latter into physical theories. And this is the category to which the Hermetic books that are extant today—among which the *Pimander* and *Asclepius* are outstanding—belong. They are full of the dreams of Alexandrian philosophy.[73]

For Brucker the entire modern Hermetic and neo-Platonic tradition was merely history—an understandable aberration of Renaissance scholars who had been fooled by forgeries.

The record suggests that Casaubon found no opponent worthy of him. Plenty of literate men ignored his arguments; plenty found them far too new and upsetting to be acceptable. The learned James Howell, for example, could not see "how they do the Church of God any good service or advantage at all, who question the truth of their [the Sibyls'] Writings (as also *Trismegistus* his *Pymandra*, and *Aristaeus*, &c), which have been handed over to posterity as incontroulable truths for so many Ages." Only "cross-grain'd People," who "read an old Author . . . to quarrel with him, and find some hole in his coat," would "prefer *John Calvin*, or a *Casaubon*" to such venerable antiques.[74] But moderately strenuous research has not turned up a sustained and cogent counterargument. To that extent, Hermes' career did reach its not inglorious end when Casaubon brought him down.

The curious case of P. E. Jablonski confirms this view. His *Pantheon Aegyptiorum* (1750-1752) reopened the case for a lost Egyptian wisdom, laid down by Hermes in books that the different priestly orders had memorized. Jablonski compared these to the Vedas of the Brahmins, "those disciples of the ancient Egyptians," and argued that they had contained sophisticated treatments of astrology, astronomy, hieroglyphs, geography, cosmography, and the behavior of the Nile. The modern writers who debunked traditional claims for Egyptian science were misguided. They did not see that the absence of any solid evidence was by itself solid proof that the Egyptians had been great sages, who had faithfully concealed their secrets from ordinary men. Had they been willing to come out with their astronomical discoveries, everyone would see that Egyptians, not Copernicus, had devised the heliocentric system.[75]

Even Jablonski, however, did not defend the *Corpus*—as opposed to the lost *Hermetica* that he constructed, his imagination untrammeled by evidence. Competent judges, he admitted, "have lost all faith in the Hermetic books. They are merely outpourings from the brains of late Pythagorean philosophers, or even Christian Gnostics."[76] In the age of *The Magic Flute*, Hermes' vision of ancient Egypt had lost most of its appeal. In any case, he had written the wrong sort of Greek to express the wrong sort of idea. If so credulous a man as Jablonski saw that, then Hermes was dead indeed.

Naturally Casaubon's treatment of Hermes had flaws. His prejudices kept him from seeing that much of what was late-antique in Hermes could as easily have been written by the pagans of the time as by Jews or Christians.[77] And his constructive arguments were far weaker than the destructive ones. Casaubon knew that some proved forgeries were historical ore that should be mined, not slag that should be discarded. He recommended the study of the *Orphic Hymns*, though he knew that Orpheus had not written them.[78] Even when he thought the *Letter of Aristeas* unauthentic, he found it interesting: "Though we do not think that little work is as old as it pretends to be, we still believe that it is quite old, and that it is the sort of book that the student of Greek language or antiquity will not regret having read."[79] But a forgery that challenged the priority in time and preeminence in clarity of the Judeo-Christian revelation—and made a poor job of the challenge—deserved contempt only. All of these defects in Casaubon's historical insight, moreover, stemmed from the convictions that impelled him to attack Hermes in the first place.

Still, Hermes had to be brought down to human size and back to historical time before the Hermetic revelation could be studied for what it was, and that Casaubon did. One can only admire the good sense of the man who wrote: "I must confess that the clearer oracles and utterances of this kind are, the more I suspect them."[80]

6

The Strange Deaths of Hermes and the Sibyls

THE VOICE of humanism never sounds more modern than when it rings out at the Renaissance high table, excising still more fakes from the canon. Historians of humanism always celebrate Lorenzo Valla's demolition of the Donation of Constantine and his suspicions about Dionysius the Areopagite.[1] More generally, the ability to distinguish between the spurious and the genuine, the modern and the antique, seems a central feature of the humanists' new sensibility. Medieval writers—we often read—produced the *Institutio Traiani,* the *De vetula,* and the *De disciplina scolarium* because they lacked a sense of individuality and historicity. Any very good book seemed to them "good enough to be by Ovid or Boethius," and they happily suited their titles and subscriptions to their naive pride in their workmanship, ignoring the anachronisms of style and content that encrust their most classical productions like Gothic ornament. Where medieval humanists accumulated, however, Renaissance ones discriminated. The revival of the classical heritage involved not only the discovery of what was lost but the expunging of what was false. This violent act of purgation deprived Cicero of his once-classic *Rhetorica ad Herennium,* Aristotle of his *Theology,* and Seneca of his correspondence with Saint Paul.[2] And in the ability to detect the corrupt and the spurious, the humanists created a critical art without a literary precedent. So we are told.[3]

Yet this story is far too clear to be correct. It suppresses, first of all, the simple fact that medieval scholars—and forgers—had a discriminating sense of the past as well as a feeling for the styles they imitated. The author of the *Institutio Traiani* set out to write a quasi-antique account of the state as a macrocosm of the human body because he knew that no ancient source offered what he needed; forgery, far from being naive, is often the tribute that the sense of history pays to the facts.[4]

More seriously, the accepted account suggests that forgery ceased to play a prominent role in Western classical scholarship after humanism arose. And that is surely false. It was not only Michelangelo who began a prominent career by faking antiques. Annius of Viterbo, as Chapter 3 showed, forged a set of complementary ancient histories of the world that satisfied most readers for two generations and were still being anthologized for young history students a century after they were published.[5] And even Joseph Scaliger, that human Geiger counter who began to tick loudly as soon as a forgery entered his presence, made some wrong diagnoses (as when he ascribed to Petrarch a supplement to Quintus Curtius that was written no later than the twelfth century) and approved enthusiastically of some fakes (like the *Hieroglyphica* of Horapollo, which he thought both lucid and genuine, and the Hermetic Corpus, which Scaliger helped to edit in the 1570s, and which he described as "even more exciting [than Philo], and very old indeed").[6] Simple maps will not guide us down the roads we need to travel.

Let us try rather to raise some complex questions, by examining a classic of Renaissance higher criticism: Isaac Casaubon's great attack on the Hermetic Corpus, that lurid and attractive body of religious and gnostic literature from the later first, second, and third centuries, written in Greek but presented as revelations associated with the ancient Egyptian sage Hermes.[7] Considered genuine in Byzantium, the Corpus enchanted its first readers and students in the West. Ficino delayed translating Plato, at the direct request of his patron Cosimo de' Medici, to put Hermes into Latin; and for the next century and more, theologians, scientists, and magicians alike saw the texts as a rich source of information about the natural theology and natural magic of the ancient Egyptians—

one both far clearer and more powerful than the later speculations of the Greeks.⁸

Casaubon read this text, perhaps for the second time, while preparing his vast attack upon the Catholic church history of Cesare Baronio. He filled his working copy of the Corpus (now in London) with acid remarks, and then transmuted these into a brilliant gem of historical criticism, a biting essay in which he proved that the text could not be genuine. Its language was Greek, not a translation from the Egyptian; otherwise it could not contain Greek puns. The Greek, moreover, was not early but late, for it was pocked with technical terms derived from Greek philosophy and Christian theology. The author mentioned the Greek sculptor, Phidias, who had lived long after the Egyptian Hermes. And he knew more about Christian theology than any pagan (or Jewish) prophet who had lived before the birth of Christ. Such a text could not come from the chronological niche to which tradition assigned it.⁹ No wonder that modern scholars have been struck by the modernity of Casaubon's method, and have reared a place of honor for him in the critical Pantheon directly between those of Valla and Mabillon.¹⁰

As we have seen, Casaubon was not the first to be struck by gross factual anachronisms in the Corpus. And even the subtler stylistic tests that Casaubon really was the first to apply to Hermes had also been in use for a long time, though in different settings. True, ancient critics' decisions about the authenticity of literary works were generally couched—or least reported—in vague language and apodictic form, and revealed little of the reasoning behind them.¹¹ But the ancient technical works so dear to the scholars of Casaubon's generation swarmed with detailed discussions of authenticity. Diogenes Laertius and the commentators on Aristotle made clear that one needed a total grasp of the Aristotelian corpus and a discriminating sense for Aristotle's style in order to tell the genuine Aristotelian texts from the spurious ones that clung to them like a layer of barnacles.¹² Galen—a connoisseur of forgery, who relished the spurious obscure work that Lucian wrote under the name of Heraclitus, fooling a famous philosopher—made the same point about Hippocrates, repeatedly and at length.¹³

Renaissance scholars knew these ancient precedents well. Consider, for example, Galen's commentary on the Hippocratic work *On Humors,* with its elaborate argument that the demands of the great Hellenistic libraries at Alexandria and Pergamum had raised the prices for rare books, and thus stimulated the Hellenistic mania for literary forgery.[14] This work not only offered localized discussions of technical points but adumbrated a history of forgery itself. And it is revealing in more than one way. The doctrines it purveys are Galen's, attested in his commentary on *De natura hominis* and elsewhere, and show how dexterously an ancient scholar could set forgeries into their context.[15] But the commentary on *Humors* is itself a Renaissance forgery, cobbled together from real bits of Galen and other sources, notably Maimonides.[16] And that shows that even a forger could know enough about the classical theories about his art form to incorporate them into a specimen of his work—and thus, paradoxically, to make a great detective help a burglar with his crime. No wonder that a learned man like Janus Wowerius could assemble a dozen stimulating examples of *antike Echtheitskritik* in his *De polymathia* of 1604.[17]

Porphyry had used even more advanced techniques in his effort to prove the inauthenticity of a very great forgery indeed, the Bible, as all readers of Jerome's commentary on Daniel knew.[18] And Porphyry, as Jerome also showed, had learned his methods at least in part from Christians, from some of the stoutest pillars of the early church. At an unknown date, perhaps around A.D. 250, Julius Africanus wrote to Origen, to urge him not to cite as authentic the story of Susanna that occurs in the Septuagint text of the book of Daniel.[19] His letter is a little masterpiece of the higher criticism. He points out that the style of the episode differs from that of the rest of the book.[20] He argues, exactly as Casaubon would 1350 years later, that a book containing Greek puns cannot be a translation from a Hebrew original.[21] He shows that the book does not fit the context it claims to come from:

> The Jews were captives in Babylonia, strangled and thrown unburied in the streets, as we are told of the first Babylonian captivity, their sons torn away from them to be made eunuchs and their daughters to be made prostitutes, as had been prophesied. Could they then pass a sentence of death, and moreover one against the wife of their king Joachim . . . ?[22]

And he goes further than Casaubon would in one crucial direction, by pointing out that the work suffered from literary as well as historical inconsistencies:[23] "Susanna having been condemned to die, the prophet, seized by the spirit, cried out that the sentence was unjust. First of all, Daniel prophesies in a different way, always by visions and dreams, and he has an angel appear to him; but never by prophetic inspiration." Africanus's arguments have been variously judged by his readers, from Origen and Porphyry on.[24] What matters for our purposes is not which side is right but that the debate took place. More than a millennium before Valla and Casaubon, pagan and Christian scholars had created what remain the standard weapons against forgery and were vigorously debating how to apply them in practice. And some Renaissance critics—notably Campanella and Heinsius—took a keen and direct interest in these ancient arguments.[25]

What we need to account for in Casaubon, then, is not the general method that he learned from others, but his ability to apply it so thoroughly and his desire to apply it to Hermes. The first question really amounts to a linguistic one. How did Casaubon know so clearly that Hermetic Greek was late? And the answer, as is often true in the history of scholarship, has less to do with method than with data. Casaubon was born in 1559. By then virtually all Greek texts that would be known before the second Renaissance of the papyrologists had reached print; and even those which had not, like the philosophical works of Sextus Empiricus, circulated in manuscript, in printed Latin versions, or at least in summaries.[26] By Casaubon's thirteenth year the greatest of all aids to Greek scholarship had also come out: the vast Greek *Thesaurus* that was the masterpiece of Casaubon's future father-in-law, Henri Estienne. This magnificent root-by-root survey of classical Greek proved so expensive, and so far beyond the needs of most scholars caught up in that terrible time of religious and civil war, that it was remaindered for almost nothing and became accessible to many who, like Casaubon himself, could never have paid the high price that the publisher demanded.[27] Late-sixteenth-century scholars, in short, could read Greek literature in the unmysterious form given it by mechanical reproduction, for the most part in the sturdy bilingual folios that became the standard format during the middle decades of the century. They could turn for help to solid manuals of usage, syntax, mythology, geography. And the very ability to read the whole tradition, end to end, had the effect that close observation always does. It changed the tradition itself. The Greek texts, seen together, shifted positions and alignments, as suddenly as iron filings obeying the commands of a magnet whose poles have switched position. One can watch this process taking place in Estienne himself. No historical thinker, he still saw that the hexameter poems attributed to Orpheus and Musaeus had less in common with the early works of Hesiod and Homer than with the late ones of grammarians like Nonnus. Accordingly, he put them at the end rather than at the beginning of the great corpus of Greek hexameter poetry that he issued in 1566.[28] And this simple editorial decision, laconically justified, did as much as many pages of Julius Caesar Scaliger to enable Renaissance scholars to follow the evolution of Greek literature as a historical process.

In this context, Casaubon's ability to deconstruct Hermetic Greek presents no mysteries. Casaubon was a master Greek scholar with a special taste for the language of the New Testament and the fathers of the church. When he saw abstract composite nouns and adjectives like *huperousiotēs, aoristia,* and *homoousios* in a text that supposedly antedated Homer and Hippocrates, Casaubon needed no time for reflection. He jotted down parallel usages in his margins, together with their late sources, Origen, Dionysius the Areopagite, and John of Damascus. And he summed up his views in one definitive statement, which derives its force from the absolute solidity of Casaubon's data: "The style of this book could not be farther from the language that the Greek contemporaries of Hermes used . . . On the contrary, there are many words here that do not belong to any Greek earlier than that of the time of Christ's birth."[29]

The question of motive is more complex. Why train such big guns on Hermes? In the ancient world forged scriptures by *prisci theologi* had sometimes encountered criticism. Porphyry wrote, at Plotinus's request, an elaborate refutation of a work attributed to Zoroaster, "which I showed to be entirely spurious and modern, made up by the sectarians to convey the impression that the doctrines which they had chosen to hold in honour were those of the ancient Zoroaster."[30] Such arguments seem immediately relevant to Hermes. And in fact, if we can trust the difficult and confusing evidence of Iamblichus *De mysteriis* 8, Porphyry did wonder why the Greek Hermetica, supposedly translated from the Egyptian, contained Greek philosophical terms.[31] We can certainly sense the embarrassment that such questions caused in Iamblichus's effort to reply (he claimed that the translators, themselves skilled in Greek philosophy, had added the offending language).[32] And we know that even less critical readers noticed seams and inconsistencies in the ancient Hermetic collections.[33] On the whole, however, Hermes suffered little damage from the critics' scalpels. Saint Augustine accused him not of forging texts but of worshiping demons.[34] And Lactantius revered Hermes, seeing him—like the Sibyl—as an ancient pagan who had clearly prophesied Christ's coming and Christian doctrine.[35]

The treatises of the Hermetic Corpus resemble a rich hot fudge sundae, the ingredients of which have flowed together so stickily as to discourage efforts at analysis. They are permeated with the contradictions and confusions natural in what began as collections of the oral teachings of leaders of small groups (and this humble social world, as Garth Fowden has shown, was the culture in which Hermetism flourished).[36] They are marked by desperate efforts to assert the validity and power of a native Egyptian culture almost extinguished by Macedonian and Roman invaders (and thus crucial aspects of the Hermetic "vision of the vanquished," like the emphasis on the divine powers of the hieroglyphs, could express themselves only in the conventional Greek terms and ideas of late antiquity). To the modern reader Hermes' emphasis on the lost Egyptian original of his writings eerily resembles Dictys's and his redactors' dithyrambs about the original Punic script—or language—of his romance.[37]

Most early readers, however, were struck above all by the clarity, originality, and power of what now strike us as turbid, derivative, and chaotic works. The Corpus did seem just the sort of thing that would have been written by Hermes, an Egyptian sage not much later than Moses; its content seemed a sober and satisfactory blend of Platonic philosophy and biblical theology. No wonder then that the blurb in the first printed edition of Ficino's Latin urged the reader, "whoever you are, grammarian or orator or philosopher or theologian," to buy Hermes "because for a small price I will enrich you with pleasure and profit."[38] No wonder either that Hermes offered especially good reading to Christians, as Angelus Vergicius assured buyers of his *editio princeps* of the Greek, since he showed that the pagans "seem to agree with our scriptures on some points, as is clear from the present text."[39] He offered rich rewards to the Platonist, as Francesco Patrizi reasoned in his account of Plato's works, since his books were the source of Plato's dialogues and were aimed at young students—and thus did not enter into polemics, as Plato had to, with men uninterested in philosophy and with sophists.[40] In short, "he blended philosophy with the divine oracle," as François de Foix de Candale assured readers of the edition he and Scaliger produced, and could thus be recommended to Christians so long as they realized that the magical sections of the Latin *Asclepius* were interpolations made by its impious translator, Apuleius.[41] Casaubon agreed with his predecessors about the nature of these symptoms, but disagreed *toto coelo* about the diagnosis that ensued. Where others read with sympathy the Hermetic accounts of creation, fall, and salvation, he read with rage. Biblical parallels proved only that Hermes had plagiarized the Bible. Otherwise one must accept, as Casaubon's predecessors did, a pagan revelation clearer than the Jewish. And that, as we have seen, he could not do. The Hermetic Corpus must be false, in short, because its content was true and its exposition clear.

Casaubon was not the first to insist that Hermes had borrowed from Revelation. In the eleventh century Michael Psellus had made the same argument: "that phrase 'And God said increase and multiply,'" he wrote in his copy of the text, "clearly comes from the Mosaic account of Creation." But he drew an inference quite different from Casaubon's—namely, that since Pimander, the narrator of the first text in the corpus, had revealed divine secrets at so early a date, he must in fact have been the devil—who is, as we all know, a great thief.[42] But Psellus's note seems to have attracted no attention in Casaubon's time; and in any event Casaubon made the author of the Corpus not a fallen angel but a deceptive man. How then can we explain that what had seemed to Ficino and Lefèvre, Champier and Patrizi holy simplicity now seemed a simple unholiness?[43] One way to do so, of course, is to recall Casaubon's own religious position. A devoted—though complex—Calvinist, he clung to his church in spite of his fellow Calvinists' ignorance and intolerance and his Catholic friends' untiring efforts to convert him. Naturally he had to reject a text that claimed to be older and clearer than the Bible.[44]

But Casaubon's Calvinism was only a necessary, not a sufficient precondition for his attack on Hermes. After all, his predecessor Génébrard was a fanatical Catholic; and Scaliger, who found the Hermetic Corpus to his taste, was a more thoroughgoing Calvinist than Casaubon himself. What bothered Casaubon in Hermes' case was less the Protestant doctrine of *sola scriptura* than the larger and more amorphous dilemma that faced all Christian humanists: how to rob the Egyptians with moral impunity. In his remarkable diary Casaubon recorded his ways of dealing with the texts that mattered most to him. In 1597 we hear him talking to himself about Seneca:

> *13 May.* Early prayers, Hebrew, and Seneca. The reading of this writer does me good because, by a new sort of burlesque, I accommodate everything that he writes about wisdom, and about the wise man that the Stoics imagined and described but never saw, to true piety. Thus I ensure that much that the philosopher wrote falsely about his wise man becomes true. For Seneca wrote much that cannot be understood or believed without the sense of true piety. Since he lacked this good, it follows that we must say that he wrote this not by certain knowledge or faith, but as it were prophesying, and seized by inspiration like the poets.[45]

Seneca, in short, did anticipate some features of Christianity, but unclearly and unconsciously; only his later Christian reader, not his pagan contemporary, could enjoy the *plērophoria* of his true sayings about the soul and the afterlife.

In February 1599, by contrast, we can listen in on Casaubon listening to Job. The contrast is sharp:

> *14 February.* Before sermon I was with Job and Mercerus, as much as I could be. O sweet food for the soul! O Job, divine example of all the virtues! For Maimonides is quite wrong to hold, in the *Guide for the Perplexed,* that the book contains not

the narration of an event but, so to speak, the image and idea of a pious man. There can be no doubt that a Job existed, the one mentioned by Ezekiel and the Apostles, and that everything contained in that book happened to him. I humbly pray, great God, that reading this may bring me as much profit as it does pious pleasure.[46]

Job—and, presumably, the rest of the Bible as well—has a simple relation to truth where the pagans have a complex one. The scholar reading Job must take care not to invent flaws in the crystalline surface of his text, and must let the fathers guide him when dark passages occur.[47] Given these simple precautions the reader of the Bible can do what readers of the pagans never can: let himself yield trustingly to the pleasures of his text.

This combination of attitudes had a specific pedigree, with the humanists of the early sixteenth century—Erasmus & co.—at its head. These men had simultaneously pleaded for the value of the pagans, the difference between paganism and Christianity, and the uniqueness of the Bible. Erasmus in particular urged Christians to read pagan texts with all due caution. His ideal Christian prince, for example, must be warned before reading a classic that "He whom you are reading is a pagan, you who read are a Christian; while much that he says is splendid, his sketch of the good prince is hardly correct."[48] Erasmus explained in detail that even Seneca and Plutarch were not Christians, the pagan historians and orators far less so. At the same time, he urged everyone, scholars, statesmen, and ordinary men and women, to see the Bible as the one perfect book, uniquely full, direct, and clear:

> The Father spoke once, and created the eternal discourse *(sermo)*. He spoke again, and created the whole machinery of this world with his omnipotent word. He spoke again through his prophets, through whom he bestowed on us the holy books, as they hid the vast treasure of divine wisdom under a few simple words . . .[49]

Casaubon looked back to Erasmus—the Erasmus of the last great programmatic works, *Lingua, Ciceronianus, Ecclesiastes*—as an intellectual model. In him he found a clear method of making classical scholarship serve Christian ends, a pure distaste for the Egyptian mysteries that fascinated so many of his friends, and a passionate commitment to scholarship as the tool that could blaze a path through pseudepigrapha and textual errors back to the early, incorrupt church. Erasmus too had denounced the early Christians, who ascribed so many spurious works to famous authors because "in those days even pious men thought it pleasing to God to use this deceit to inspire the people with eagerness to read." Erasmus attacked Dionysius the Areopagite and Seneca the correspondent of St. Paul; his colleague Vives did the same to Aristeas.[50] Casaubon—and his intellectual conscience, Scaliger—paid tribute of many kinds to Erasmus's memory. But their homage to him was nowhere manifested more profoundly than in their systematic effort to purge the early church of the work of those who "thought the word of God so feeble that they feared the kingdom of Christ could not be furthered without lies"—an effort that led Scaliger to attack the Jewish forgeries of "Aristeas" and "Phocylides" as well as the "Pseudosibyllina oracula" that he attributed to the Christians.[51]

One of these parallel traditions—that of the Sibyls—deserves our attention here. For no pagan anticipator of the sixteenth-century True Churches found more respect—or more readers—in all of them than the various Sibyls, those powerful ladies who recounted the Creation and the history of the world and predicted the coming of Christ and the downfall of Rome in pedestrian but popular Greek hexameters. After all, they had the best imaginable letters of recommendation from the ancient world. On the one hand Cicero had denounced at least one of them as a fraud in *De divinatione*.[52] He argued that the author of the acrostic oracle he saw had disguised an account of the past as a prediction of the future. She had been deliberately obscure in order to preempt criticism, and in any case she had given her game away by the genre in which she chose to write. Acrostics, Cicero stiffly explained, must have been the work "scriptoris . . . non furentis, adhibentis diligentiam, non insani." On the other hand better men than Cicero loved the Sibyls dearly. Virgil wove them into his poems. Constantine himself had quoted the verses of the Erythraean Sibyl, repeatedly praised their clarity, and used Cicero's testimony to prove the antiquity of the *Oracula* against those unnamed critics who saw them as Christian forgeries.[53] Even Augustine sided with Constantine and speculated that the Sibyl must inhabit the *civitas Dei*.

Most sixteenth-century readers threaded their way without anxiety through what now seems a minefield of contradictory opinions. They found guidance in the late Greek prologue to the collection of the *Oracula*. This repeatedly praised the clarity with which the Sibyl expounded the Mosaic history of the Creation and the Fall, the Evangelists' account of the career and passion of Jesus and the basic doctrines of Christianity: "In manifold ways they tell of certain past history, and equally, foretell future events, and to speak simply, they can profit those who read them in no small way" (tr. J. J. Collins). Those most committed to the Sibyl—like the first editor—usually found further reassurance outside the text, in their own comfortable humanist belief in the partial goodness of all men, pagans as well as Christians. The teacher and dramatist Xystus Betuleius argued in his *editio princeps* of 1545 that the pagan belief that the oracles were obscure proved only that they were unintelligible to those not meant to grasp them ... a mark of all true prophecies.[54] Cicero's proof that the oracles were written with deliberation, not from inspiration, amounted only to a recognition of the oracles' sanity and holiness.[55] And Cicero's argument that very accurate prophecies must really be histories written *post eventum* quite escaped Betuleius's notice. Instead he concentrated on listing parallels between the Sibyls and the Hebrew prophets: "Do we not clearly hear Zacharias here ... ? Does she not say exactly what Hosea says ... ? And she describes the same celestial city of which a fine *ekphrasis* is to be found in the Apocalypse."[56]

As Johannes Geffcken remarked, even in the early sixteenth century "Freilich gab es schon ... kritische, uns sehr sympathische Leute."[57] Sebastian Castellio, in the dedication to his Latin translation of 1546, admitted that "some people may find these oracles too clear *(nimis aperta)* and therefore think some Christian forged them."[58] He accepted their description of the text but drew the opposite conclusion from it. The Gentiles, he argued, "had to have clearer predictions because they did not have Moses or any other preparation for Christ."[59] And in ancient Greece, in any case, the oracles would not have seemed so clear to the few who read them as they did in 1546. In fact, a forger would have made them more obscure. "Their clarity," he concluded, "makes them seem genuine, not forged."[60]

This argument did not convince everyone. The Basle scholar and Erasmian J. J. Grynaeus argued in 1569, on exactly the ground Castellio tried to refute, that the oracles were "not very old."[61] But while everyone knew that the Sibyls could be challenged, few saw any reason to do so. A nice case in point comes from Princeton, where an anonymous early reader has left copious traces in a copy of Castellio's edition. He often shows alertness, as when he remarks that the sixth oracle, the shortest and most obviously Christian one, "seems to be by a different Sibyl."[62] But he offers no criticism, expresses no reservations, and summarizes at eloquent length the Sibyl's effusions about divine inspiration and lamentations about her Jugendsünden.[63] Ambivalence and credulity appear still more tightly mingled in Ioannes Temporarius. In the first edition of his *Chronologicae demonstrationes* (1596) he rehearsed and rebutted the objection that the Sibyl spoke too clearly, and concluded that though a pious Christian had compiled the extant *Oracula*, he had used real pagan sources. In the second edition (1600) he treated the *Oracula* as genuine pagan texts marred only by "nonnulla a Christiano quopiam intermixta." And in both editions, the cautious first and the bold second, he quoted the Sibyl as a prime authority on the degeneration of mankind before the Flood.[64]

The best Hellenists in sixteenth-century Europe agreed with Castellio and the anonymous in Princeton against Grynaeus. When Johannes Opsopoeus (known to his friends as Koch) came to Paris in the 1580s to work on the Oracula, he received a wealth of help from members of the same Parisian circle that Casaubon moved in during the 1590s—and from their masters, dead and alive, as well. The Royal Library supplied him with a late-fifteenth-century manuscript of the Oracles, François Pithou with another; Claude Dupuy lent him a ninth-century manuscript of Lactantius. Turnèbe's emendations proved helpful, as did Dorat's—some of which the old man gave Opsopoeus *viva voce*.[65] And Turnèbe's son provided Opsopoeus with an especially rich lode of relevant material: Aimar Ranconet's working copy of Castellio's edition, richly annotated and accompanied by a manuscript "quaternio."[66] This last batch of material, now in the Gennadius Library, Athens, enables us to meet Ranconet not merely in his familiar capacity as collector and patron but as a serious scholar, working through the Oracula, establishing book divisions and connecting related passages—and recording his conclusions in fluent Greek.[67] The Parisian Greek scholars had devoted themselves to the Sibyls. And they had done so, so far as I can tell, from the same motives that had inspired Castellio: the hope of finding in these texts the remnants of an ancient theology. This, at least, was how Dorat presented the matter in the 1560s to his eager student Willem Canter.[68]

As Opsopoeus attacked this mass of filings, the magnet finally shifted its poles—and for the same reasons that operated when Casaubon confronted Hermes. Opsopoeus too inherited both the learning of the mid-century Hellenists and the prejudices of humanist theology. He too knew that clarity meant fraudulence: "Isaiah predicted vaguely: Behold a virgin will bear a boy. But the Sibyl does so by name: Behold a virgin named Mary will bear a boy Jesus in Bethlehem. As though the Prophets predicted the future with less divine inspiration than the Sibyls . . ."[69] Like Casaubon, Opsopoeus piled up arguments from diction and content to prove the lateness of his text. And as he did so, he proved able to frame a larger argument—or rather to revive the larger argument that Cicero had advanced against his pagan Sibyl in 45/44 B.C. Opsopoeus too now saw that exact prophecy must really be disguised history: "many of the things the Sibyllines describe are portrayed so fully, copiously, and clearly as to make it seem plain that these things were not predicted."[70] And he too now saw as Cicero had that these texts were the product "of a calm mind, not a raging one."[71] An ancient model of higher criticism once again provided the armature on which modern prejudices arranged newly discovered facts.

We have seen two pagan sages executed, in public and with crowd-pleasing elegance. Method played little part in either execution, and such critical techniques as were applied were more or less as old as the texts they were applied to. In each case, clarity condemned, and scholarship carried out the sentence. Neither execution could have taken place unless prejudices and facts had interacted as they did. And neither corpse proved entirely willing to lie down. It would take a heart of stone to feel no sympathy with those seventeenth-century lovers of the Discarded Image who defended the sages rather than their learned hangmen. James Howell, for example, saw the rise of higher criticism as clear evidence of the decline in general morality. "Madam," he wrote—appropriately, to Lady Sibylla Brown—"in these peevish times, which may be call'd the *Rust* of the *Iron* Age, there is a race of crossgrain'd People, who are malevolent to all Antiquity. If they read an old Author, it is to quarrel with him, and find some hole in his coat . . ."[72] Following the authority of the Fathers, he did not hesitate to declare his belief in the Sibyls, who "spake not like the ambiguous *Pagan* Oracles in riddles, but so clearly, that they sometimes go beyond the *Jewish* Prophets."[73] He even managed to cite on their behalf the testimony of their great enemy Cicero. While Howell and other defenders could do nothing to diminish the power of Casaubon's and Opsopoeus's criticisms, the very existence of their views proves that the criticisms could not take shape save in a very specific context—and one, moreover, by no means coextensive with the whole learned world of the time.

This complex story has three simple morals. First, Casaubon's famous attack on Hermes employed principles that had been in use for more than a thousand years (though not continuously, of course). Casaubon specifically imitated Opsopoeus's recent attack on the Sibyls and indirectly followed Opsopoeus's ancient model, Cicero. Second, the higher criticism was the object as well as the instrument of a classical revival. The New Scholarship applied to Hermes and the Sibyls was not the invention of late medieval jurists and Renaissance humanists, but a very old scholarship indeed (though none the less impressive for that).[74] Third, the old methods Casaubon and Opsopoeus applied were made new because the data relevant to them had been enhanced by an order of richness. New data do not generate new questions. But they do, apparently, transmute old answers into something rich and strange. Porphyry, Cicero, Africanus, Erasmus—none of them could have framed arguments from diction as Casaubon and Opsopoeus did. They checked their texts against the new lexica they could read and the even more impressive mental lexica they had compiled by their own hard work, which divided the history of the Greek language into periods as hard and well-defined as geological strata. And this act of checking transformed their criticisms of texts from disputable hypotheses into what still read out as irrefutable statements of fact—the foundations of the dull but solid edifice of modern literary history.

At the end of the seventeenth century, Richard Bentley's enemies would make fun of him for dating the letters of Phalaris by the history of the Greek language: "He knows the age of any Greek word, unless it be in the Greek Testament, and can tell you the time a man lived in by reading a page of his book, as easily as I could have told an oyster-woman's fortune when my hand was crossed with a piece of silver."[75] So the late astrologer William Lilly was made to ridicule Bentley in a dialogue of the dead by William King. We, however, can see that Bentley's historical lexicon was even more profoundly modern than his other polemical weapons. We can see that the attacks on the Sibyl and ps.-Hecataeus that worried early readers of his *Epistola ad Millium* were implicit statements of his own debt to Scaliger and Casaubon.[76] And we can see that though he was no dwarf—indeed, he was so fat that he had to take two seats in the post coach—in his scholarship as in other areas he managed to stand on the shoulders of giants.

7

Humanism and Science in Rudolphine Prague: Kepler in Context

IN THE WINTER of 1612 humanism and science confronted one another in Prague. Humanism took the solid and threatening form of Melchior Goldast, Saxon *Rat,* and envoy; science took that of the imperial mathematician Johann Kepler, once called a pretty boy by the Greek scholar Martin Crusius in Tübingen but now bowed by financial trouble, racked by bad eyesight, and tormented by his inability to complete his great Rudolphine Tables of planetary motion and to discover all the harmonic ratios that governed the motions of the planets. The confrontation went exactly as one would expect, as Goldast recorded in his diary:

> Kepler the mathematician boasts that he has discovered a new world in the moon, which is supposed to be much bigger than this habitable world of ours. He thinks we will be put there after the Resurrection. But I quoted Scripture to him: "Heaven and earth shall pass away etc." He showed me an instrument with which I was supposed to look at the moon. The moon appears to be formed so that it is higher on one spot than on another. He wanted to convince me that these were mountains and valleys. I would rather believe him than climb up there and investigate.[1]

Here we see scientist and literary scholar trying to talk about a vital matter but kept apart like Pyramus and Thisbe by the thick intellectual brick and mortar that already separated the two cultures. Kepler is fired with enthusiasm by Galileo's discovery of the moons of Jupiter—a discovery first announced to him by his philosopher friend Johann Matthias Wacker, who was so excited that he shouted the news from his coach in the street outside Kepler's house.[2] Kepler imitates his Italian correspondent and turns a telescope toward the moon. He finds, as Galileo had, that the moon is not the regular and perfect sphere of Aristotelian cosmology but a bumpy and imperfect planet like the earth. Immediately he begins to see visions and dream dreams of a new heaven and a new earth. Goldast, jurist, historian, book thief, and literary scholar, reacts to new discoveries about the universe by looking for enlightenment in old texts. The editor, among other things, of a collection of treatises on the power and precedence of the Holy Roman Empire, volume 2 of which alone weighs in at ten pounds on my bathroom scale, Goldast kept his vision fixed firmly on the past and on his books. Even a look through Kepler's telescope inspired in him nothing more profound than a feeble joke.[3]

The encounter seems freighted with meaning; it seems, indeed, to reveal two world views in collision. One is empirical, turned toward the direct study of nature, open to imaginative speculations that could go wherever the facts might lead. The other is literary, bounded by vast authoritative texts that made speculation difficult and deluded Rudolf II and many others into thinking that the old world of humanism and empire that they knew was not slipping into dissolution. It took more modern men than Goldast, we think, to appreciate Kepler as we do—men like that English traveler, diplomat, and dilettante Henry Wotton, now remembered for his famous and injudicious remark that an ambassador "is a good man sent to lie abroad for his country." When Wotton met Kepler in Linz in 1620, Kepler showed him a landscape drawing he had executed—"me thought," says Wotton, "masterly done." Kepler smiled enigmatically and explained that he had drawn it "non tanquam pictor, sed tanquam mathematicus" (not as a painter but as a scientist). "This set me on fire," Wotton wrote; and he gave Kepler no peace until he had explained and demonstrated the camera obscura he had devised, which Wotton promptly described in detail to an even more famous and modern correspondent in England: Francis Bacon. With characteristic initiative Wotton invited Kepler to come back to England with him.[4] Such incidents—and others could be cited—seem to point up the backward and literary character of imperial culture in those transitional years around 1600, as well as the isolation of an innovator like Kepler in his German setting, and the peculiarly modern qualities of his mind and interests. We regret that he refused to abandon Linz for London because, as he quaintly explained to Bernegger in Strasbourg, a German like him, someone who loved to have a whole continent around him and feared the narrowness and isolation of an island, could not possibly accept the invitation Wotton offered—far less drag along his *uxorcula* and their *grex* of children.[5] In doing so, however, we wrongly accept modern disciplinary divisions as eternally valid. The twentieth-century historian of literary culture has normally felt justified in leaving on the shelf the stately gray and white volumes of Kepler's *Gesammelte Werke*, abandoning the historian of science to decode their highly technical diagrams and tables—and their highly elaborate Latin—unaided. And the historian of science in his turn has happily ignored backward fellows like Goldast and Herwart von Hohenburg, with whom Kepler spent time and exchanged letters, but who had no new data or theses to offer him. The result, wished for by no one but brought about by many, has been a distortion in our vision of the past. We have allowed the divergent forms of scholarship that we now recognize and practice delude us into reconstructing a past culture as fragmented as our own.

In recent years, to be sure, the walls between these two separate histories have begun to crumble in a few strategic spots. Robert Evans's and Thomas DaCosta Kaufmann's powerful books on the cultural history of the Habsburg lands have shown that the separation between scientists and humanists was not nearly so sharp as Kepler's confrontation with Goldast would suggest. The two men in fact formed part of a larger but coherent social world, a Prague province of the *Respublica litterarum,* which included scientists, scholars, brilliant, obsessive painters, and bold universal philosophers among its citizens. Readers of Raymund Lull and practitioners of artificial memory like Wacker and Hans von Nostitz tried simultaneously to explore the details of the mundane world, to botanize and to observe, and to enfold the new data they obtained into comprehensive systems as inclusive as the inherited one of Aristotle but more up-to-date in their factual content.[6] Indeed, we know that Goldast, Kepler, Wacker, and von Nostitz all had lunch together one day in February 1612, though sadly we do not know what they talked about.[7] Friedrich Seck has taught us to appreciate Kepler's devotion to literary and humanistic enterprises like the writing of Latin verse; Kepler, we now know, covered his scrap paper not just with thousands of computations and successive discarded models of planetary movements but also with successive drafts of Latin poems. He expressed his distaste for efforts to censor the theories of Copernicus in a bold epigram:

> Ne lasciviret, poterant castrare poetam,
>   Testiculis demptis vita superstes erat.
> Vae tibi Pythagora, cerebro qui ferris abusus,
>   Vitam concedunt, ante sed excerebrant.
>
> [They wished to keep the poet away from whores
>   So they castrated him, the awful bores.
> Thus of his testicles bereft of force
>   The poet could live, tormented by remorse.
> Poor you, Pythagoras, to feel worse pain,
>   You whose genius wore out iron chain.
> They took your brain out with their surgeon's knife
>   And left you what it's wrong to call a life.][8]

Not Martial or John Owen, certainly, but a sincere addition to the treasury of abusive Latin epigrams, which draws its tribute from the fifteenth-century Italian humanist Panormita, the sixteenth-century English humanist More, and the seventeenth-century German humanist Kepler.

To move from high style to high science, Nicholas Jardine has recently focused attention on Kepler's efforts to reconstruct the history of mathematics and astronomy in the ancient world. Kepler, he shows, both recreated specific ancient innovations with a lens grinder's meticulous attention to detail and rooted these in their wider social and cultural settings with a historian's bold flair for generalization.[9]

I hope to push these pioneering investigations a little further. We will see that the confrontation I began from actually stands out for its idiosyncrasy in Kepler's life and world. The tall volumes of his works reveal his intense and life-long devotion to the humanist enterprises of eloquence and exegesis. More surprisingly, Kepler's contributions to these fields show so high a level of creativity and learning as to establish him as one of the most distinguished humanist scholars of his time—one whose work demands the attention of humanistic scholars now. The journey will be a difficult one, through the dusty Faustian studies of scholars grappling with forgotten scholarly disciplines, across rebarbative pages of that macaronic language, half-German and half-Latin in vocabulary, half-Gothic and half-Roman in script, in which the scholars of the old Empire debated. But I hope that we will see enough curious sights and win enough enlightenment to make this strenuous effort worthwhile.

WE BEGIN with some parameters—and first, Kepler himself. He was born in 1571 and educated at Tübingen, where he learned astronomy from an expert Copernican, Michael Maestlin. He engaged in all the standard practices of the late-Renaissance arts student. He wrote mannered Latin on the Mannerist themes of physical curiosities and obscure emblems. He described this phase of his life in the explication that he drew up for his own horoscope, one of the most revealing autobiographical statements by any Renaissance humanist:

> This man was fated from birth to spend his time on difficult things that every one else shies away from. In his boyhood he precociously attacked the problems of metrical composition. He tried to write comedies, and chose the longest psalms to memorize. He tried to learn every example in Crusius's grammar by heart. In his poems at first he worked on acrostics, riddles, anagrams; when his judgment became more mature and he could esteem these at their own small value he tried various difficult genres of lyric poetry; he wrote a song in Pindaric metre, he wrote dithyrambs. He treated unusual subjects: the sun at rest, the origin of rivers, the view from Atlas over the clouds. He took delight in enigmas, looked for the saltiest jokes, played with allegories in

such a way that he followed out every minutest detail and dragged them along by their hair.[10]

Kepler thus portrayed himself as the normal graduate of the late Renaissance arts course, a lover of obscurity and erudition, of emblems and hieroglyphics, rather like those Altdorf students whose graduation exercises, including long speeches decoding the elaborate medals coined for the occasion, have been brilliantly studied by F. J. Stopp.[11] And he long kept his delight in such pursuits. He never gave up his search for a sufficiently ingenious anagram for his name, even though his efforts included such euphonious and elegant pseudonyms as Kleopas Herennius, alias Phalaris von Nee-sek. And in writing his Pindaric poem on a friend's wedding, following the meters of the first Olympian *(ariston men hudōr)* line by line, he showed how fully he shared the fascination of contemporaries like the younger Joachim Camerarius and Erasmus Schmid with Pindar's difficult but dazzling rhetoric and meter.[12]

Kepler, then, was by the 1590s a practicing humanist of the most up-to-date imperial style. He was also, thanks to Maestlin's influence, a practicing scientist, as the provincial astrologer in Graz, where he published yearly predictions and taught mathematics and astronomy. And after 1596 he became a famous and influential member of the German scientific community, thanks to his first remarkable book, the *Mysterium cosmographicum*. Here he tried to show that the Copernican world system was not only true in itself but also the key to a still deeper revelation: the very logic of geometrical proportion that had guided the Creator's hand. Kepler proved, so he and many readers thought, that God had used basic principles of geometry in laying out the planetary spheres. These could be shown to be separated by varying distances, in Copernicus's system, which in turn were exactly those which would have separated them if God had taken the five regular Platonic solids, arranged them in an aesthetically pleasing order, and interposed the spheres. Since there were five and only five such solids and six and only six planetary spheres, and since the correspondences were very close, Kepler felt certain that he had unlocked the Pythagorean logic that underpinned the process of Creation itself. And though such bold explanations and novel world systems were not uncommon in Kepler's time, the elegance of his geometry and the mastery of planetary theory that supported his philosophical and aesthetic arguments were so palpable that he found receptive readers across Europe. Even Galileo, though not very responsive, doodled some calculations modeled on Kepler's. Tycho Brahe, the great Danish observer of the heavens who was soon to take his vast collection of empirical data about the stars to Rudolf's Prague, was very impressed. Eventually he invited Kepler to join him in Prague, and thus made the astronomical revolution happen.[13]

Kepler must also be set into his environment—the encyclopedic intellectual world of the early-seventeenth-century empire that has found its sympathetic chronicler in Robert Evans. The old Empire—especially its Bohemian heart—was the *locus classicus* for the powerful agglomerative impulse that motivated so many late-Renaissance scholars. Some of the monumental products of that world still inspire awe and attract attention: for example, Athanasius Kircher's stately volumes on the monuments and hieroglyphs of Egypt. Kircher characteristically also produced profusely illustrated tomes on other subjects as varied as the route followed by Noah's Ark and the early history of China.[14] His interests—and those of his contemporaries—sprawled across centuries and continents, genres and disciplines with what now seems terrifying abandon. The ability of seventeenth-century scholars to combine scientific and humanistic interests, to use Near Eastern languages as well as Western ones, to move with obvious intellectual comfort from history to law to moral philosophy, is more likely to inspire bewilderment than admiration in the modern reader. True, efforts were made toward the end of the seventeenth century to map this vast and inaccessible intellectual country. Morhof's *Polyhistor* and its savage parody, Mencke's *Orations on the Charlatanry of the Learned,* offer a vivid panoramic introduction to the mental world of the polyhistors. But no modern scholar has retained full control over this dizzying baroque plethora of theories and information.[15]

Yet it seems clear that Kepler—student of texts, music, perspective, astronomy, and mathematics, measurer of barrels and writer of Neo-Latin poems—fits naturally, if not neatly, into this variegated intellectual scene. The comprehensive impulse of the encyclopedists appears in Kepler's desperate, life-long effort to distill neat geometrical models—or at least neat algebraic formulae—from the apparent chaos of the astronomical data. And the scattershot omnivorous quality of the polyhistors' interests characterizes many of his major and minor productions. Consider, for example, the tiny book that he dedicated to Wacker in 1611: his *Strena seu de nive sexangula*. Here Kepler describes himself as hurrying across the Karlsbrücke in Prague, desolate at his lack of an appropriate New Year's gift for Wacker, when snow begins to fall. Noticing that the drops are all hexagonal, Kepler wonders what secret logic of geometrical form or physical function can account for uniformity in so impermanent a material as snow. Two-dimensional hexagons lead to hexagonal solids, and soon Kepler is off onto a brilliant study of the advantages of hexagonal cells, their ability to be packed together without wasted space and their immense structural stability—both qualities preeminently visible in the beehive. Snow brings Kepler to geometry, geometry to bees, the combination of questions to the world's first essay on what might now be called crystallography and what was in its day a pioneering inquiry about natural processes that result spontaneously in geometrically regular products. All of this is wrapped, moreover, in a fine covering of Latin rhetoric, as Kepler divagates feverishly about the appropriateness of this gift, an essay on insignificant snowdrops, to Wacker whom he calls a "lover of nothing"— that is, a reader of the Epicurean atomist Giordano Bruno: "Eia strenam exoptatissimam Nihil amanti, et dignam quam det mathematicus, Nihil habens, Nihil accipiens, quia et de caelo descendit et stellarum gerit similitudinem"—"Here was the ideal New Year's gift for the devotee of Nothing, the very thing for a mathematician to give, who has Nothing and receives Nothing, since it comes down from heaven and looks like a star."[16] Kepler's birdlike hopping from subject to subject, his effort to find God's logic in the smallest and most evanescent of His creations, his strenuous efforts to cloak cogent and original argument about structures in the traditional strained conceits of Neo-Latin wit—all these characteristics mark him out as one of that breed of humanist-encyclopedists whose last and noblest representative was Leibniz, and whose brutal parody was Dr. Pangloss.

Yet we can, I think, go still further in teasing out the interrelationships between science and scholarship in Kepler's years in Prague, 1600–1612. In fact, much closer and more profound connections ran between his scholarly and his scientific pursuits than I have yet suggested; and by pursuing some of these threads we will find ourselves drawn deep into the dark heart of the culture of the old Empire. As an astronomer, Kepler had to interact with humanists and humanistic studies in three very precise ways. He had to interpret references to celestial phenomena in classical texts. He had to use his astronomical expertise to date events in ancient history. And he had to apply the humanists' methods of exegesis to the classical sources of his own discipline—above all the greatest ancient astronomical work, Ptolemy's *Almagest,* and many collateral sources. In each field his work resulted in triumphant applications of—and remarkable improvements on—the philological *Wissenschaft* of the humanists.

The interpretation of astronomical bits in literary texts had preoccupied philologists since ancient times, and some of the problems Kepler was asked to solve were traditional. In 1599, for example, Maestlin asked Kepler to do a little job for Crusius. The dean had interpreted the encounters of the gods in Homer as favorable and unfavorable conjunctions of the planets named after them, and wanted Maestlin to work out the technical details. Maestlin in turn claimed to find the suggestion reasonable, but urged that it be carried out by an astrologer, not an astronomer— and that Kepler was just the astrologer to do it.[17] In this case the basic problem Kepler confronted went back to the Hellenistic origins of literary scholarship. Maestlin cited an ancient commentator on Aratus to the effect that Homer had been an astronomer. And an ancient he did not cite, Heraclitus, recorded—and refuted—the suggestion of an unknown critic that the battle of the gods in *Iliad* 20 and 21 in fact represented a conjunction of all seven planets, of the sort that would occur at the end of the world.[18] Kepler replied by making fun of the whole enterprise. He urged Maestlin to take on the enormous and impractical job of computaton: "Why don't you read all of Homer, assign his story the firm chronological place it still lacks, fix the individual colloquies (of gods) to their calendrical dates, do the computation, and produce

Ephemerides for twenty years?"[19] And he promised that if Maestlin did the astronomy, he, Kepler, would happily do the astrologer's proper job of interpreting the planetary positions and predicting their effects. Evidently he was as unconvinced as Heraclitus or Plutarch that Homer had described precise conjunctions, and like them saw no need for elaborate counterarguments. Kepler's attitude was individual enough; in the baroque Empire most scholars thought Homer a learned authority on everything from history to husbandry.[20] But it was clearly inspired by—and did not transcend—the mild skepticism of the Greek students of Homer.

In other cases problems and solutions alike were both more original in conception and far sharper in definition. Johann Herwart von Hohenburg, chancellor of Bavaria and a scholar of great energy—if little judgment—asked Kepler for enlightenment on what he considered a vital source for the early history of the Roman empire. The text in question was not historical but literary; it formed the end of book 1 of *De bello civili,* Lucan's epic on the Roman civil wars, that vast poem which employs the meter of Virgil and the artistic sensibility of Roger Corman to give the fall of the Republic punch and drama. At the end of book 1, Caesar has crossed the Rubicon and Pompey has left Rome. Terrible omens appear: animals speak, women give birth to creatures monstrous in the sizes and number of their limbs, and urns full of the ashes of dead men let forth groans. The Etruscan seer Aruns has a bull killed, but it leaks slimy liquid instead of blood, and its flabby liver, streaked and growing a monstrous extra lobe, fills him with horror. Then a more reputable prophet comes on stage: Nigidius Figulus, Pythagorean astrologer and friend of Cicero. He too prophesies doom, but he uses up-to-date Chaldean astrology to do so:

> ... Extremi multorum tempus in unum
> Convenere dies. Summo si frigida caelo
> Stella nocens nigros Saturni accenderet ignes,
> Deucalioneos fudisset Aquarius imbres,
> Totaque diffuso latuisset in aequore tellus.
> Si saevum radiis Nemeaeum, Phoebe, Leonem
> Nunc premeres, toto fluerent incendia mundo
> Succensusque tuis flagrasset curribus aether.
> Hi cessant ignes. Tu, qui flagrante minacem

> Scorpion incendis cauda chelasque peruris,
> Quid tantum, Gradive, paras? nam mitis in alto
> Iuppiter occasu premitur, Venerisque salubre
> Sidus hebet, motuque celer Cyllenius haeret,
> Et caelum Mars solus habet. Cur signa meatus
> Deseruere suos mundoque obscura feruntur . . . ?

[The lives of multitudes are doomed to end together. If Saturn, that cold, baleful planet, were now kindling his black fires in the zenith, then Aquarius would have poured down such rains as Deucalion saw, and the whole earth would have been hidden under the waste of waters. Or if the sun's rays were now passing over the fierce Lion of Nemea, then fire would stream over all the world, and the upper air would be kindled and consumed by the sun's chariot. These heavenly bodies are not active now. But Mars—what dreadful purpose has he, when he kindles the Scorpion menacing with fiery tail and scorches its claws? For the benign star of Jupiter is hidden deep in the West, the healthful planet Venus is dim, and Mercury's swift motion is stayed; Mars alone lords it in heaven. Why have the constellations fled from their courses, to move darkling through the sky?] (1.650–664, tr. J. D. Duff.)

Herwart read these lines as a description of the configuration of the skies at a given time. He fixed this on general grounds as between 50 and 38 B.C. But how to gain greater precision? The humanist commentators, Giovanni Sulpizio and Ognibene da Lonigo, had applied their normal dull-edged tools. They explained the names of the planets, paying special attention to the title of Mercury, Cyllenius. They named the signs of the zodiac and listed those which enhance the power of each planet. They tabulated the periods in which the planets make their way around the Zodiac and found references to these in the text. And then they went on their way rejoicing, not having explained in more than the vaguest, most qualitative way what Lucan—or Nigidius—was actually saying about the position of the planets, or indeed when he said it.[21] Herwart did his best. He took Lucan as placing Saturn in Aquarius, the Sun in Leo, Mars near the end of Libra, and Jupiter in Scorpio. He found that they had indeed been in those positions toward the middle of 39 B.C., and thus could have presaged Augustus's victory in the civil wars. But he could not make sense of the positions of the inferior planets, Venus and Mercury, and turned to Kepler—as he also had to Tycho and Maestlin on similar problems—for help: "To what date between 50 and 38 B.C. does the celestial configuration that Lucan describes precisely correspond?"[22]

Kepler answered with a meticulous essay. In it he showed the philologist how to do philology:

> Before we try calculations, and let these be tossed about to no avail in this uncertain ocean of twelve whole years, let us first examine the poet's description of the constellation word by word. If Saturn, he says, "were in the zenith," that is, in Cancer, and there "kindled"—that is, brought into operation and aroused by his conjunction—"black fires"—that is, the mist-wrapped stars Aselli and Praesepe—then a flood would be portended. He says "Aquarius would have poured out rains" either for the benefit of his poetic fiction and because Aquarius is the zodiacal sign most appropriate to describing a flood (or because the Sun was eclipsed in Aquarius or because it rains most heavily when the sun is in Aquarius . . .). But it follows from this description that the poet is *not* putting Saturn either in Cancer or Aquarius. For what the poet feigns is—as I interpret it—that if it were in Cancer, there would be rain . . . But neither is the poet putting it in Aquarius, so as to say that Saturn is in Aquarius, in the Mid-Heaven; for again the poet would be using a contrary-to-fact figure; if Saturn were in Aquarius, then Aquarius would rain. These words imply negation.[23]

Kepler saw, as Herwart did not, that Lucan's constructions were contrary to fact. Accordingly they gave positions that the planets did not occupy, not those they did. Moreover, the text as a whole made clear the general date of the prophecy, and here too philology, not astronomy, by itself showed Herwart to be wildly wrong. "Setting out to describe the civil war between Pompey and Caesar, Lucan starts at the very beginning, the crossing of the Rubicon and the fall of Ariminum . . . There is no doubt, then, that this constellation must be sought in 49, 50 or at the earliest 51 B.C."[24] Kepler then offered a possible solution to the problem

Herwart had set him by determining when Mars entered Scorpio and setting out a possible version of Nigidius's horoscope for a date in January 50 B.C.

Herwart was not satisfied. He insisted that Nigidius must have drawn his figure to predict the civil wars of Augustus, not those of Caesar. He insisted that the text did describe Saturn in Aquarius and the Sun in Leo. And he explained the contrary-to-fact character of the description as referring not to the planets' position but to their effects—which would be neither a flood nor a fire, but war. That was what Lucan meant when he said that Mars alone lorded it in heaven.[25] But Kepler stuck to his guns. He held that a close reading of Lucan's text revealed that he was offering either a horoscope for 51 B.C. or a purely imaginary one. And he now found the latter explanation likelier. After all, he pointed out, what Nigidius offered was actually a description of astrological theory of a very elementary kind. In each case, he predicted what would happen if the planet were in the zodiacal sign of its *domus,* where it exerted its greatest power. "If Saturn is in its house, which is Aquarius, he says there will be a flood. If the Sun is in its house, Leo, there will be a fire. If Mars is in its house, Scorpio, there will be war."[26] This was no description of the heavens but a set of elementary astrological doctrines, cobbled together from a manual without understanding. Competent astrologers know that no single planet produces overwhelming effects on its own. What brings about floods and fires and wars is conjunctions, oppositions, and other meaningful configurations of two or more planets at once, not the appearance of one planet in one place. And in any event planets had their benevolent and malevolent effects when in various signs, not only their own *domus.* Kepler's conclusion was lapidary: "Sed tyronem aliter loqui non decet." Accordingly, his "facies Caeli," his configuration of the stars, was not to be sought for in the heavens or in human time.

Herwart did not drop his bad counterarguments. Indeed, two years later Kepler complained that Herwart still exhausted him with his continued inquiries. But he held firmly that Nigidius's horoscope of the Roman civil wars was as imaginary as the horoscopes of Plato, Paris, and other ancient heroes contained in the ancient astrological manual of Firmicus Maternus.[27] For our purposes, what matters is first of all that Kepler was right. Twentieth-century commentators confirm his interpretation of Lucan's dramatic date for the horoscope, his contrary-to-fact reading of Lucan's verses on Saturn and the sun, and his conclusion that the horoscope was imaginary and Lucan incompetent.[28] What matters even more is the way Kepler reached his conclusions. He did so, as he rightly said, not by computing expertly but by reading carefully, by sticking *mordicus,* with his teeth, to the words of his text. He practiced not astronomy here but hermeneutics, and he did this so expertly as to prove himself the master of both cultures— or else, perhaps, to prove their basic unity. Kepler's works are strewn with less extensive but equally provocative discussions of ancient texts; it seems not only a pity but an injustice that they have not earned him a place in modern histories of classical scholarship.

EVEN MORE powerful are the interventions that Kepler made in another area of study where philology and science intersected: technical chronology. Kepler would not have agreed with his acquaintance Helisaeus Röslin that chronology was "ultimus finis astronomiae"—far less with Röslin's belief that he himself had, thanks to divine aid, made such progress "in sacro et prophano calculo mit Hülff Astronomici calculi . . . das mir nit ein scrupulus pleiben soll."[29] But he studied chronology from early on in his academic career: "In history he gave a different explication of Daniel's weeks. He wrote a new history of the Assyrian monarchy; he studied the Roman calendar"—so runs his horoscope of 1597.[30] On Maestlin's recommendation he read Joseph Scaliger's *De emendatione temporum* as soon as it became available in the pirated Frankfurt edition of 1593—though, as he confessed in a sentence that he wisely deleted from his draft of a letter to Scaliger, the difficulty and idiosyncrasy of book 1 made him so sleepy that he found it impossible to work through the book as a whole. Kepler and his teacher corresponded as eagerly about chronological as about astronomical matters. It is engaging to hear Kepler confess that he and Maestlin had often spent days, and even weeks, worrying about the chronology of the book of Judges. And it is instructive to watch

Maestlin instruct Kepler about a major chronological problem: the eclipse that supposedly heralded Xerxes' crossing into Greece in Ol.75, 1 (480–479 B.C.):

> The eclipse must have taken place in Ol.74, 4 in the spring, in the year 268 of Nabonassar = 480 B.C. But there was no eclipse then. The notion that some have that it was the eclipse of Ol.74, 3 is absurd. The sun was eclipsed for only one digit; a soldier in a military formation couldn't even have noticed it. In Ol.75, 2 = Nabonassar 270 there was a solar eclipse of some 10 digits...[31]

And this, Maestlin argued, must have been confused by historians with a prodigy that actually preceded Xerxes' expedition. This last thesis is perhaps fanciful. But the rest of Maestlin's comment shows that he could have given lessons to many of the twentieth-century scholars who still try to deal with the same data.[32] No doubt Kepler owed him much.

By his Prague years, however, Kepler had gone beyond his teacher. In a splendid letter to Scaliger, the monarch of the discipline, Kepler amended central tenets of the *De emendatione temporum*. Scaliger had laid great stress on Herodotus's report (1.32) of a conversation between Solon and Croesus, in which the gloomy Solon proved that a man had many chances of suffering ill fortune, given a life span of 70 years with 360 days per normal year (or 25,200 days—or, counting every other year as an embolismic one of 390 days, 26,250 days). Scaliger had both used the passage in his own reconstruction of a nonlunar Greek calendar and accused Herodotus of inventing a nonexistent calendar.[33] Kepler's response was devastatingly insightful: "No precise computation of days is undertaken here; Solon is using rounded numbers to persuade Croesus how easy it is, given this great sum of days, for one of them to be unfortunate."[34] And yet this passage may have stung less than a later one. Scaliger had cited, among other texts that seemed to support the existence of a nonlunar calendar, Plutarch's *Camillus* 19.5. This seemed to set the battle of Naxos on the fifth day before the end of Boedromion *at full moon* (which, naturally, should fall at the middle, not near the end, of a lunar month).[35] Kepler simply advised Scaliger "Velim tamen in emendato Plutarchi contextu requiras"—"please look up the passage in a correct text of Plutarch"[36]—for as he saw, and Scaliger failed to, Scaliger's theory rested solely on two independent clauses wrongly conflated in the Aldine edition of Plutarch. For want of a comma Scaliger had committed himself to a wrong thesis. Kepler thus triumphantly anticipated Dionysius Petavius's full-scale attack on Scaliger a few years later. To be sure, Kepler's positive efforts were less innovative and less successful than his critical ones; he shared with Scaliger the erroneous belief that the Attic months in Plutarch were not strictly lunar.

Though quantitative and technical enough for most of us, however, the chronology practiced in Habsburg imperial circles was anything but a simply empirical or technical discipline. The Habsburgs liked universal claims and programs (like the claim to domination of the natural world advanced by the painter Arcimboldo in his famous portrait of Rudolf II as Vertumnus, Roman god of the changing year, where the vegetable ingredients of Rudolf's face powerfully express the analogy between divine rule of the cosmos and Habsburg rule of the Empire).[37] The chronologers they patronized had to devise original and impressive general theses (like Lazius's proof that the Austrians were directly descended from the Jews who settled Europe after the flood).[38] Chronologers also had to impose neat patterns on events, showing that these lined up in numerically elegant ways, were closely connected to celestial and other omens, and pointed explicitly toward imperial hegemony in general and Habsburg power in particular.

Thus the Austrian Freiherr Michael von Aitzing—later to win fame as the first author of political newsletters or Zeitungen—made his *Pentaplus regnorum mundi* (1579), which he dedicated to Rudolf, a key not only to the dating of events but to their inner providential logic.[39] His fabulously profuse and detailed tables illustrate both the variety of the causal systems he invoked and the inventiveness with which he fused them into a whole. One of them listed the so-called great conjunctions of Jupiter and Saturn that occurred every twenty years through all of history, providing the celestial omens that occurred most regularly. Another by contrast divided history up into the reigns of seven angels, each in charge for 792 years, the first five before Christ and the sixth and seventh after him; these in turn are divided into twelve subsections corresponding to the signs of the Zodiac, twenty-four subsections corresponding to the hours of the day, forty-two corresponding to the places where the Jews stopped between Egypt and Palestine, and so on. The last conjunction falls in 1583, and the last angel runs out in 1584; all this to show that history will soon play the starring role in a crowd-pleasing death scene. Meanwhile, and for good measure, a splendid emblematic illustration (see Figure 1) summed up the most important lesson of all. The letters above the two columns headed by the sun and moon are the initials of the biblical patriarchs from Adam to Noah; these, when rearranged by von Aitzing's instructions—and when Noah's name is given in its alternative Greek form, Ianus—spell out the name of Rudolf's father, Maximilian. Meanwhile four beasts—the traditional ones of the Book of Daniel—represent the four Empires, the last of which is the Holy Roman Empire represented by the Habsburg double eagle. Here was a powerful skeleton key to history itself, and the door thus opened revealed the Habsburgs at the very heart of things. Many others tried to forge and apply similar keys. Indeed, doing so was part of Kepler's job. His predecessor as imperial mathematician, Ursus, had written, supposedly in 1596, a work that conflated chronology with eschatology to show that the world would end "innerhalb 77 Jaren."[40] Kepler, accordingly, had to comment when, as in 1603, a particularly vivid great conjunction took place—especially as this one was followed by the appearance of a nova, a brilliant new star that seemed to portend some special change in human affairs.

Kepler did not drag chronology off the traditional rails, as he did astronomy. He took great interest in the old doctrine of the great conjunctions, compiling his own table of the correspondence between the beginning of the great conjunction cycle every 800 years and major changes in events. It runs from the first great conjunction, that of Adam in 4000 B.C., to that of 1603, which portended something for Rudolf, and as Kepler said, for "our life, our fate, and our prayers." A final line pointed out that in A.D. 2400 still another cycle would begin. Kepler commented: "Where will we be then, and our Germany which is so prosperous now? And who will come after us? And will they remember us? If the

DE INTELLECTV
PENTAPLI,
PARS PRIMA.

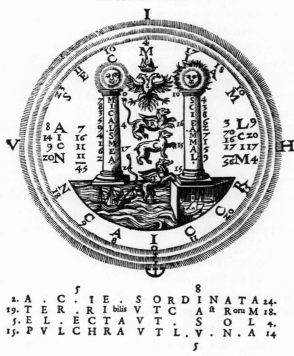

Figure 1. World history in emblematic form. From Michael von Aitzing, *Pentaplus regnorum mundi* (Antwerp: Plantin, 1579). Courtesy of the Rare book and Manuscript Division, Princeton University Library.

world lasts so long."[41] So far there was nothing novel in Kepler's sensibility or method, even if he did insist that he loved the neat series of great conjunctions largely for the prosaic reason that it served as such a splendid mnemonic device for historical dates.

In other respects, however, Kepler's approach to chronology was as novel and elegant as his approach to Lucan. He argued, in the first place, that the great conjunctions were a useful tool for finding order in the past but not a valid guide to the immediate or distant future. The only prediction he felt able to make on the basis of the new star was that it portended "den Buchdruckhern grosse unrhu und zimlichen gewin darbey," since every theologian, philosopher, doctor, and mathematician in Germany would write about it.[42] He argued again and again that only God knew the date of the end of the world, and urged chronologers to study only the past. He thus separated the discipline's two traditionally related functions and deprived one of value or interest.

In the second place, and more important still, Kepler used even the traditional doctrine of the great conjunctions in a novel way. Most chronologers packaged history neatly. They ignored structural differences between people and events in order to make recurrent celestial events line up as neatly as possible with similar earthly ones. Chronology was for them not a means of discovering new facts or connections but a way of imposing an order on facts already known. And it constantly threatened to degenerate into a mindless repetitive series of lists of names and numbers signifying nothing.

Kepler by contrast applied the theory of the great conjunctions heuristically rather than rhetorically. Did the conjunction of 1603 and the nova of 1604 portend an especially radical change? He was not sure. And so he examined other great conjunctions of the last two centuries. He argued that more prominent conjunctions than that of 1603 had occurred in the sixteenth century. He also argued that these had cumulatively had a special effect, one visible from the historical record. They had stirred up men's minds in uniquely powerful ways. Above all, the invention of printing had transformed the intellectual world, creating a new community of scholars not confined to monastic orders: "After the birth of printing books became widespread. Hence everyone throughout Europe devoted himself to the study of letters. Hence many universities came into existence, and at once so many learned men appeared that the authority of those who clung to barbarism soon declined."[43] Kepler connected this rise of a new "public domain of knowledge"[44] with the discovery of the new world, the growth of rapid means for communicating goods and knowledge, and the development of modern techniques in every area from warfare to textual criticism. And he explicitly contrasted the limited achievements of the ancients with the greater ones of the moderns: "What did they have in more recent antiquity that resembled present-day knowledge in the art of war?"[45] The great conjunction and nova could not portend any very radical change, for the world had already been radically transformed by the stars and men working together for the last century and a half.

Kepler never abandoned his belief that the configurations of the stars provided a constant and vital thread in the tapestry of the past; he redated Christ's birth to a few years before the outset of the Christian era in part because he could thereby bring it nearer to a great conjunction that might have been related to the Star of the Magi.[46] But in arguments like those I have rehearsed he used this traditional doctrine in a novel way: to show that society and culture changed by accretion and development, not by sudden seismic shifts engendered from on high. And he developed a penetrating insight into the fabric of society as well as into the movements of the stars.

In attacking the question of cultural development in this broad-gauged way Kepler once again drew on the classical and humanist tradition. The Roman historian Velleius Paterculus (first century B.C.) had inquired, in a long digression, why all the arts from grammar to sculpture had reached perfection so quickly both in ancient Athens and in his own modern Rome. While he did not claim to have a fully satisfactory explanation, he suggested that competition, both malevolent and friendly, was the most probable one: "alit aemulatio ingenia" (1.16–17). Lorenzo Valla, trying both to prove the necessity of reviving classical Latin and to explain the flourishing of the same group of *artes* in Rome and in his own day, accepted—or reinvented—and embroidered on Velleius's thesis. He agreed that "aemulatio" explained artistic and intellectual progress. And he insisted that the existence of a single perfect language, classical Latin, in its turn explained "aemulatio." The universality of Latin had enabled all to communicate with and thus to compete with one another. Without the common medium of Latin, civilization could never have grown to perfection, for "Any of the arts is quite as difficult to perfect as a city is. Therefore just as no city, so also no art can be established by a single man, nor indeed by a few men; it needs many, very many men, and these men must not be unknown to each other—how otherwise could they vie with each other and contend for glory."[47]

If Valla attributed explanatory power to language and Kepler to printing, they strikingly agreed to invoking the medium of communication as the central factor in cultural change. And Kepler's adaptation of this humanistic approach, with its emphasis on human volition and the cumulative efforts of individuals, could not have contrasted more sharply with the practice of most astrologers and chronologers. Girolamo Cardano, for example, had also discussed Velleius's thesis in his influential commentary on Ptolemy's manual of astrology, the *Tetrabiblos*. He agreed that Velleius had asked a good question, but denied that the historian had offered a good answer. "Aemulatio" might explain the growth of the arts but could not explain their equally rapid decline. Hence it must be rejected in favor of the stars, whose "constitutiones" lay behind all multiple simultaneous changes. The sun's return, for example, made plants grow, trees flower, birds build nests, and animals make love. Since the arts flowered and withered simultaneously, Cardano concluded, "it is obvious that they are principally the result of the general *constitutiones* of the heavens"—even if the causes invoked by Velleius might "aliquid in rem facere."[48] The greater depth and density of Kepler's attack on the subject are manifest; so too its closer connection to the scholarly than to the scientific tradition.

Chronologers and astrologers had always seen stars (and angels) as the decisive factors, and human beings and society as largely passive. Yet Kepler the chronologer and astrologer cut the role of stars to mere stimulation. He constantly adjusted the celestial history of conjunctions to fit a messy human history that he studied field by field and decade by decade. And he thus raised chronology to the level of a powerful tool for studying social and cultural processes, one capable of using and enriching the most mature historical ideas of the humanists instead of ignoring these in favor of neat numbers.

KEPLER'S third transformation of the study of the past lay in his own special discipline of astronomy. Renaissance astronomers had always had to confront the work of their classic predecessors: the preserved work of Ptolemy, the *Almagest;* the lost works of Hipparchus, which Ptolemy had used and quoted; and the large corpus of ancient anecdotes and shorter texts about astronomers. They used these materials as their primary sources of data, models, and techniques. And they also used them as something more: as the classical foundation, or pedigree, that gave astronomy legitimacy and dignity in their own time. But the very need for legitimacy that made astronomers study their ancient predecessors also made them depart from objective truth in their interpretations. After all, a substantial body of ancient anecdotage treated astronomy as the oldest and purest of the sciences—developed by Jewish patriarchs, virtuous Near Eastern priests, and Gallic Druids; preserved during the Flood by being engraved on stone tablets; gradually lost in later, less pure times by Greeks and Romans.[49] When a late-Renaissance scholar wrote the history of astronomy—as Henry Savile did in 1570, preparing for his Oxford lectures on Ptolemy—he tended to insist that the primeval astronomy of the patriarchs had been both simpler and more accurate than the developed astronomy of Ptolemy. He divagated at length about the astronomical achievements of such dubious luminaries in the field as Enoch, Seth, and Hermes Trismegistus, which could be reconstructed with great confidence primarily because not a scrap of evidence about them survived. And he described the purpose of astronomy in his own time as being to recover the lost wisdom of pre-Socratics, patriarchs, and other sages.[50] Such views were widely shared and long persisted; Newton himself was inspired by them to form his own unhistorical view of the history of astronomy, according to which Chiron the Centaur, no less, laid out the first constellations on a sphere for the use of the Argonauts.[51]

Kepler himself took a very different view. From early in his career he devoted attention to the ancient astronomers. Just arrived in Prague, he spent time unwillingly on what he considered a philological, not a mathematical, task: showing that his predecessor Ursus had misrepresented the history of ancient astronomy in order to claim that Tycho had derived his compromise system of the universe, in which the planets revolved about the sun and the sun in turn about a stationary earth, from classical sources. Already in his *Apologia* against Ursus, Kepler insisted on the primitive conditions in which the ancients had worked. In the teeth of a tradition that arranged all ancient sages genealogically into schools, Kepler pointed out that ancient astronomy had not really had an institutional base and that its chief developments had come about through the discontinuous efforts of a few individuals.[52] After leaving Prague, in the great *Tabulae Rudolphinae* of planetary motion that he compiled from Tycho's data, Kepler went much further. He began the work with a history of astronomy, and astronomical tables in particular, in which Seth, Enoch, and other early sages played no part. The first real developments he could trace fell in the third century B.C., when the Greeks gained access, under the Seleucid kings of Babylon, to a not very old corpus of Babylonian observations. Hipparchus, in the second century B.C., offered the first adumbrations of tables of the motions of the planets, which made rough predictions of their future positions possible; and Ptolemy, three hundred years later, perfected the exact science of sciences and offered the world its first full tables. Other sections traced the later and even more sophisticated work of Alfonso of Aragon, Johannes Regiomontanus, and Erasmus Reinhold. The message of the whole was clearly that astronomy had grown, by uneven increments, from primitive beginnings in the ancient world to modern perfection.[53] And Kepler gave this thesis pictorial and poetic as well as technical and prosaic form. The title page of the Tables (see Figure 2) represents an "Astro-poecilo-pyrgium," a "variegated-star-tower," or temple of astronomy. Here architectural orders drive home historical lessons. The Chaldean page appears in the very back, sighting through his fingers at the stars and standing by a rough wood column, almost still a tree. Aratus and Hipparchus on the left, Meton and Ptolemy on the right, the heroes of Greek astronomy, have their names attached to plain brick columns. But the heroes of modern times, Copernicus and Tycho, dominate the foreground, and they receive the compliment of classical ornament. Corpernicus sits by an Ionic column. Tycho, who is arguing with Copernicus—he points upward to a diagram of his own system on the roof and asks "Quid si sic?"—has the most glamorous column of them all, a fine Corinthian. Thus modern culture is revealed as older—that is, more experienced and

Figure 2. The *Astro-poecilo-pyrgium.* From Johann Kepler, *Tabulae Rudolphinae* (Ulm: Saurius, 1627). Courtesy of the Rare Book and Manuscript Division, Princeton University Library.

sophisticated—than that of the so-called ancients, which was ignorant and primitive. And just in case any reader failed in his duty to decode this rich historical emblem, Kepler had the Ulm *Gymnasialrektor* Johann Baptist Hebenstreit provide a pedestrian *Idyl-*

lion to explicate it, imaginary brick by brick.

In advancing this thesis Kepler made his own a modern and perceptive view of the history of ancient astronomy. It was not entirely his own invention, to be sure. He owed parts of it at least to Pico della Mirandola, whose *Disputationes contra astrologiam divinatricem* of a century before provoked and fascinated Kepler as did few other modern texts. Pico had incorporated similar arguments, similarly based on direct study of the sources, into books 11 and 12 of his great work. He had argued there, as Kepler would, that astronomy and astrology had no very ancient pedigree, and that the ancients' boasts of possessing records of observations stretching back for hundreds of thousands of years were not borne out by the facts: "When the founders of astronomy, Hipparchus and Ptolemy, produce the ancients' observations in order to lay the foundations of their own doctrines, they cite none older than those made under Nabuchodonosor [Nabonassar] in Egypt or Babylon. Hipparchus flourished around the six hundredth year after his reign."[54] Pico, like Kepler, located the beginning of mathematical astronomy near Nabonassar's accession in 747 B.C., and its perfection in the time of Hipparchus. He thus adumbrated Kepler's polemical history, and to the same end of puncturing the pretensions of astrologers and believers in the *prisca theologia*. And Kepler actually cited Pico's book in a letter in which he sketched a genealogy of ancient astronomy, though in a different context— a long letter to Herwart in which he said that he condemned astrology "tantum . . . quantum Picus."[55]

For all Kepler's traditional interests and beliefs, in following and developing Pico's ideas he challenged one of the deepest convictions of late Renaissance intellectuals in general and his Prague friends like Wacker and von Nostitz in particular. They saw truth as residing in a primeval revelation, handed down in its purest form by God at the outset of human history and degraded ever since by contact with mere humans.[56] Pico and Kepler saw truth as the product of human effort, untidy and inconsistent, but gradually able to reach perfection over time. Kepler's *Tabulae Rudolphinae*, the last and most technical product of his Prague years, not only replaced ancient astronomy but also attacked ancient myths about the nature and origins of human culture.

The case of Kepler, then, offers by itself enough dynamite to explode any notion that the scientific and the humanistic cultures of the Empire existed in isolation from each other. The scientist could not perform his function without being enough of a scholar to decode the classical texts that still contained his richest sets of data. The scholar could not read his poems without having recourse to scientific concepts and methods. And in some of the most fashionable and attractive studies—like astronomy and chronology—scholarship and science were necessarily fused into a single pursuit not identifiable with any modern discipline. Both cultures, in other words, formed parts of the same vast Mannerist garden, and a single wind could send pollen from each side to fertilize the other. Like a German court garden of the time—like those at Heidelberg, soon to be destroyed—they make a lurid, variegated and alien spectacle to the modern onlooker. And yet, if we limit our explorations to the familiar we must fail to understand the principles of order and the connecting links of method that bound science to scholarship and mathematics to letters in this singularly fascinating lost intellectual world.

8

Isaac La Peyrère and the Old Testament

IN 1582 the schoolmaster Noël Journet was burned for heresy at Metz. His errors were grave, and Protestant and Catholic divines joined in condemning them. He had found inconsistencies and mistakes in the Bible. He could not understand how Sarah provoked the lust of the Egyptians when she was an old lady, or how the Egyptian magicians could turn water into blood when Moses had already transformed all the water in Egypt. He found clear evidence in the description of Moses' death and elsewhere that Moses had not written Deuteronomy. And he used these considerations to attack not only the Old Testament but Christianity as well. No wonder that he had to die in the flames, and his books with him.[1]

In 1692 Pierre Bayle published the first specimen of his *Historical and Critical Dictionary*. He too discussed Sarah's long-lived beauty. He speculated amusingly about the brand with which God protected Cain from strangers who might murder him, in a world supposedly still unpopulated. And he worried about God's hardening of Pharaoh's heart—and His allowing Pharaoh's magicians to work their magic. He too encountered opposition, and had his difficulties at Rotterdam. But his critics did far more barking than biting. Bayle lived to fill the huge volumes of the *Dictionary*, and his books outlived him, to become the much-thumbed favorites of Voltaire, Jefferson, and Herman Melville.

Evidently something had happened in the intervening century. An exegetical as well as a scientific revolution had taken place. The sharp-edged criticism of Spinoza and Simon had deprived the Bible of some of its luster of sacredness and perfection. A Bayle could wonder about it and live; soon Jean Astruc would begin to chop the Pentateuch into its underlying Jahwist and Elohist sources. No one did more to make this revolution happen than the little-remembered French Calvinist Isaac La Peyrère.[2]

La Peyrère came from a Calvinist family in Bordeaux. After an early career as a soldier, he served the Prince of Condé as a secretary and traveled in Holland and Scandinavia. Around 1640 he wrote out an elaborate treatment of an idea he had first conceived as a child: that the two Creation stories of Genesis 1 and 2 in fact told of two different Creations, that of mankind as a whole and that of the Jews, which came much later. Men had lived before Adam, for millennia. By accepting this simple principle one could explain many puzzling inconsistencies within the Bible. One could reconcile its short chronology of world history with the much longer ones of the ancient pagans, the American Indians, and the Chinese. One could understand how Greenland and the Americas had been populated without assuming nonexistent land bridges between Asia or Europe and the New World. In short, one could know where Cain's wife came from.

These theories found sympathetic listeners at first. La Peyrère's ideas circulated clandestinely, in conversation and through peeks at his manuscript, to which he gave the catchy title "A southerner's dream of men before Adam." And in the Paris salons, where the wits gossiped about arcane heresies and wondered where they could lay hands on that enticing, elusive—in fact, as yet nonexistent—classic, "On the Three Impostors" (Moses, Jesus, and Muhammad), his views seemed mild in tone and constructive in content.[3] Gui Patin began by making fun of La Peyrère's "chimerical" ideas, but came to find the theory "belle" and wish it true. Marin Mersenne thought that it made "several passages in Scripture easier to understand"—though he too had his doubts.[4]

Wider communication of these ideas turned partial, private sympathy into a storm of virulent, public abuse. Hugo Grotius, a man of strong views about the Bible and strong loyalty to his Swedish patrons, wanted to prove that the Indians reached North America from Scandinavia, via Greenland (thereby giving Swedish claims to colonies in the New World a basis in *jus primae occupationis*). He borrowed the manuscript, disliked it, and criticized it in print as a "danger to piety." This surprised and annoyed La Peyrère.[5] And when Christina of Sweden and other early readers seemed more sympathetic, he allowed them to persuade him to publish his work, in revised form, anonymously in Holland. The text was reprinted four times, translated into English and Dutch, and caused a sensation. But the authorities condemned it, in Calvinist Holland and Catholic France alike. Nineteen refutations appeared in 1656 alone. La Peyrère was arrested by the Inquisition in Brussels. He escaped the worst only because his master Condé remained faithful, and secured his release at the price of his conversion to Catholicism—which absolved him of errors committed as a Protestant. La Peyrère had to publish a retraction of his theory; promised pensions never materialized, and he ended his days a pauper in an Oratorian seminary outside Paris.

For the next half-century every critic who ventured farther down the paths La Peyrère had cleared used his theories—often while denouncing their author, quite unfairly, as impious. Spinoza owned La Peyrère's book and drew on it. Simon, who produced the first critical history of the Old Testament text—which he treated as an imperfect compilation drawn from lost earlier sources—knew La Peyrère well, corresponded with him, and wrote a curious memoir of him in the form of a letter. To this extent La Peyrère posthumously won out against the orthodox critics, whose opposition he consistently ascribed to ignorance and bigotry. And in centuries to come his ideas flourished and he himself came to be seen as a heroic martyr, the Galileo of the exegetes. His visions of the human past—and future—became grist for the varied mills of anthropologists explaining racial diversity, Zionists demanding a Jewish homeland, and experts in "Niggerology" who tried to prove the separate origins of whites and Negroes, and the inferiority of the latter, by interpreting Scripture as well as by measuring skulls.[6]

Richard Popkin, whose *Isaac La Peyrère (1596–1676)* sums up earlier scholarship—much of it his own—has anchored La Peyrère firmly to his intellectual context. In particular, he has shown that La Peyrère was even more concerned with the future than with the past. In his prophetic *Du rappel des juifs* and elsewhere he elaborated a rich vision of the future of mankind. The Jews had the starring role in his scenario. They would soon be converted to a painlessly simplified form of Christianity. Led by a French holy ruler, they would return to the Holy Land, and the restoration of Israel, cleansed of its rejection of Jesus, would mark the beginning of the millennium. Then lions would lie down with lambs, and one race with another, in harmony. La Peyrère's obsessive belief in the central importance of the Jews generated and ruled his explication of Scripture. He needed to believe that the Old Testament told only the story of the Jews, God's incomprehensibly chosen people, not that of the larger human race. To give the Jews their proper place in history—and, he hoped, to convince both them and the Christians of the rightness of his plan to convert them—he had to devise his iconoclastic exegesis. Popkin suspects—as others have before him—that La Peyrère was himself a Marrano, and that his origin explains his devotion to the Jews. But no decisive evidence supports this view. What Popkin has shown is that, in La Peyrère's as in many other cases in the seventeenth century, prophecy dominated historiography. And that is a result of great importance.

But if the target La Peyrère aimed at is now starkly clear, the quality of his equipment remains obscure. He tried to connect pagan sources and scholarship with the Bible; to show that, on his assumption, the Bible "is wonderfully reconciled with all prophane Records whether ancient or new, to wit, those of the Caldeans, Egyptians, Scythians and Chinensians."[7] In particular, he believed the claims of Diodorus Siculus and others about the vast age of the Babylonian and Egyptian kingdoms, and of the hundreds of thousands of years during which the astronomers of those nations had perfected their art. These data gave the external, as prophecy gave the internal, stimulus that provoked him to write. But in undertaking to reconcile pagan and Jewish records, astronomy and history, La Peyrère ventured into one of the most complex and sharply contested territories in the map of early modern learning. His progress in this area remains to be charted.

La Peyrère's enterprise was not wholly new. From the generation of Joseph Scaliger and Isaac Casaubon on, as we have seen, many early modern scholars comfortably combined classical with biblical studies, Latin and Greek with Hebrew and Arabic, philology with mathematics. Scientists like Kepler did profound historical work on classical sources; scholars like Scaliger tried to do systematic work in the exact sciences. La Peyrère was only one of many polymaths who freely crossed and recrossed what now seem high and forbidding barriers between disciplines. Isaac Vossius—the Dutch-born canon of Windsor who wrote on ancient prostitution and modern China, tried to solve the same chronological puzzles that absorbed La Peyrère, and was reputed to believe anything so long as it was not in the Bible—is an equally typical, and more familiar, citizen of this world.

La Peyrère, moreover, not only dealt with fashionable, cutting-edge intellectual issues but moved with confidence in fashionable intellectual circles. In Paris he discussed astronomical problems with Gassendi and cartographical methods with Roberval. During his travels in the north of Europe he became a friend and correspondent of the wonderfully learned Danish doctor Ole Worm, who collected a vast range of natural and historical objects of interest, studied runes and Icelandic, and went out into the countryside to question the peasants about their customs.[8] La Peyrère too took a deep interest in both human and natural history, examined narwhal horns as eagerly as early texts, and wrote charming descriptions of Iceland and Greenland that remained standard until the nineteenth century. Around 1660 he knew the members of the informal academy of science that met under the patronage of Montmor. Christiaan Huygens recorded in his Paris diary some fascinating encounters with "M de la Peirere the Pre-Adamite."[9] La Peyrère, in short, swam in the same waters as Hobbes and Pascal.

It seems only natural, then, to connect La Peyrère's boldest contribution to learning with the learned world he lived in. After all, the polymaths whom he admired had published ancient materials that supported his belief in the longevity of pagan history—and even, perhaps, in the incompleteness of the Bible. Scaliger, as we have seen, had brought to light Berosus's troubling account of the Babylonian *Urgeschichte*. He had also published and defended Manetho's shocking lists of the kings of ancient Egypt, which began—by Scaliger's computation—before time itself did. And he had thus provoked a controversy that engaged, among others, Dionysius Petavius, Gerardus and Isaac Vossius, and John Marsham. Gulielmus Sossus, a generation later, defended the pagans' knowledge of the earliest times even more sharply in his dialogues *De numine historiae.* Josephus described the two columns of stone and brick on which the "antenoëmici mortales" had inscribed their wisdom. The one of these that survived the Flood gave later pagans as well as Jews a "floridum & certum . . . historiae antelucanae testimonium."[10] Moreover, Greek writers from Xanthus of Lydia down to Hermippus and Hermodorus had dated the earliest pagan prophet, Zoroaster, to very early times indeed; Eudoxus put him 6000 years before Plato's death. He could then have been Adam himself, whom the Septuagint put 5200 years before the birth of Christ. More likely he was Mesraim, who lived just after the Flood, and 6000 was simply an error for the true interval of 3000 years between him and Plato. "If you accept this opinion," one of Bossus's interlocutors told the other, "only think how many things Zoroaster knew about the beginnings of the universe and set out in his countless verses."[11] Thus the tradition of the *prisca theologia* had paradoxically taken on new strength in the same learned milieus in which its basic texts, like the Hermetic Corpus, were condemned.

La Peyrère did cite some of the great works of sixteenth- and seventeenth-century scholarship. He admitted that he had drawn information not only from Diodorus and Cicero but also from *"Salmasius . . .* in his Climactericall years";[12] that is, in more modern terms, from that "inimitable monument . . . of erudition and untidiness," the *Diatribae de annis climactericis* of the Protestant scholar Claude Saumaise, Scaliger's successor at Leiden, who attacked Milton but befriended La Peyrère.[13] But if Saumaise had cited Diodorus and Cicero on the claims of Chaldean astronomers to thousands of years of observation, he had clearly done so with disapproval. He took Berosus, following a long tradition, as a proficient Chaldean astronomer and a transmitter of Eastern wisdom to the West. He pointed out that Berosus had thought that the moon shone with a light of its own. "It is quite extraordinary," he commented, "that in a space of 470,000 years the Chaldeans could not determine by their continuous observation of the stars whether the moon shone by its own or by a borrowed light."[14] And he found their astrological beliefs equally absurd, and equally out of keeping with the supposed antiquity and authority of Near Eastern culture. Evidently Saumaise's views were the oppposite of La Peyrère's.

The same is true of the other learned source on which La Peyrère drew. He never mentioned the Egyptian dynasty lists that Scaliger had published in the *Thesaurus temporum*, presumably because he never explored that vast and difficult work. But he did mention that Scaliger had recorded in the *De emendatione temporum* the claim of the Chinese that their empire had begun 880,073 years before A.D. 1594. La Peyrère found this claim—which he slightly misquoted—plausible enough, since it matched those of the Phoenicians and the original inhabitants of the Americas. But even he had to admit that Scaliger had considered it "monstrous," *"prodigiosa,"* and had attributed it to the Chinese scholars' great distance from the West and consequent ignorance of the Bible.[15]

No wonder then that La Peyrère's learned readers had little trouble in deconstructing the classical evidence he deployed. Philippe le Prieur quoted Pico della Mirandola against him.[16] Richard Simon argued that only Ptolemy's testimony, according to which the Chaldean records stretched back only to the accession of Nabonassar, deserved credence.[17] These men knew that the scholarly tradition was firmly in their corner against the bumptious biblical critic. Paradoxically, La Peyrère's critical attitude toward the Bible derived from his uncritical and unironic reading of the advanced classical scholarship of his time. He exploited some of the information provided by Scaliger and Saumaise, but discounted or ignored the arguments in which the data were imbedded. Simon was right to describe La Peyrère as a man of very limited learning.[18] As a scholar he was dwarfed by the great erudite dinosaurs of his time, though he did filch vital factoids from them.

Yet if La Peyrère's erudition disappears on inspection, his intelligence seems all the more impressive. In particular, his radical view of the Bible—his insistence that "those Five Books are not the Originals, but copied out by another"—seems all the more incisive when one realizes that he had no grasp of the similar analytical techniques that professional scholars had applied to pagan texts like the Hermetic Corpus.[19] Neither Italian materialists nor Marlowe, not even Hobbes, analyzed the text with such pertinacity, or showed so fully that it was not a seamless divine garment but a human product full of rents and patches.

La Peyrère surely did not know the precedents that some scholars have cited for his views—like the analytical insights that crop up in Ibn Ezra's commentary on the Bible. He was too ignorant. And in fact the closest parallel to his thought comes not from the world of learning but from that of semipopular heresy. Journet, that poor schoolmaster who knew no foreign language save a little German, anticipated some of La Peyrère's individual objections to the Bible and adumbrated his general rejection of its authenticity and coherence. The two men differed in many ways; Journet was an explicit and radical critic of Christianity. But their resemblances repay study. Both were innocent of professional scholarly training—the training that would have shown why the text really made sense even where it did not seem to. Both ruminated the text itself, over and over, seeking coherence and authenticity and finding the errors and confusions natural in a collection of ancient holy books written at different times. And both ended by tearing the text to pieces, such was the power of their need for truth and the depth of their conviction that a truly divine work must be orderly in every respect.

It was precisely the amateur quality of Journet's and La Peyrère's reading of the Bible that so infuriated the professional theologians—even analytical and critical professionals like Simon. Jean Chassanion, the minister from Metz whose report is our source for Journet's views, drew a profound hermeneutical moral from Journet's death: "This should warn everyone how and in what manner one ought to devote oneself to reading the Holy Scriptures. First of all one must be persuaded that the infallible truth of God is set out for us here. One must not contradict this even with the least thought that one's mind contains."[20] Simon regarded La Peyrère with affectionate irritation. But even he found it remarkable that "He knew neither Greek nor Hebrew and yet insisted on involving himself in devising new senses for several passages in the Bible."[21] An educated La Peyrère would have seen that many of the problems the Bible seemed to pose were illusory. He would never have challenged the tyrannical authority of the text, much less have provoked others to do so, as Simon did, with far more powerful weapons.

Two centuries later, La Peyrère might have stood at the back of a revival meeting, shouting "Hey, mister, where *did* Cain's wife come from?"—or haunted public libraries, combing the plays of Shakespeare for evidence of Bacon's authorship. A century before, he would have been burned with Journet and Domenico Scandella, called Menocchio.[22] As it was, he fused for a poignant moment the popular skepticism that could call any text into question and a few powerful fragments of learned scholarship. The combination was potent enough to stimulate, to split, and eventually to transform the world of learning to which he never really belonged.

One facet of La Peyrère's character may have been crucial to his survival. He was very funny. In fact, he provoked as much amusement as consternation in some of his critics. Even the General of the Jesuit order said that he and the Pope had laughed themselves silly when they read the *Pre-Adamites*. Even on his deathbed he made an equivocal joke rather than a full confession. When the confessor said that his soul was free to depart this world, La Peyrère asked, "Where do you want it to go?"[23] Did La Peyrère serve as a learned jester to some of the scholars of his time? Were any of his works comic or ironic in intent? His world is hard for any modern scholar to penetrate, and jokes are notoriously the hardest of all past utterances to interpret. It will take a historian with a keen sense of humor to feel his way past La Peyrère's stream of baroque paradoxes to whatever reality may lie beneath. But no one will expect a deep analysis of humor from a study of the humanist tradition. It is enough for now to have shown that the humanists sometimes had to engage in dialogue with men and works from a low rather than a high cultural tradition—and that humanism was enriched as well as challenged when they did so.

9

Prolegomena to Friedrich August Wolf

IN 1908 classical philology was still important enough to provoke a satire. Ludwig Hatvany provided it. His *Die Wissenschaft des nicht Wissenswerten* is a slashing parody of—of all things—a Berlin student's notes on a year's work in classics. The professors whose lectures it records are nightmare figures, academic hobgoblins. The Latinist, Woepke, buries Catullus's Lesbia poems under heaps of unnecessary lexical distinctions: *"Vivamus* here means not only 'let us live' but 'let us enjoy life'; we find this expression used in the same sense in an inscription in the C.I.L. . . ."[1] The Greek professor spews out over Plato's *Protagoras* a flood of details about the daily trivia of ancient life: "As you see, gentlemen, the porter shut the gate [on Socrates and his companions]. At this passage anyone would be struck by the question of how this gate was constructed, and also by the important, still unsolved problem of door-shutting in antiquity."[2] Hatvany's fellow students are no better than their masters. One barges into Hatvany's room on an April morning, waking him, to introduce himself: "My name is Meier, student of philology. I've come to explain why I'm so serious."[3] Hatvany's own mock-seminar report uses the fragments of Sappho to argue that she was the headmistress of a *Mädchenpensionat*.[4]

Like most elaborate satires, Hatvany's had a serious message. It was an attack on the main tenets of *Altertumswissenschaft*—above all, on the demand that the student pay equal attention to every aspect of the Greek and Roman worlds and fit every text into an elaborate political, social, and material context. For Hatvany this approach destroyed the literature it was meant to explain. The student was too bogged down in a swamp of facts to appreciate any single work of art, too worried about the details of Catullus's hypothetical biography—and the place of pet sparrows in Roman society—to take fire from his poems.

Hatvany admitted that the eighteenth-century founders of *Altertumswissenschaft* had not been misguided. In the Enlightenment, Homer and Plato had ceased to be thought of as real men. They had become colorless counters in the abstract intellectual games of the *philosophes*. The only way to breathe life back into them had been to learn—as F. A. Wolf and his contemporaries had—to see them vividly, in all their colors and dimensions. And the only way to do that was the one Wolf had taken: to master the apparently inconsequential details of public institutions and private life, architecture and mythology, etiquette and coinage—and so learn to imagine the ancient world as it had been.[5] Wolf's successors, however, had made the means into the end. They did not read *(lesen)* ancient literature, they read it to pieces *(zerlesen)* in their frenzied search for raw materials from which to make new lexica and handbooks—but never a new vision of the past. This perversion of Wolf's discoveries Hatvany sought to expose.

Unlike his dissections of Mommsen and Wilamowitz, Hatvany's praise of Wolf was conventional in tone and content. But the estimate of Wolf's work and worth that he accepted needs more examination than he gave it. Claims for Wolf's originality as thinker and scholar rest on the work he did as a professor at Halle. He arrived in 1783 as a stiff-necked but promising young product of Göttingen. There—so he later told the story—he had learned much from the matchless library but little from the lectures of his teacher C. G. Heyne. He had been a great success at Gymnasium teaching and had published very little. That was all. By 1806–07, when the closing of the university forced Wolf to leave Halle, he had become the dominant scholar in north Germany. His lectures and seminars had made him so famous that Goethe hid behind a curtain to hear him teach. He had created a school of original scholars and competent Gymnasium teachers. He had won and declined the offer of a chair at Leiden, still the center of Greek studies in Europe. Throughout this time—so the histories tell us—his teaching, his research, his mental life aimed at one end: to replace the sterile polymathy of the seventeenth and eighteenth centuries with a new, historical brand of philology.[6]

To judge this account we must venture outside the orderly, well-mapped intellectual landscape found in most histories of scholarship. We must confront Wolf's ideas and results with those of his immediate predecessors and contemporaries. These are buried in hundreds of forgotten books, whose neoclassical title pages unfairly raise the hopes of the reader, soon to be discouraged by foxed paper and ugly type. Most of them are too technical to attract historians and too obsolete to interest classicists. No wonder, then, that the dusty sectors of eighteenth-century culture bounded by their covers have become a *terra* largely *incognita*. But two forays into this intellectual heart of darkness may prove rewarding. The spoil will amount to a new background for Wolf's thought; set before it, both his general ideas and his specific results will look quite different.

THE CASE for Wolf's originality as a thinker rests largely on the general program for *Altertumswissenschaft* which he set out in his lectures on "The Encyclopedia of Philology" and applied in his teaching to a wide variety of texts and problems. In Halle, a newish university, educational theorists had called for years for the abandonment of ancient languages in favor of modern, useful subjects.[7] Wolf had to prove that classical studies offered a form of knowledge still worth having. He did so by identifying the central subject of philology as the Greek national character. Each subdiscipline had value insofar as it contributed to "the knowledge of human nature in antiquity, which comes from the observation of an organically developed, significant national culture, founded on study of the ancient remains."[8] Greek culture, of course, was more organically developed and significant than any other. To know it the student must master all twenty-four disciplines that inform us about the ancient world—"introductory" like grammar and textual criticism and "material" like geography and mythology.[9] Thus he could control all the evidence. More important, he could read literary documents historically, in the light of the situation, needs, and values of their original audience. Above all, he could follow the evolution of the Greek spirit, which was faithfully reflected in each period by language, art, and social and political life. By watching the uniquely independent and creative Greeks learn to exercise in harmony all the powers of their souls, modern men could wake and harness the powers of their own souls. True, no modern man could know everything about the Greeks. But a serious effort to make their world and culture one's own would ennoble the mind and soul. And the modern student could understand those aspects of the Greek mind which were represented by several forms of evidence "in some respects more profoundly than the ancients themselves."[10]

Wolf's program was elegantly thought out, powerfully described, and rooted in deep knowledge of the sources. Novel it was not. As a research program it differed little from Heyne's practice. As Menze and Mettler have shown, Heyne too believed that the scholar must employ a historical method and exploit all relevant disciplines and forms of evidence.[11] In his lectures on Homer, for example, he carefully treated what was known of the poet's life, the culture of his Ionian homeland, the mythological background he had reworked, and the geographical setting of the Trojan wars. He envisioned Homer's world as a whole, comparing Ionian "dialect, trade, elegance, refinement of customs and languages" with those of Attica.[12] And he used the details of the texts as clues to the general history of Greek culture. At *Iliad* 1.599, for example, Homer describes the gods as bursting into "unquenchable laughter" at the sight of Hephaestus hopping about to pour nectar. Heyne explained: *"Gelōs* (laughter) here merely means joy. Because his age was uncultivated, Homer can only express 'joy' through 'laughter.' "[13]

Homer, in other words, wanted to say not that the gods were amused but that they were joyful. But the Greek language of his time—like other primitive languages, as Heyne knew from contemporary anthropology—lacked abstract terms, especially for anything so rarefied as a mental state. Hence Homer had had to use the existing, concrete term and image. However leaden its execution, Heyne's approach was clearly historical, not grammatical, in conception.[14]

But Heyne too was less original than most of his modern students have argued. He was at Göttingen, after all, *the* university for noblemen in eighteenth-century Germany. Founded in 1734, it was meant to train public servants. Its curriculum stressed the fields of social philosophy most closely related to the needs of government: history, statistics, political economy. In the history school of Achenwall and Gatterer, training in diplomatics, sigillography, and numismatics gave the future civil servant the tools to sift his information until only facts were left. The analytical tools of "statistics" enabled him to see his country's history not as a succession of reigns but as the natural fruits of its permanent traits: population, natural resources, climate, and institutions.[15] Heyne's official address on the opening of Gatterer's Historical Institute admirably summed up the school's principles:

> History cannot be studied properly unless you know the geniuses of peoples, their customs, rites, institutions, laws, arts, crafts, and all products of the human intellect. You must also know the causes that lie behind these things: above all the quality, nature, gifts, and benefits of the soil and climate. From these, unless they are corrected or corrupted by deliberate action, all those other phenomena necessarily proceed, as it were from their seeds.[16]

If Wolf heard little about historical method in his few visits to Heyne's classroom, he certainly knew the essays in which Heyne put such principles to work. In his description of "The Genius of the Age of the Ptolemies," for instance, Heyne portrayed the Hellenistic mind through key characteristics—subtlety, love of detail, lust for the strange and marvelous—that ran through all fields of thought. He explained these habits of mind by reference to the geographical, political, and cultural situation of Greek intellectuals in Egypt: their contact with Asian cultures, isolation from politics, and excessive learning.[17] And Wolf used Heyne's analytical scheme, giving due credit, in his *Prolegomena ad Homerum* of 1795, to explain why Hellenistic intellectuals had devoted themselves so heartily to minute pieces of grammatical and philological work.[18]

The Göttingen scholars in their turn had codified principles well known elsewhere. In treating the *Iliad* and the *Odyssey* as primitive poetry, Heyne owed much to the English traveler Robert Wood, as his colleague Michaelis suggested in a letter to Wood: "It seems to the three of us, Heyne, Beckmann, and me, that Homer ought to be read as you have read him; and we would prefer that no one in Germany read him differently."[19] Wood was in part refining arguments advanced by the Neapolitan jurist Gravina years before, in a famous letter to Scipione Maffei.[20]

To take a more general case, Heyne's Leipzig teacher J. F. Christ had argued for the combined use of separate disciplines as early as the 1720s:

> Students of Roman Law should take considerable care to keep in mind the clearest possible images of ancient life. But it is almost impossible to give a proper account of visual matters unless they are before one's eyes. Hence antiquaries should collect coins, vases, marbles, and the like, which preserve the images of ancient use . . .[21]

Christ's fascination with visual evidence led him to reproduce ancient works of art in his own endearingly naive engravings. His desire to bring complementary runs of evidence together sometimes led to coherent analyses of aspects of ancient society very alien from his own. Take his discussion of the Venus of the type known as *aux belles fesses*. He took her to be pulling her clothes aside to engage in what he modestly called—in Greek—"congress *more ferarum*."[22] He remarked that sodomy and similar goings-on were not common in Germany, "thanks more to the northern cold than to our morals," "save that as I recall simple shepherds who debauch their beasts must now and then be restrained by the law."[23] Hence he refused to shock German readers with an engraving of the statue. But he admitted that such things were managed differently in France, where a Venus *aux belles fesses* had been set up at Versailles. He continued:

> In different ages customs were different. The books of the Greeks and Romans show that they not only did not detest this sort of statue as we do, but loved and almost adored them . . . In those (Roman) days to love boys—if not considered praiseworthy, as it was by the Greeks—was not considered blameworthy. In those

days Martial's wife, doubtless a respectable lady, dared to protest to her husband that if he was so mad for boys, "she too had a bottom."[24]

Here was historicism indeed.

Christ stood out in Germany, but not in Europe. His French contemporary Caylus put a similar range of questions to a similar array of material remains.[25] Indeed, sixteenth- and seventeenth-century students of philology and antiquities had long known how to set literary documents into a rich context. A polymath like Lucas Holstenius, who learned Greek in Leiden and archaeological field technique in Italy, knew perfectly well that the *Passio Perpetuae* was the record of a tragedy played out in a late-antique legal, religious, and cultural setting.[26] He brilliantly used the art of the catacombs to illustrate the manifest content of Perpetua's vision:

> *In habitu pastoris oves mulgentem* ("A man dressed as a shepherd, milking his sheep"). Christ appeared to Perpetua in the form in which he was presented in the oratories and sacred vessels of that time. For since the Gospels assign the title and character of a shepherd to Christ, the early Christians loved to show him in this most amiable of forms to the faithful, who were regularly called his sheep. In Bosio's *Roma sotterranea* there are many scenes taken from funeral monuments, in almost all of which Christ is depicted in the form of a shepherd with his flock—in the first, in the very form shown here of a shepherd milking his sheep.[27]

Holstenius did here just what he had been taught to do by his own master Cluverius—and what had been done by his master's master, Joseph Scaliger.[28]

To watch Wolf applying his general programme to a specific document is to confirm the view that much of his work was traditional in character. His 1789 edition of Demosthenes' *Against Leptines*—which he passed out to his students sheet by sheet as it was printed—makes a case in point. Superficially this looks like a brilliant specimen of the new *Altertumswissenschaft*. Wolf explicitly connected his work on the speech with his general hermeneutics. He dealt with "both what can serve to explicate the speech and what can be learned from it," and he promised that the reader would "so far as possible understand the speech in the same way as those who originally heard the orator deliver it."[29] His *Prolegomena* not only summarized the speech and analyzed its style but set out in economical, lucid prose the Athenian liturgies that were at issue and the form of legal action in which the *Leptinea* fell.[30] They inspired his students—notably Böckh, who developed Wolf's sketch into his own massive *Public Economy of Athens*—and remained standard until late in the nineteenth century.[31]

Though he did not say so, however, Wolf had a model—one neither contemporary nor German. Jacques de Tourreil, *robin* and member of the Académie des Inscriptions et Belles Lettres, had argued in his own Demosthenes of 1721 that the serious reader must "act and think like the Athenians of that time, take on their feelings and prejudices, embrace their interests, their quarrels, their envies, their fears, their hopes. Otherwise Demosthenes would hardly find his audience."[32] He repeatedly explained that facts or events that seemed bizarre "par rapport à nos usages" were "not at all surprising in the context of ancient customs."[33] Like Wolf, he covered a wide range of points in Athenian law and institutions in deft and concise notes.[34] And Wolf borrowed more than one might infer from his admiring reference in a footnote to Tourreil's Gallic lucidity.[35] Wolf argued elaborately that the many conflicting ancient descriptions of the trierarchy could be reconciled only by scholars aware that the institution itself had "repeatedly been changed from top to bottom."[36] Tourreil had commented: "What makes this subject so obscure are the continual changes that took place in the Trierarchy, as a result of which the ancient authors, each describing it in the state in which it was in his time, have almost all disagreed."[37] Again, Wolf tried to show that the repetitions that cropped up in the speech would have been effective, not annoying, in oral delivery to a mixed audience: "a speech affects a reader and a listener quite differently."[38] This seems a characteristic sample of Wolf's sensitivity to conditions of performance in an oral or partly oral culture. But it is also an updated version of Tourreil's neat remark: "one fails to see the need for these repetitions unless one puts oneself in the place of those to whom they were originally addressed."[39] Tourreil, of course, was no exception to French norms, but a competent practitioner of the interdisciplinary approach seen at its best in the medieval scholarship of his confrère La Curne de Sainte-Palaye.[40] No wonder then, that German students and observers met Wolf's pedagogical program with so much sympathy.[41] Eloquently and incisively, he was telling them what well-read scholars already knew.

It could be argued in favor of Wolf's originality that he wished to reform contemporary culture as a whole as well as classical scholarship, while Heyne, an unpolitical German *avant la lettre*, cultivated his detachment from the life of his time. This view, which has been advanced, implies a misunderstanding of Heyne.[42] He loved and exploited Göttingen's privilege of free speech. He discussed the possibility of an American revolution with Benjamin Franklin in the 1760s. And he clearly saw classical studies as a key to understanding pressing modern problems. His defense of the Athenian condemnation of Phocion and his essay on the agrarian law both commented by indirection on political crises of his day.[43] Yet there is something to this point of view. For Heyne did not anticipate Wolf's thesis that a broadly historical training was per se the best discipline for mind and soul. His own arguments for the value of the classics were the established ones of the humanist tradition, and very lame they sounded in the 1770s.[44]

But Wolf's thesis was not his own. He owed it to his close friend Wilhelm von Humboldt, whose self-obsession—unusually intense even for a member of Werther's generation—had led him to devote much of his life to thinking about the best way to form his mental powers and sensibility. In the 1790s, impressed by Wolf, he decided that Greek culture held the key, and sent Wolf an essay "On the Study of the Ancients."[45] This Wolf drew on in his lectures and in his published *Darstellung der Altertumswissenschaft* of 1807.[46] Wolf also learned from letters and conversation. Naturally he coined formulas of his own. Humboldt admitted that the definition of *Altertumswissenschaft* as "knowledge of human nature in antiquity" was Wolf's.[47] But Wolf's most striking general tenets were borrowed. His description of language as the bearer of "what is highest and deepest" in human nature, for instance, came directly from one of Humboldt's letters to him.[48] And the excerpts from Humboldt's draft essay that Wolf quoted in the footnotes of his

Darstellung put the main points with a sharpness and generality that his own work did not attain:

> The treatment of ancient works is most rewarding when one considers less the works themselves than their authors and the period from which they come. Only this approach can lead to a true, philosophical knowledge of man. For it forces us to work out the character and complete situation of a nation, and to grasp all aspects of it in their correlation. The effort to attain such knowledge—for no one can hope to complete the search—can be called necessary for any man . . .[49]

Wolf's importance as a thinker, then, lay less in his originality than in his ability—something like Gibbon's—to fuse materials from the divergent realms of philosophy and erudition. Heyne could not provide a cogent program for educational reform. Humboldt did not know in detail what the sort of education he called for should look like. Wolf was less original than either, but his capacious mind enabled him to compose a forceful manifesto. And even the deficiencies of his *Darstellung* had one happy result. They helped to stimulate Böckh to make his own incomparably more subtle and discerning effort to solve the same set of problems.[50]

THE CLAIM for Wolf's originality as a scholar rests not on his general theories but on his technical research—above all on the notorious *Prolegomena ad Homerum,* which he published in 1795 after spending almost twenty years framing and solving and reframing the same set of problems. Here, as is well known, he argued that the standard text of Homer barely resembled the original poems. In the first place, Homer had not known how to write; in fact, the Greeks of his day had been illiterate. Hence he could not have composed epics of the size of the *Iliad* and *Odyssey.* His poems—like the ballads of the Druids and the early Germans—must have been short enough for professionals, the rhapsodes, to memorize them and recite them in public. The rhapsodes freely altered their material. And even the altered poems were not written down and arranged into coherent, unified epics until the time of

Peisistratus. In the second place, the Athenian written text had been further altered, emended, cut, and extended by early revisers, or *diaskeuastai,* and above all by such Hellenistic critics as Zenodotus, Aristophanes of Byzantium, and Aristarchus. The extant manuscripts preserved a corrupt form of this final Alexandrian revision. A modern editor could hope only to restore the Alexandrian vulgate. He could never know which sections really went back to Homer.[51]

The *Prolegomena* evoked violent reactions. Wolf presented his thesis as novel and daring (one long, defensive footnote begins "Iacta est alea..."). Others agreed. J. H. Voss, who had made Homer come alive in modern German, denounced it. Herder, the passionate student of folk poetry, annexed it. Goethe took it as encouragement to carry on with an epic of his own, now that he need no longer fear to be overshadowed. Then he rejected it in favor of his old belief in Homer's unity and perfection. Heyne reviewed it favorably, suggesting that some of Wolf's central arguments were well known. Hermann took it as a brilliant solution of an unusually intractable problem of editorial method.[52]

To argue that Homer had not written his poems, and that they had undergone radical changes as they were preserved only in performers' memories, was a commonplace of eighteenth-century literary theory. Robert Wood had argued that Homer's poems, like Ossian's, only took on their coherent form thanks to the deliberate intervention of learned collectors, after centuries of oral transmission as separate ballads:

> Just as some curious fragments of ancient poetry have been lately collected in the northern parts of this island, their reduction to order in Greece was a work of taste and judgment: and those great names which we have mentioned [Lycurgus, Solon, Peisistratus, Hipparchus] might claim the same merit in regard to Homer, that the ingenious Editor of Fingal is entitled to from Ossian.[53]

Wolf used the evidence Wood had collected to show that the Greek alphabet was a late invention. Though he did not yet know the related, much more complex ideas of Vico on the true Homer, he did know and use other ideas first advanced by contemporary critics and philosophers, such as Rousseau's argument that writing came late to the Greeks.[54]

Scholars had raised related problems even before literary critics took up the Homeric question. After all, reputable authors—Cicero, Aelian, Plutarch—suggested that the Homeric poems were only put into the accepted order after centuries of transmission in a rough and confused form.[55] Obertus Giphanius wove these hints into a short but suggestive history of Homer's text in his edition of the *Iliad* and *Odyssey* (1572). He knew from Josephus that "traces of the earlier confusion are still to be found in Homer . . . since some passages contain contradictions."[56] He also knew that Peisistratus was said to have interpolated verses in the text.[57] Soon after Giphanius, Isaac Casaubon drew much more radical conclusions. In his commentary on Diogenes Laertius (1583), he remarked that

> If what Josephus says is true, that Homer did not leave his poems in written form, but they were preserved by memorization and written down much later, then I do not see how we can ever have them in a correct form, even if we have the oldest MSS. For it is likely that they were written down in a form quite different from that in which they were first composed.

Ménage, commenting on the same passage, agreed. Richard Bentley, whom Wolf admired greatly, went even further in a passing remark, and denied the original unity of the epics:

> Homer wrote a sequel of Songs and Rhapsodies, to be sung by himself for small earnings and good cheer, at Festivals and other days of Merriment; the Ilias he made for the Men, and the Odysseis for the other Sex. These loose Songs were not collected together in the Form of an Epic Poem, till about 500 years after.[58]

Wolf knew all of these remarks and quoted many of them in his footnotes. Accordingly, his claim to be advancing novel and audacious views was at least partly insincere. And his critics' attacks on his "literary impiety" applied less to his ideas per se than to his having dared to state them so directly and to make them the foundation of true Homeric studies. Wolf himself paradoxically complained in 1804 that his views had been far less revolutionary than his opponents had claimed.[59]

Since the late nineteenth century, competent historians of scholarship have accepted these points. Some have reacted to them by denying that Wolf did anything of interest, or at least anything original.[60] Others, however, have tried to save the originality of the *Prolegomena*. They have done so by drawing attention to the more technical—and less popular—concluding sections of the book, where Wolf used the Venice scholia on the *Iliad*, published by Villoison in 1788, to trace in detail what each of the Alexandrian critics had done to the text of Homer.[61]

These chapters reveal impressive technical dexterity and attention to detail. Wolf worked again and again through the Venice scholia, collecting in enormous footnotes every reading and interpretation that they attributed to each of the major ancient critics and recensions. He paid careful attention to the language of the scholiasts and the nature of the ancient critical signs, and took pains to work out the limits of what can be known about the work of each ancient critic. Every assertion, however uncontentious, rested on a solid base of close-packed references and quotations. In support of his thesis that the ancient vulgate was based largely on the recension of Aristarchus, for example, Wolf wrote: "Relevant to this are the words *epeisthē hē paradosis Aristarchōi* in the scholia, as in 4.138, 5.289, 20.357, or *epeisthēsan autōi hoi grammatikoi*, 16.415, or *epekratēsen hē anagnōsis autou*, 1.572, 5.69, 6.150, 7.289, 22.67, or *houtōs echei ta tēs anagnōseōs*, 11.651 (652), 23.387 . . ."[62]

Wolf's work was not only thorough but full of insight. Anyone used to nineteenth-century treatments of ancient scholarship will be refreshed by a dip into the *Prolegomena*. Wolf's successors, Lehrs and—to a lesser extent—Ludwich, set out to refute Wolf. They described Aristarchus et al. as professional scholars like themselves, who had collated manuscripts, established critical editions of texts, and presumably published them with the Alexandrian equivalent of B. G. Teubner.[63] Wolf's historical vision was far clearer. He argued that improvement of the text as a literary masterpiece, not preservation of the text as a historical document, had been the aim of Hellenistic criticism:

> We must completely rid ourselves of the notion that makes us imagine the critics of that age in the terms of the modern state

of the art ... Some of them perhaps tried to represent Homer as accurately as they could. But they worked harder to ensure that he never seemed inconsistent or weak, often removing verses, elsewhere adding polish where none belonged ... This whole art had its origin in what we Germans call the aesthetic rather than the critical faculty—or, to put it another way, relied on poetic, not diplomatic standards of argument.[64]

Wolf did not blame the Alexandrians for having aims different from his. He merely insisted that their ends be understood—that modern scholars realize that, even if an Alexandrian critic did consult manuscripts while seeking the "genuine" form of the text, "the genuine form was the one that seemed most appropriate to the poet; and that obviously depended on the arbitrary judgment of the Alexandrians."[65]

It was not unusual for a major eighteenth-century edition to begin with a history of its text. Heyne's great Tibullus began with one.[66] And even the bare texts of the Bipontine Press started with elaborate bibliographies of earlier editions and commentaries. But as one might expect from editors with rich libraries of printed books, no travel grants, and a strong interest in the history of scholarship, Heyne and his colleagues concentrated on the history of their texts in the age of printed editions. They investigated this field with exemplary bibliographical skills and historical insight, carefully evaluating the work of Renaissance editors in the light of the more primitive critical methods they had known.[67] The manuscript traditions of their texts they usually treated in less detail, and with less hope of arriving at definitive results. Ernesti and Heyne turned up important textual witnesses and had their insights into the genealogy of manuscripts. But for the most part they lacked access to the evidence.[68] As to ancient editions and variants, they knew that such things had existed but made little serious effort to reconstruct them.[69]

Wolf defied readers' expectations and warned them *not* to expect a detailed account of modern Homeric scholarship: "We will examine ancient sources above all, not these modern ones of printed books. The changes that the latter have produced in the text, compared with those produced by the former, hardly deserve the attention of a busy man."[70] He explicitly pointed out that this limitation was unusual. And in the announcement of his edition that he printed in the Jena *Literatur-Zeitung,* he claimed that his work on Homer was novel precisely because he was the first editor to master the sources for the ancient history of the text.[71] Within the genre of eighteenth-century prefaces, then, the *Prolegomena* were something of an exception.[72]

But they were a predictable sort of exception. Scholars had long seen that Homer's text could not be established except by investigation of what the Alexandrians had done to it. Giphanius knew that the Hellenistic scholars had divided Homer's continuous poems into books. He also knew that Aristarchus had "noticed many spurious verses in the text of Homer, and marked them with the obelus": "This is why many verses are cited by the ancients, such as Aristotle and others, which are not found in our texts today. For they used other recensions, we that of Aristarchus."[73] J. R. Wettstein, speaking in 1684 "de fato scriptorum Homeri per omnia secula," seems to have agreed.[74] Ludolf Küster—whose *Historia critica Homeri* of 1696 Wolf reprinted—disagreed. Eustathius's comments showed that the Homeric vulgate contained passages and lines that Aristarchus had condemned; hence it could not be the text he had established.[75] In two careful dissertations defended in 1732, the Wittenberg scholar J. M. Chladenius outdid both Giphanius and Küster in subtlety. Assembling cases in point from almost all the Greek scholia in print, he showed how to study the ancient critic "completing a job in his workshop."[76] His results anticipated Wolf's. Ancient critics had had other aims in mind than reconstructing a text *wie es eigentlich gewesen:* "The ancients will give us the example of a criticism that serves piety rather than the integrity of poets' works. For they allowed nothing to remain that did not accord with their religion—or rather superstition."[77] Unfortunately he drew no detailed conclusions about the history and state of the text of Homer.

By the 1780s even very ordinary scholars were taking positions on the Alexandrians' work. Thomas Burgess—later a marvelously credulous theologian, who defended the *Comma Johanneum* and argued that the *De doctrina Christiana* could not be by Milton, since it was not orthodox—in 1781 described the text of Homer as a *rifacimento* "not so radically altered as Berni's Boiardo or

Dryden's or Pope's Chaucer, but more so than John Hughes's Spenser." He explained that

> many ancient critics set out to emend Homer. And though the extant scholia often call our attention to changes Aristarchus and others made, we have only a small part of the commentaries—especially the older ones—on the poet. Hence we probably fail to recognize many of their emendations, which are today believed to be by Homer.[78]

Wolf himself had argued a parallel case as early as 1783. In his edition of Hesiod's *Theogony,* he argued that the extant scholia revealed "only the names of those who commented on" Hesiod. To produce a really satisfactory edition one would have to know from sources more solid than conjecture "exactly what the Alexandrian critics contributed to it, and from whose recension our form of the text derives."[79]

Villoison quoted this body of scholarship lavishly in the *Prolegomena* to his edition of the Venice manuscript. He saw that the new material he was publishing provided just the information whose absence Burgess and Wolf had lamented: "Our great collection of ancient commentaries and scholia will remedy this confusion."[80] And he set out to assemble what the scholia revealed about the ancient critics.[81] In short, it was well known before 1795 that a critical edition of Homer would have to be based on a critical history of Alexandrian learning, itself to be based on the Venice scholia.

Indeed, Wolf knew—and told his students—that he was hardly the first scholar to anatomize in detail a set of scholia on the *Iliad.* He regarded L. C. Valckenaer's *Dissertatio* on Leiden manuscript Voss. gr. fol. 64, which appeared in 1747, as the first important essay: "after that time scholars became more interested in bringing scholia to light from libraries."[82] Valckenaer had argued that all scholia are mixtures of old and new, useful and useless materials. Those in his Vossianus included Homeric questions and answers apparently derived from Porphyry, excerpts from Eustathius that helped to date them, and many notes by a gentleman oddly named Senacherim—who, Valckenaer rightly guessed, was a grammarian of the twelfth or thirteenth century.[83] With minute care and copious examples he dissected marginal and even interlinear notes. He reproduced and defined the technical terms and signs used in his manuscript and other Homeric commentaries. And he took an informed interest in the general history of Alexandrian scholarship:

> I believe that a great many *Poetic Questions* had their origin in the Alexandrian Museum. Scholars spent their spare time in devising them and finding clever solutions for them. Thus they passed their holidays and refreshed minds tired out by dealing with more serious problems. If this is true, the custom will not be cause for criticism. That it is likely will be clear from the scholium of Porphyry given at the end of *Iliad* 9. I will give only its beginning, from the Leiden MS. "PORPHYRY: In the Alexandrian Museum it was the rule for questions to be proposed and their answers recorded . . ."[84]

Valckenaer made clear that his method was not original: "Hemsterhusius disclosed the true method of dealing with scholia in his notes on the scholia to Aristophanes' *Plutus* (1744)."[85] Hemsterhusius had argued there that the historical analysis of scholia was central to Greek studies. He had also laid down specific rules for carrying it out. All scholia should be treated as conglomerates. And the occurrence of divergent or contradictory glosses in one note should be taken as evidence that two different sets of scholia had been conflated.[86] Drawing on Estienne, Casaubon, and Bentley, he had also shown how to understand and classify the special language of scholiasts: "Grammarians often use *hoti* ('that') in such a way that one must understand *sēmeiou* ('Note' [imperative]) to complete the sentence."[87]

Hemsterhusius's views were shared by most of his Dutch followers. Valckenaer published scholia on Euripides as well as on Homer, and drew on scholia as well as on Seneca and Stobaeus to reconstruct the *Hippolytus Kalyptomenos*—the original, unsuccessful form of Euripides' *Hippolytus*.[88] David Ruhnken, the dominant Greek scholar in Holland in Wolf's time, drew evidence from the scholia about the first version of Apollonius Rhodius's *Argonautica*, and devoted more than forty years to editing the Plato scholia.[89] Wolf certainly knew that he was entering Dutch territory in the *Prolegomena*. That may help to explain why he dedicated the book "Davidi Ruhnkenio principi criticorum."[90]

Wolf also knew that he was not the first German to encroach on this Dutch specialty. Ernesti, setting out to make German Greek scholarship as rigorous as Dutch, included the scholia in his 1753 edition of Aristophanes' *Clouds*. He explained the proper point of view from which to examine them:

> I must warn at the outset that these scholia—I mean the Aldine—are not the work of one man, even when they are presented as one man's, and are not separated by the normal term *allōs* ["An alternative explanation is . . ."]. One must imagine a farrago of excerpts from the commentaries of grammarians on Aristophanes, Apollonius Rhodius, and other poets, from ancient glossaries, even from writers of quite a different sort: Thucydides, Plutarch, Lucian, Stephanus *de Vrbibus,* and many others, sometimes cited under their names, often anonymously. Nor did the farrago itself have a single author.[91]

In his lectures he used the confused and corrupt text of the scholia as a body of cases from which his students could learn the elements of textual and historical criticism. His successor Reiz continued to lecture on scholia.[92] Heyne planned an elaborate study of the ancient commentators on Virgil, which he abandoned on learning that Ruhnken was already at work in the field.[93] His pupil J. P. Siebenkees published a long account of the A scholia to the *Iliad* in the Göttingen *Bibliothek der alten Litteratur und Kunst* in 1786, two years before Villoison's edition came out. This careful essay dwelt on the ancient critical signs and important, puzzling subscriptions that appeared at the end of each book in A: "Why does the subscription never mention Zenodotus, whose recension seems to be used very heavily in the MS? This seems surprising . . . Perhaps the scholia which cite and criticize Zenodotus's emendations—and usually reject them—are the notes of Aristonicus [who was mentioned in the subscription]?"[94] And if one looks—as any German scholar of the 1780s and 1790s regularly did—outside the Dutch and German world, one finds that English and Italian scholars had had their say as well—notably A. Bongiovanni, who had published the B scholia to *Iliad* 1 in 1740, with an exemplary preface.[95] No wonder then that C. D. Beck, writing his inaugural dissertation *De ratione qua scholiastae poetarum Graecorum veteres, inprimisque Homeri, ad sensum elegantiae et venustatis acuendum ad-*

hiberi recte possint (1785), could cite a rich body of monographic work to support his perverse belief that no extant set of scholia antedated Eustathius: "rivalry led others to follow his example, and even to try to surpass Eustathius's diligence."[96] No wonder, too, that a far less conventional man like J. J. Reiske denounced the study of scholia as a fad as early as 1770:

> Had the decision lain wholly with me, I would have added no scholia at all to my Demosthenes. Most of them are emptier than a nut with no meat, futile, footling, trivial, childish. The language is full of unbearable solecisms. But who can stand against a flood? I therefore decided to leave the commentaries commonly attributed to Ulpian to others to edit—if anyone wants to do so—and give here only what my Bavarian and Augsburg codices afforded me. It is not that they are better than the common ones—they are about the same—but that they will give my edition the attraction of novelty and make it more saleable . . .[97]

By 1788, then, German scholars were well prepared to see the historical importance of the A and B scholia. Prepublication announcements had made them curious as well.[98] Hence several of them—including Wolf and Heyne—skimmed the cream from this material in elaborate reviews.[99] The anonymous reviewer in the *Bibliothek der alten Litteratur und Kunst,* for instance, took Villoison's edition as the starting-point for a long investigation of all the Homer scholia in print, which he traced back to two distinct original sources.[100] G. C. Harles assembled the views of Villoison and his critics in staggering detail in the first volume of his new edition of Fabricius's *Bibliotheca Graeca* (1790). He dealt with the scholia in all their forms, assembled the information they yielded about the lives and lost works of the Alexandrian scholars, and, imitating Fabricius's treatment of Eustathius, had indices to the new scholia drawn up.[101]

Wolf was as excited as anyone by the Venice scholia, but his treatment of them differed markedly from those summarized by Harles. As C. G. Schütz put it in 1796:

> Anyone who compares Harles's edition of Fabricius and Villoison's *Prolegomena* will easily see that Mr. Wolf has done more than emulate in his own futher investigations their *Sammlerfleiss*—

a quality really valuable in the light of the intention behind their works. By historical reasoning and ingenious rapprochements [of facts] he has established new results, and thus greatly surpassed them.[102]

A more precise way of expressing the difference would be to say that Wolf wrote the history of scholarship rather than the history of scholia. He knew that the scholia needed analysis, that each Alexandrian needed reliable dates and bibliography. But he devoted most of his time and effort to reconstructing the technical method each ancient critic had used—and, through them, the story of Homer's fate. Villoison and others, as we saw, had known that the scholia should be made to tell this tale. But they did not make them tell it.[103] Paradoxically, Wolf learned more than others had from the scholia precisely because the scholia themselves did not interest him as much.

THERE WAS a model for the "reasoning" and "rapprochements" that set Wolf's work apart. Since the Reformation, the textual history of the Bible—especially that of the Old Testament—had received much passionate attention. It posed intractable problems. The Hebrew text of the extant manuscripts and the printed editions had undergone radical changes since the days of Moses and David. The square-charactered alphabet in which it was written was not the original one, as Jerome and others pointed out. The word divisions, the vowel points, the accents, the marginal apparatus of variant readings had apparently been introduced by the Masoretes, the Hebrew grammarians of the Near East, during the first millennium A.D. Some Protestants inferred from these facts that the Septuagint was more reliable than the Hebrew. After all, it had been translated with divine help from good manuscripts. The Greek and Hebrew texts disagreed because perfidious Jews had deliberately altered their manuscripts. Others denied patristic and Jewish evidence alike and claimed that the extant Masoretic text of the Pentateuch went back to Moses, vowel points, variants, and all. Catholics on the other hand tended to claim that the Latin vulgate surpassed both the Hebrew and the Septuagint.[104]

Fueled by *odium theologicum* as well as *philologicum*, the debate ran long and hot. Scaliger, Casaubon, the Buxtorfs, the Cappells, Morin, and the Vossiuses took sides. By 1678 the Oratorian Richard Simon had produced a *Critical History of the Old Testament*, which used the patristic evidence, the Talmud, and the Jewish scholarly apparatus and commentaries to reconstruct the state of the biblical text century by century from its origins to his own time. He hoped to show that no version was complete and reliable. Hence one had to rely on Mother Church, which was guided by infallible Tradition, to compensate for faulty Scripture.[105]

This tradition of biblical research and controversy was still active in the eighteenth century; and Wolf had it very much in mind as he worked on Homer. In calling for more collations of manuscripts, he wrote that "Homer still awaits his Kennicott."[106] He referred to Benjamin Kennicott, whose critical edition of the Hebrew Bible, based on obsessively minute collations, had appeared at Oxford in 1776–1780. In reviewing Villoison, Wolf explicitly compared the Alexandrian scholars to the Masoretes.[107] In the *Prolegomena*, he likened the Venice scholia to the apparatus of the Masoretic Bible: "Let the Oriental masters, proud of their Masorah, cease to deplore the bad fortune that makes us rely on such late manuscripts for the text of Homer . . . We too now have a sort of Greek Masorah." And he made clear that his efforts to reconstruct the methods of Alexandrian scholars were exactly similar to what the orientalists had already done for ancient Jewish scholars (though they had been hindered by the comparative poverty of their sources):

> The Oriental masters would, I think, be delighted if they knew for certain what Gamaliel or any other early Jewish doctor read in three passages of Moses and the Prophets. We know what Zenodotus read in Homer in some four hundred passages, what Aristophanes read in two hundred, what Aristarchus read in more than a thousand.[108]

Wolf even compared those who tried to defend the authenticity of the standard text of Homer to the Buxtorfs, who had defended the *theopneustia* of the Masoretic Old Testament.[109]

It is not surprising that Wolf knew this body of work. Many of his best friends were theologians who used a historical method.

One of the few Göttingen teachers he had liked, Michaelis, had been a pioneer in applying a critical technique to the Old Testament. Wolf's closest ally in the early days at Halle, Semler, was a radical theologian and textual critic of the New Testament. And he knew and greatly respected Semler's pupil Griesbach, whose edition of the New Testament was the model for Wolf's own final deluxe Homer.[110]

But there is more than general analogy and biographical anecdote to connect Wolf's work to that of the biblical scholars. The evidence suggests that the *Prolegomena* were directly modeled on one of the most controversial products of the German biblical scholarship of Wolf's time: J. G. Eichhorn's *Einleitung ins Alte Testament,* which began to appear in 1780. Like Wolf, Eichhorn studied at Göttingen under Heyne and Michaelis. He returned there as a professor in 1788.[111] His works on the Old and New Testaments fascinated literati of widely different stripe. Coleridge filled the margins of his copies with approving, detailed notes.[112] Wolf thought the volumes on the New Testament exemplary. He cited those on the Old Testament in passing in the *Prolegomena,* and recommended them to his friends and students.[113] But the connection between his work and Eichhorn's is far closer than his explicit remarks suggest.

Like Wolf, Eichhorn treated his text as a historical and anthropological document, the much altered remnant of an early stage in the development of human culture.[114] Like Wolf, he held that the original work had undergone radical changes, so that the serious biblical scholar must reconstruct "the history of the text."[115] Like Wolf, too, he saw the true history of the text as its ancient history, before the standardized manuscripts now extant had been prepared. With the work of the Masoretes, he wrote, "properly ends the history of the written text; for the chief work was accomplished, and the Hebrew text continued now, some insignificant changes excepted, true in all its copies, to its once-for-all established pattern, as is clear from Kennicott's 'Collection of Variations.' "[116] Unlike Kennicott, Eichhorn spent hardly any space on the printed editions of the Hebrew text.

Eichhorn ransacked the Masorah for evidence about the methods of its creators as ruthlessly as Wolf later attacked the Venice scholia.

His conclusions, set out point by point in heavily documented chapters, resembled Wolf's far more closely than anything I have been able to find in strictly classical philology. If Wolf showed great resource in classifying the critical signs by which Alexandrian scholars expressed their opinions of the received text of Homer, Eichhorn had already done the same for the critical remarks that filled the margins of the Masoretic text (and which, like the Alexandrian signs, assumed that the received text should be respected). Here, for example, is his chapter on the Masoretic marginal direction *Qerē velo Cethib* (literally "read and not written"), which instructs the reader to insert a word into a given passage of the Bible when reading it aloud or analyzing it, but not when copying the text proper:

> The Talmud already knows seven instances of the Qerē velo Cethib; the Masorah lists ten of them at the beginning of the fifth book of Moses. Our editions, finally, note a still greater number, but they deviate from one another in the passages where they omit the Qerē velo Cethib. The passages in the Masorah are: Judges 20.13; Ruth 3.5, 17; II Sam. 8.3; 16.23; 18.20; II Kings 19.31, 37; Jerem. 31.38; 50.29. In our editions cf. II Kings 20.13; Ezech. 9.11; Is. 53.4, 9; Ps. 46.2; Jos. 22.24 . . .
>
> The Qerē velo Cethib are not variants, as they have been previously presented, but exegetical glosses . . . For all the words added in the Qerē could be omitted without harming the sense. An interpreter presumably added them for the sake of clarity in meaning. Further, in the passages that I checked by way of trial, there is no evidence that the ancient translators read them. Finally, they are simply exegetical glosses . . .[117]

Even the Dutch had not made such systematic use of evidence from scholia.

These parallels, though striking, do not prove that Wolf imitated Eichhorn. For Eichhorn was not the first to classify and analyze Masoretic critical methods. The Masoretes themselves sometimes listed the passages at which a given marginal direction occurred; the *Masora magna* to Deuteronomy 1.1, for instance, gives ten cases of *Qerē velo Cethib*.[118] Jacob ben Chajim and Elias Levita discussed in the sixteenth century the origins and nature of the different forms of variant found in the Masorah.[119] Louis Cappell reworked the evidence in his *Critica sacra* of 1650; in the case of the *Qerē velo Cethib* as elsewhere, he provided the bulk of Eichhorn's collections of cases in point.[120] The younger Buxtorf went over the same ground from the opposite point of view in his reply to Cappell, the *Anticritica*.[121] Brian Walton summarized Buxtorfs and Cappells in his *Prolegomena* to the London Polyglot of 1657. Those who referred the *Qerē* "to the *traditio* of Moses on Sinai," he wrote, "deserve laughter rather than serious refutation." It was clearly the work of later scholars, and consisted partly of variant readings drawn from manuscripts, partly of rabbinical conjectures designed to replace "voces obscoeniores."[122] The most active eighteenth-century textual critics of the Hebrew Bible, Kennicott and de Rossi, naturally also raised and gave answers to what had become a canonical set of problems about the extent to which the *Qerē* came from collation or conjecture.[123] And Simon had shown long before Eichhorn how to insert a systematic analysis of the Masorah into a diachronic account of the history of the biblical text. Wolf did not have the Hebrew to study the Masorah on his own; but the scholarship of the Republic of Letters could obviously have offered him a plethora of sources and partial models for the *Prolegomena* even if he had not read the *Einleitung*.

Yet it seems all but certain that Eichhorn was Wolf's source. For Eichhorn provided both formulations and solutions for technical problems that the scholia posed Wolf. I will give two examples. First, Wolf argued that the Venice scholia did not fully explain the methods Alexandrian scholars had used in collating manuscripts and assessing the worth of variants: "The most important facts—that is, what innovations Aristarchus made in the poems as a whole, how scrupulously he dealt with old MSS, how he used the recensions of Zenodotus, Aristophanes, and the rest—these cannot now be ascertained by certain or even probable arguments."[124] Wolf explained this gap in the information offered by the scholiasts historically, in terms of the cultural situation of the ancient grammarians. Exact details of a critic's reasoning about variants simply had not interested them: "The ancients apparently never worried about *our* problem."[125] Eichhorn had already solved a parallel problem in a parallel way: namely, why the Masoretes did not bother to preserve the older, unpunctuated manuscripts of the Bible from which they copied their own more usable texts:

> Truly the manuscripts of the ancient pattern were worthy of preservation for the sake of their critical value. But at that time criticism was regarded in a different point of view from that of our days. The Jews believed themselves to have furnished manuscripts better, and containing more information, than those of their ancestors, and to have imparted to the former all that was valuable in the latter; they were flattered also at beholding the manuscripts of their creation adopted and preferred. How probable then, under such circumstances, is the disuse and neglect of the older manuscripts.[126]

Second, Wolf argued that in the Venice scholia the opinions of individual scholars were irretrievably mingled and confused, since the Alexandrians had not considered it worthwhile to keep them separate: "Once Aristarchus's *anagnōsis* became the vulgate... new emendations and observations were added to it; the names of the original authors of the readings were generally omitted, except perhaps where they disagreed."[127] Hence no single recension—even the best-attested, that of Aristarchus himself—could be reconstructed line by line or word for word. Eichhorn had made exactly the same point about the Masorah: "We must regret that in the Masorah the earlier and later recensions of the Jews are mixed together, that each Jew did not publish the results of his critical work separately or designate his own contribution more precisely—in short, that we can no longer distinguish the old Masoretic recension from the new one."[128] In each case, Eichhorn's argument combined existing methods in a novel way. Michaelis had given him the techniques for studying Masoretic scholarship, and the conviction that the Masoretes had introduced changes and errors in their text.[129] But Michaelis had freed himself from the established Lutheran view that the Masoretic text was perfect only at the cost of refusing any sympathy to the Masoretes "whose names we do not even know," and whose arguments were often "jüdische Grillen."[130] Eichhorn's other teacher, Heyne, had consistently taught that one must bring imaginative sympathy to every past person and phenomenon—that even early textual critics deserved admiration for their learning and brilliance "if one considers the times in which [they] lived."[131] Eichhorn applied Heyne's historicism to the material Michaelis had made available. Only in the *Einleitung* could Wolf have found a method so apt to fit his own prejudices and interests.

It will be clear by now that Wolf's treatment and Eichhorn's both led, like so many modern *Textgeschichten,* to results of great interest to historians but minimal utility to editors. In both cases, close study of the evidence showed that the work of the ancient scholars could not be reconstructed fully—and thus that a really critical text could not be produced. It would be hard for two treatments of different bodies of evidence to resemble one another more closely.

Three further bits of evidence confirm the view that Wolf imitated Eichhorn. In a brilliant essay in the *Repertorium für Biblische und Morgenländische Litteratur,* Eichhorn argued that the stories about Ptolemy Philadelphus and the LXXII translators of the Bible had been conflated from two divergent accounts, both intended to win credit for the Septuagint. He compared them explicitly with the story transmitted by a scholiast to Dionysius Thrax: that Peisistratus had reassembled the poems of Homer with the help of LXXII grammarians.[132] Wolf, discussing the story about Peisistratus, made the same analogy in reverse. The LXXII Homeric grammarians were a Hellenistic invention, and should be eliminated from history just as "viri docti" had already eliminated the "Jewish invention of the LXX interpreters." Wolf's footnote made his reference clear: Eichhorn had treated this problem "acutissime omnium."[133] In this one case Wolf admitted how much his historical insight owed to Eichhorn's.

Second, at least one contemporary observer saw the resemblance between Wolf and Eichhorn. In 1827 H. K. A. Eichstädt remarked that Eichhorn had set out "to achieve with equal success for biblical criticism what Wolf did for Homeric criticism."[134] Both, he reflected, had destroyed far more than they had built, but that was only to be expected in critics. Eichstädt's chronology was backward—but such an error is not surprising in a man so cavalier with details.[135]

Third, Wolf himself hinted in a variety of ways that biblical scholarship was the model that anyone interested in the history of ancient scholarship must imitate. In Chapter 4 of the *Prolegomena* he wrote that the true nature of ancient criticism would become clear "through comparison of both of these farragoes, Greek and Hebrew."[136] Of the two chapters of Part II of the *Prolegomena* that he completed, one compared at length the textual histories of the Old Testament and Homer, and the origins of the Masorah and the Venice scholia:

> Our Hebrew text derived from a *paradosis;* so, clearly, did our vulgate Homer. In each *paradosis* a choice was made among readings, which we may nonetheless rework. In each text, the *paradosis* itself has undergone some mutilation and corruption . . . The Masorah is full of trifles and feeble, superstitious inventions; this mass of scholia does not lack similar things. True, Greeks rave in one way, Jews in another . . .[137]

And he explicitly advised his students, if they hoped to understand Greek textual criticism, to approach it through that of the Jews:

> One who wants to penetrate more deeply must concern himself with the history of the Masoretic MSS. True, these came into being much later than Greek criticism. But there is much similarity with the earliest Greek criticism. It is at least clear that they were not so bold as the Greek scholars. But one must not think that we have the Old Testament in its original form.[138]

I do not think that it is stretching a point to take this passage as autobiographical. Wolf himself had learned from the Masorah, as analyzed by Eichhorn, to read the Venice scholia.

Moreover, Wolf stated in the preface to his Homer of 1804 that he used critical techniques "established by Griesbach, that outstanding practitioner of *Critica sacra.*"[139] Griesbach had argued that the extant manuscripts of the New Testament fell into three great recensions and laid down rules for choosing among the readings they offered.[140] Wolf held that these rules should guide the Homeric scholar as well. He too had to choose not merely among manuscripts but among distinct recensions—the ancient ones that Wolf had reconstructed. This passage shows that Wolf could set out deliberately to borrow tools from another discipline. And it suggests that his grand plan was to forge a Homeric criticism that drew on both Old and New Testament scholarship. The former would show how to reconstruct ancient recensions, the latter how to use them.

Wolf, then, did not create a new model of philological criticism. Rather he annexed for classical studies the most sophisticated methods of contemporary biblical scholarship—just as Lachmann was soon to annex with greater success the methods of New Testament textual criticism.[141] True, the biblical scholarship that Wolf used had received powerful help from eighteenth-century classical scholarship; but Heyne, not Wolf, deserves the credit for that.

WHAT REMAINS of the *Prolegomena* after all this is subracted? A great deal. Wolf combined detailed work on textual criticism with a general contribution to broad and fashionable literary problems. He thus made philology more intellectually respectable and interesting than it had been since the late Renaissance. As Humboldt put it in another context, he showed that the study of grammatical details could be *geistvoll*.[142] This was the best possible response to the *philosophes'* scoffing about pedantry.

Wolf also achieved a masterpiece of presentation. He could state the commonplaces of eighteenth-century scholarship in a style that gleamed with polish. No one could put the insights of Bengel and Griesbach more pithily:

> Newness in MSS is no more a vice than youth in men. For just as age does not always make men wise, so insofar as each MS closely follows an old and good authority, it is a good witness.[143]

No one could admit with a prettier pretence of humility that poets understand the growth of an epic more profoundly than scholars:

> I have done what I could . . . But this is a subject worthy of the study of many, and of men who will follow a different path; above all, of those who can measure the poetic abilities of the human mind by the standard of their own intellect, and possess powers of judgment trained by classical literature: the Klopstocks, Wielands, Vosses.[144]

No one could explain more powerfully just what it meant to think historically, to recognize the pastness of the past:

Here we must completely forget the archives and libraries that nowadays preserve our studies, and transport ourselves into a different age and a different world. There a great many devices that we think necessary for good living were unknown to wise men and fools alike.[145]

If the ideas were well known, Wolf's axioms and images were his own. To that extent at least—as a literary masterpiece composed from already existing sources—the *Prolegomena* deserved the enthusiastic reception they received from such connoisseurs of style as Flaxman, Goethe, and Leopardi.

The ultimate fate of the *Prolegomena* had little to do with the work's real origin and merits. In a sense, it is emblematic of Wolf's own larger fate as a writer. The *Prolegomena* became *the* model of philological criticism. Even those who lived through the origins of the *Historische Schule*, men like Niebuhr and Varnhagen von Ense, came to see the *Prolegomena* as the work "in which the higher criticism attained completion"; as the work that inspired a revolution in biblical as well as classical scholarship.[146] Newcomers to German philology, like the American George Bancroft, were told to master "Wolf & yet Wolf & yet Wolf."[147] The *Prolegomena* were reprinted several times in the nineteenth century so that students could have direct access to them.

Like some other classics of German scholarship—F. Schlegel's book *Ueber die Sprache und Weisheit der Indier* is a case in point—Wolf's work was seldom criticized in detail.[148] Even those who denied the novelty of his general thesis applauded the rigor and originality of his technical work.[149] This need not surprise us. The admission that Wolf could have had predecessors in his technical discoveries could have suggested that the *Historische Schule* itself did not spring full-grown from the *Sturm und Drang*.

What is more surprising, at first, is that even those who continued Wolf's work did not mention or follow up his comparisons between Alexandrians and Masoretes. Karl Lehrs, a converted Jew, set out in the 1830s to show that Wolf had greatly underestimated the consistency and seriousness of Alexandrian textual criticism. Lehrs did not use the Masorah and made clear in passing that he had no knowledge of things or languages oriental.[150] Parthey, the historian of the Alexandrian Museum, followed Wolf rather than Lehrs in his discussion of textual criticism. He too failed to refer to or use Jewish material. Parthey's critic Ritschl showed no greater comparative interest.[151] Even Gräfenhan, whose comprehensive *Geschichte der klassischen Philologie im Alterthum* summed up all that was known of the subject by the mid-1840s, used strictly Greek sources to reconstruct Greek methods.[152] Only in very recent times have historians of Judaism started to raise again Wolf's questions—apparently in ignorance of Wolf's work.[153] The "two farragoes" have not yet been compared in detail.

The reasons why Wolf's program was not carried out are not far to seek. Jews and Judaism seemed far less attractive in Restoration Germany than they had in the late Enlightenment. To use Jewish material to clear up Homeric problems would perhaps have seemed an oxymoron in 1840. More generally, the ready cross-cultural comparisons that had been Wolf's normal practice seemed suspect in the age of pure historicism. Classical scholars were professional specialists, who neither expected nor were expected to deal with Semitic philology. Wolf was not condemned for his excursions into Judaica; rather, they were not noticed. The founder of German *Philologie* simply could not have done something so uncharacteristic of the true *Philologe*. Generations of scholars apparently read and digested the *Prolegomena* without even noticing—far less building on—the remarks about the Masorah.

Nineteenth-century readers failed to see that Wolf, for all his self-conscious modernity, remained a good citizen of the old-fashioned, Latin-speaking Republic of Letters—that rich continent of the mind, whose baroque contours have not yet received their map. Wieland, a fellow citizen of the Republic, sketched Wolf's true historical location far more vividly than Wolf's professional successors could: "I have always wanted to see the vast learning of Salmasius, which I know from his *Exercitationes Plinianae,* united in one mind with the elegance of my idol Hemsterhuys. I have now found them truly united in Wolf."[154]

NOTES

ACKNOWLEDGMENTS

INDEX

NOTES

Introduction: The Humanists Reassessed

1. The *loci classici* for this line of argument are Bacon, *Novum Organum,* bk. 1, aphorisms 63–98; *Works* (London, 1879), 2:438–449; Descartes, *Discourse on Method,* section 1.
2. Bacon, *Novum Organum, Parasceve ad historiam naturalem et experimentalem,* aph. 3; *Works,* 2:505. This translation—like all others not explicitly attributed to a translator—is mine.
3. Many scientists did not share these views. For one case see W. Pagel, "The Place of Harvey and Van Helmont in the History of European Thought," *Journal of the History of Medicine and Allied Sciences* 13 (1958), 186–199; for another, see Chapter 7 below. In general, too, it is important to bear in mind that all assessments of the state of scholarship were bulletins from a complex and protean front line— and often shaped more by their authors' prejudices than by the facts. For this important methodological point see F. Waquet, *Le modèle français et l'Italie savante (1660–1750)* (Rome, 1989).
4. Pascal, "Préface sur le Traité du vide," *Oeuvres complètes,* ed. L. Lafuma (Paris, 1963), 230–231.
5. J. Kepler, *Gesammelte Werke* (Munich, 1937–), 16:329. Cf. Chapter 7 below; and for a more nuanced account of these matters see H. Blumenberg, *Die Lesbarkeit der Welt,* 2d ed. (Frankfurt, 1983), and P. Rossi, *La scienza e la filosofia dei moderni* (Turin, 1989), chs. 3–4.
6. Boulliau is quoted by F. F. Blok, *Nicolaas Heinsius in dienst van Christina van Zweden* (Delft, 1949), 111–112; for Casaubon see M. Casaubon, *Epistolae,* in I. Casaubon, *Epistolae,* ed. T. Janson van Almeloveen (Rotterdam, 1709), 23–24; cf. ibid., 17, and M. R. G.

Spiller, *"Concerning Natural Experimental Philosophie": Meric Casaubon and the Royal Society* (The Hague, 1980), 146.
7. See e.g. E. W. Cochrane, *Tradition and Enlightenment in the Tuscan Academies, 1690–1800* (Chicago, 1961), ch. 5; F. Venturi, "History and Reform in the Middle of the Eighteenth Century," *The Diversity of History: Essays in Honour of Sir Herbert Butterfield,* ed. J. H. Elliott and H. G. Koenigsberger (Ithaca, N.Y., 1970); J. B. Knudsen, *Justus Möser and the German Enlightenment* (Cambridge, 1986); *Aufklärung und Geschichte,* ed. H. E. Bödeker et al. (Göttingen, 1986).
8. N. Swerdlow and O. Neugebauer, *Mathematical Astronomy in Copernicus's De Revolutionibus* (New York, 1984), 1:48–54, 92–93, 182–190; L. Joy, *Gassendi the Atomist* (Cambridge, 1987); V. Nutton, *John Caius and the Manuscripts of Galen* (Cambridge, 1987); P. Dear, *Mersenne and the Learning of the Schools* (Cambridge, 1988); Rossi, *La scienza e la filosofia dei moderni.*
9. See the important essay by V. Nutton, " 'Prisci dissectionum professores': Greek Texts and Renaissance Anatomists," *The Uses of Greek and Latin: Historical Essays,* ed. A. C. Dionisotti et al. (London, 1988), 111–126, rightly criticizing me and others for ignoring this point.
10. This situation has since changed; see e.g. J. Monfasani, *George of Trebizond* (Leiden, 1976); M. Allen, *The Platonism of Marsilio Ficino* (Berkeley, 1984); A. Field, *The Origins of the Platonic Academy of Florence* (Princeton, 1988); and the forthcoming book by J. Hankins.
11. See e.g. E. Garin, "Le favole antiche," *Medioevo e Rinascimento* (Bari and Rome, 1980), 63–84; and the essays by V. Branca collected in *Poliziano e l'umanesimo della parola* (Turin, 1983).
12. See Chapter 2 below for the results of this encounter and for an introduction to this body of scholarship.
13. See A. Grafton, *Joseph Scaliger: A Study in the History of Classical Scholarship,* I: *Textual Criticism and Exegesis* (Oxford, 1983).
14. For further information on these scholarly traditions, see Chapters 1 and 9 below.
15. M. Casaubon, "On Learning," in Spiller, *"Concerning Natural Experimental Philosophie,"* 214.
16. The best general account of both Bentley and the *Epistola* remains R. C. Jebb's *Bentley* (London, 1882). For the general context see J. M. Levine, *Humanism and History* (Ithaca, N.Y., and London, 1987).
17. See G. P. Goold's introduction to his reprint edition of the *Epistola* (Toronto, 1962); L. D. Reynolds and N. G. Wilson, *Scribes and Scholars,* 2d ed. (Oxford, 1974), 166–170; C. O. Brink, *English Classical Scholarship* (Cambridge and New York, 1986), 41–49.

18. Sophocles frag. 1126 Pearson. For the patristic sources and the context see also the treatment in *Fragments from Hellenistic Jewish Authors*, 1: *Historians*, ed. C. R. Holladay (Chico, Calif., 1983), 279, 318–319, 335, with bibliography. For earlier discussions see J. Freudenthal, *Alexander Polyhistor* (Breslau, 1875), 166–169.
19. Bentley, *Epistola*, ed. Goold, 42–44.
20. Ibid., 31–32.
21. Ibid., 35.
22. R. Bentley, *Correspondence*, ed. C. Wordsworth (London, 1842; repr. Hildesheim and New York, 1977), 1:13. The politeness maintained by both sides is all the more remarkable given that Bernard was deeply offended by Bentley's refusal to accept his views, as his unpublished letters show (private communication from J. M. Levine).
23. Ibid. 23; *Epistola*, ed. Goold, 144. Curiously, Pearson, who calls this parallel "decisive," failed to notice that Bentley had already adduced it in his addenda, and criticizes him for not supplying strong enough verbal evidence.
24. Bentley, *Correspondence*, 1:17; *Epistola*, ed. Goold, 145.
25. Ibid.; cf. Chapter 6 below.
26. See in general R. L. Colie, *Light and Enlightenment* (Cambridge, 1957); G. Aspelin, *Ralph Cudworth's Interpretation of Greek Philosophy* (Göteborg, 1943); D. C. Allen, *Mysteriously Meant* (Baltimore, 1970); D. P. Walker, *The Ancient Theology* (London, 1972); S. K. Heninger, Jr., *Touches of Sweet Harmony* (San Marino, Calif., 1974).
27. E. Stillingfleet, *Origines Sacrae* (London, 1701), bk. 1, ch. 6, 70.
28. T. Gale, *The Court of the Gentiles* (Oxford, 1669). Cf. the Latin Argumentum, beginning:
   Censeri Veteres puerili ardore laborant
   Stulti homines, gaudentque Vetusti Sordibus Aevi;
   and the even more curious English verse one, reading in part:
   Phenicia must with Palmes no longer crown
   Sanchoniathon, falling down,
   Like *Dagon*, to the Ark, who there adores
   Diviner stores.
29. R. Cudworth, *The True Intellectual System of the Universe* (London, 1678), 363.
30. *Conway Letters*, ed. M. H. Nicolson (New Haven and London, 1930), 83. The deepest study of More's use of the Cabala is B. P. Copenhaver, "Jewish Theologies of Space in the Scientific Revolution: Henry More, Joseph Raphson, Isaac Newton and their Predecessors," *Annals of Science* 37 (1980), 489–548.
31. *Conway Letters*, 82.

32. See the essays by S. Hutton and J. M. Levine and the introduction by R. Kroll in *Philosophy, Religion and Science in England, 1640–1700*, ed. R. Kroll et al. (Cambridge, forthcoming); also the important review essay by M. Hunter, "Ancients, Moderns, Philologists, and Scientists," *Annals of Science* 39 (1982), 187–192.
33. *Conway Letters*, 304.
34. S. Hutin, *Henry More* (Hildesheim, 1966); D. P. Walker, *Il concetto di spirito o anima in Henry More e Ralph Cudworth* (Naples, 1986).
35. *Conway Letters*, 294.
36. Cf. M. C. Jacob, "John Toland and the Newtonian Ideology," *Journal of the Warburg and Courtauld Institutes* 32 (1969), 307–331.
37. The best recent studies of Bentley have reached divergent conclusions about the extent to which his thought was rooted in late-seventeenth-century tradition or strongly anticipated the Enlightenment. Cf. Reynolds and Wilson, *Scribes and Scholars*, 166–170; L. Gossman, *Medievalism and the Ideologies of the Enlightenment* (Baltimore, 1968), 223–228; S. Timpanaro, *La genesi del metodo del Lachmann*, repr. of 2d ed. (Padua, 1985), 13–16; and J. M. Levine's forthcoming study of the Battle of the Books. For the wider European context see the broad-gauged study by C. Borghero, *La certezza e la storia* (Milan, 1983).
38. Reprinted in I. Newton, *Papers and Letters on Natural Philosophy*, ed. I. B. Cohen (Cambridge, Mass., 1958), 358.

1. Renaissance Readers and Ancient Texts

1. B. Massari to L. Guidetti, 14 September 1465, ed. (with the rest of their correspondence) by R. Cardini, *La critica del Landino* (Florence, 1973), 267. For the circumstances see ibid., 39–41; for the debate itself see Cardini's masterly analysis, which I generally follow, ibid., 41–62.
2. Guidetti to Massari, 18 September 1465; ibid., 268.
3. Ibid., 268.
4. Massari to Guidetti, 26 September 1465; ibid., 270.
5. Ibid., 271.
6. Guidetti to Massari, 25 October 1465; ibid., 273–274.
7. Ibid., 274.
8. Massari to Guidetti, 31 October 1465; ibid., 279.
9. See Chapter 2 below and the works cited there.
10. For Machiavelli see above all the preface to his *Discourses on Livy;* for Guicciardini see *Ricordi* C 110, *Opere*, ed. V. de Caprariis

(Milan and Naples, 1953), 120; I use the translation by M. Domandi, *Maxims and Reflections of a Renaissance Statesman* (New York, 1965), 69.

11. A. Moss, *Ovid in Renaissance France* (London, 1982), 23–36. Rabelais's attack on some allegorical readings needs to be interpreted in the light of his own use of Lavinius's allegories; see D. Quint, *Origin and Originality in Renaissance Literature* (New Haven and London, 1983), 170.

12. For Lipsius see G. Oestreich, *Neostoicism and the Early Modern State,* ed. H. G. Koenigsberger (Cambridge, 1982). Scaliger's comment appears in his table-talk, the *Scaligerana* (Cologne, 1695), 245: ". . . neque est Politicus, nec potest quicquam in Politia: nihil possunt pedantes in illis rebus; nec ego nec alius doctus possemus scribere in Politicis." Yet Scaliger was not completely consistent. In 1606 he produced an edition of Caesar for Plantin, the anonymous preface of which (in fact by Lipsius) insisted that good scholars would seek not textual corruptions but political lessons in the text; *C. Iulii Caesaris quae extant ex emendatione Ios. Scaligeri* (Leiden, 1635), sig. ** recto.

13. P. de Nolhac, *Pétrarque et l'humanisme,* 2d ed. (Paris, 1907); R. Sabbadini, *Il metodo degli umanisti* (Florence, 1922); B. L. Ullman, *The Humanism of Coluccio Salutati* (Padua, 1963); R. Weiss, *The Renaissance Discovery of Classical Antiquity* (Oxford, 1969); G. Billanovich, "Petrarch and the Textual Tradition of Livy," *Journal of the Warburg and Courtauld Institutes* 14 (1951), 137–208; S. Timpanaro, *La genesi del metodo del Lachmann,* repr. of 2d ed. (Padua, 1985); S. Rizzo, *Il lessico filologico degli umanisti* (Rome, 1973).

14. E.g. L. D. Reynolds and N. G. Wilson, *Scribes and Scholars,* 2d ed. (Oxford, 1974); P. Burke, *The Renaissance Sense of the Past* (New York, 1969); D. R. Kelley, *Foundations of Modern Historical Scholarship* (New York, 1970); J. Bentley, *Humanists and Holy Writ* (Princeton, 1983).

15. M. Lowry, *The World of Aldus Manutius* (Oxford, 1979), 238.

16. M. Mund-Dopchie, *La survie d'Eschyle à la Renaissance* (Louvain, 1984), 8, 96–97.

17. L. Panizza, "Textual Interpretation in Italy, 1350–1450: Seneca's Letter I to Lucilius," *Journal of the Warburg and Courtauld Institutes* 46 (1983), 40–62; S. Timpanaro, "Atlas cum compare gibbo," *Rinascimento* 2 (1951), 311–318; J. Dunston, "Studies in Domizio Calderini," *Italia Medioevale e Umanistica* 11 (1968), 71–150; D. Coppini, "Filologi del Quattrocento al lavoro su due passi di Properzio," *Ri-*

nascimento 16 (1976), 219–221; J. Kraye, "Cicero, Stoicism and Textual Criticism: Poliziano on *katorthōma*," *Rinascimento* 23 (1983), 79–110; A. C. Dionisotti, "Polybius and the Royal Professor," *Tria Corda: Scritti in onore di Arnaldo Momigliano*, ed. E. Gabba (Como, 1983), 179–199; cf. A. Porro, "Pier Vettori editore di testi greci: la 'Poetica' di Aristotele," *Italia Medioevale e Umanistica* 26 (1983), 307–358; J. D'Amico, *Theory and Practice in Renaissance Textual Criticism* (Berkeley and Los Angeles, 1988).

18. Mund-Dopchie, *La survie d'Eschyle*, ch. 5; cf. the even more pointed account by J. A. Gruys, *The Early Printed Editions (1518–1664) of Aeschylus* (Nieuwkoop, 1981), 77–96.
19. J. Franklin, *Jean Bodin and the Sixteenth-Century Revolution in the Methodology of Law and History* (New York and London, 1963), 140–141.
20. J. Bodin, *Methodus ad facilem historiarum cognitionem* (Paris, 1572), ch. 4, 60: "Eorum vero minus probandae sunt narrationes qui nihil aliud habent quam quod ab aliis audierunt ... nec publica monumenta viderunt. Itaque optimi scriptores quo maior fides scriptis haberetur e publicis monumentis ea se collegisse aiunt."
21. Bodin did subscribe to many conventional notions about human nature and historical writing. See in general G. Nadel, "Philosophy of History before Historicism," *History and Theory* 3 (1964), 291–315; R. Koselleck, "Historia magistra vitae: Über die Auflösung des Topos im Horizont neuzeitlich bewegter Geschichte," *Vergangene Zukunft* (Frankfurt, 1984), 38–66; E. Kessler, "Das rhetorische Modell der Historiographie," *Formen der Geschichtsschreibung*, ed. R. Koselleck et al. (Munich, 1982), 37–85.
22. *Commentaria fratris Ioannis Annii Viterbensis ... super opera diversorum auctorum de antiquitatibus loquentium conficta* (Rome, 1498) (unpaginated): "Et declarat quod sacerdotes olim erant publici notarii rerum gestarum et temporum: qui presentes essent aut ex antiquioribus copiarent: sicut nunc instrumentum publicum et probatum dicitur: quod a notario presente publicatur et scribitur: aut ex antiquiore notario per presentem notarium traducitur."
23. B. Guenée, *Histoire et culture historique dans l'Occident médiéval* (Paris, 1980), 133–140.
24. See J. M. Levine, "Reginald Pecock and Lorenzo Valla on the Donation of Constantine," *Studies in the Renaissance* 20 (1973), 118–143; W. Setz, *Lorenzo Vallas Schrift gegen die Konstantinische Schenkung* (Tübingen, 1975); J. C. H. Lebram, "Ein Streit um die Hebräische Bibel und die Septuaginta," *Leiden University in the Seventeenth Century*, ed. T. H. Lunsingh Scheurleer (Leiden, 1975),

37; H. J. de Jonge, "Die Patriarchentestamente von Roger Bacon bis Richard Simon," *Studies on the Testaments of the Twelve Patriarchs,* ed. M. de Jonge (Leiden, 1975), 3–42; and Chapters 5 and 6 below, with further citations.
25. See A. Momigliano, "Ancient History and the Antiquarian," *Studies in Historiography* (New York, 1966), 1–39; E. H. Waterbolk, "Reacties op het historisch pyrrhonisme," *Bijdragen voor de Geschiedenis der Nederlanden* 15 (1960), 81–102; H. J. Erasmus, *The Origins of Rome in Historiography from Petrarch to Perizonius* (Assen, 1962); F. Wagner, *Die Anfänge der modernen Geschichtswissenschaft im 17. Jahrhundert,* SB Bayer. Akad., phil. -hist. Kl., 1979, Heft 2 (Munich, 1979); W. Hardtwig, "Die Verwissenschaftlichung der Geschichtsschreibung und die Ästhetisierung der Darstellung," *Formen der Geschichtsschreibung,* ed. Koselleck et al., 147–191; A. Krauss, "Grundzüge barocker Geschichtsschreibung," *Bayerische Geschichtswissenschaft in drei Jahrhunderten* (Munich, 1979), 11–33; *Historische Kritik in der Theologie,* ed. G. Schwaiger (Göttingen, 1980).
26. See Chapter 5 below.
27. F. Bacon, *De augmentis scientiarum* 2.4; *The Works of Lord Bacon* (London, 1879), 2:317.
28. E. Hassinger, *Empirisch-rationaler Historismus* (Bern and Munich, 1978), as qualified and extended by U. Muhlack, "Empirisch-rationaler Historismus," *Historische Zeitschrift* 232 (1981), 605–616.
29. J. J. Rambach, "Entwurf der künftig auszuarbeitenden Litterairhistorie," *Versuch einer pragmatischen Litterairhistorie* (Halle, 1770), 182–183; for Heyne and other eighteenth-century continuators of the humanist tradition see Chapter 9 below.
30. J. Seznec, *The Survival of the Pagan Gods,* tr. B. F. Sessions (New York, 1953); D. C. Allen, *Mysteriously Meant* (Baltimore, 1970); I. Maïer, "Une page inédite de Politien: la note du Vat. lat. 3617 sur Démétrius Triclinius, commentateur d'Homère," *Bibliothèque d'Humanisme et Renaissance* 16 (1954), 7–17; A. Levine [Rubinstein], "The Notes to Poliziano's 'Iliad,'" *Italia Medioevale e Umanistica* 25 (1982), 205–239; Moss, *Ovid in Renaissance France,* 44–53; M. Allen, *Marsilio Ficino and the Phaedran Charioteer* (Berkeley, 1981); Allen, *The Platonism of Marsilio Ficino* (Berkeley, 1984); M. Murrin, *The Veil of Allegory* (Chicago and London, 1969); M. Murrin, *The Allegorical Epic* (Chicago and London, 1980); Quint, *Origin and Originality.*
31. C. Lemmi, *The Classic Deities in Bacon* (Baltimore, 1933); P. Rossi, *Francis Bacon: From Magic to Science,* tr. S. Rabinovitch (Chicago, 1968), ch. 3.
32. Bacon, *De augmentis* 2.13; *Works,* 2:323.

33. J. E. McGuire and P. M. Rattansi, "Newton and the 'Pipes of Pan,'" *Notes and Records of the Royal Society of London* 21 (1966); D. P. Walker, *The Ancient Theology* (London, 1972), chs. 6–7.
34. M. W. Croll, *Style, Rhetoric and Rhythm* (Princeton, 1966); J. D'Amico, "The Progress of Renaissance Latin Prose: The Case of Apuleianism," *Renaissance Quarterly* 37 (1984), 351–392; M. Fumaroli, *L'Age de l'éloquence* (Geneva, 1980); W. Kühlmann, *Gelehrtenrepublik und Fürstenstaat* (Tübingen, 1982).
35. E. Garin, "Le favole antiche," *Medioevo e Rinascimento* (Bari, 1954), 66–89.
36. Cardini, *La critica del Landino*, 62–65. For a challenging case study that cuts across these distinctions—and to which this essay is much indebted—see E. F. Rice, Jr., "Humanist Aristotelianism in France: Jacques Lefèvre d'Etaples and His Circle," *Humanism in France at the End of the Middle Ages and in the Early Renaissance*, ed. A. H. T. Levi (Manchester, 1970), 132–149.
37. A. Poliziano, "Oratio super Fabio Quintiliano et Statii Sylvis," *Prosatori latini del Quattrocento*, ed. E. Garin (Milan and Naples, 1952), 878, drawing on Tacitus, *Dialogus de oratoribus* 18.3. See Quint, *Origin and Originality*, 223–224n15.
38. J. Scaliger to C. Saumaise, 20 November (Julian) 1607; Scaliger, *Epistolae*, ed. D. Heinsius (Leiden, 1627), ep. 247, 530. This influential periodization implies the necessity of a decline at some point, since it rests on an organic analogy; hence it undercuts its own defense of the products of a later period. See W. Rehm, *Der Untergang Roms im abendländischen Denken* (Leipzig, 1930), 76, 112, 154, 162; R. Pfeiffer, *Die Klassische Philologie von Petrarca bis Mommsen* (Munich, 1982), 150–151; J. H. Meter, *The Literary Theories of Daniel Heinsius* (Assen, 1984), 21.
39. A. Alciato, ep. ded. to *In tres libros posteriores Codicis Iustiniani annotatiunculae* (1513); *Le lettere di Andrea Alciato giureconsulto*, ed. G. L. Barni (Florence, 1953), 219–220.
40. F. Bacon, *De sapientia veterum*, praefatio; *Works*, 2:704. See Lemmi, *The Classic Deities in Bacon*, 41–45, and L. Jardine, *Francis Bacon: Discovery and the Art of Discourse* (Cambridge, 1974), 192–193n2.
41. C. Dionisotti, "Calderini, Poliziano e altri," *Italia Medioevale e Umanistica* 11 (1968), 151–179.
42. A. Poliziano, *Commento inedito alle Selve di Stazio*, ed. L. Cesarini Martinelli (Florence, 1978), 7; cf. L. Cesarini Martinelli, "In margine al commento di Angelo Poliziano alle *'Selve'* di Stazio," *Interpres* 1 (1978), 96–145, and "Un ritrovamento polizianesco: il fascicolo per-

duto del commento alle Selve di Stazio," *Rinascimento* 22 (1982), 183–212. In this passage "consideraremus" poses a real interpretative and textual problem, which Poliziano does not discuss directly.

43. Poliziano, *Commento alle Selve*, 51.
44. Ibid., 430–431.
45. Ibid., 431–437.
46. See de Nolhac, *Pétrarque et l'humanisme*, 1:123–161; V. Kahn, "The Figure of the Reader in Petrarch's *Secretum*," *PMLA* 100 (1985), 154–166; A. von Martin, *Coluccio Salutati's Traktat "Vom Tyrannen"* (Berlin and Leipzig, 1913), 77–98; Ullman, *Humanism of Salutati*, 21–26, 95–97; R. G. Witt, *Hercules at the Crossroads* (Durham, N.C., 1983), esp. ch. 8.
47. Mund-Dopchie, *La survie d'Eschyle*, 377–378.
48. *Scaligerana*, 37; cf. G. Demerson, "Dorat, commentateur d'Homère," *Etudes seiziémistes offertes à V.-L. Saulnier* (Geneva, 1980), 223–234, and P. Ford, "Conrad Gesner et le fabuleux manteau," *Bibliothèque d'Humanisme et Renaissance* 47 (1985), 305–320.
49. J. J. Scaliger, *Elenchus utriusque orationis chronologicae D. Davidis Parei* (Leiden, 1607), 80–81. Admittedly euhemerism seeks to substitute a more modern notion of factual plausibility rather than philosophical rigor for the lost truth concealed by a mythical narrative; but it is as much a part of the allegorical tradition as the theological interpretations Scaliger rejected.
50. J. Le Clerc, *Ars critica*, 4th ed. (Amsterdam, 1712), 1:304–305 (criticizing Hugo Grotius for taking Seneca's *Anima mundi* as the Christian soul); 1:288–290 (criticizing ancient scholars for failing to see that Homer had a deep knowledge of astronomy). In the first passage Le Clerc insists on historical distance, in the second he seeks to eliminate it; both form part of the first great manual of hermeneutics and higher and lower criticism.
51. See Chapter 2 below, n29; K. Krautter, *Philologische Methode und humanistische Existenz* (Munich, 1971); M. T. Casella, "Il metodo dei commentatori umanistici esemplato sul Beroaldo," *Studi medievali*, ser. 3, 16 (1975), 627–701; Coppini, "Filologi del Quattrocento," 221–229.
52. *Opus epistolarum Des. Erasmi Roterodami*, ed. P. S. Allen et al. (Oxford, 1906–1958), 3: 328; quoted by Moss, *Ovid in Renaissance France*, and tr. R. A. B. Mynors and D. F. S. Thomson in *Collected Works of Erasmus* (Toronto, 1982), 6: 23–24.
53. Panizza, "Textual Interpretation in Italy," 59–60.

54. Erasmus, *Enchiridion militis Christiani* (1503); *Ausgewählte Werke*, ed. H. Holborn and A. Holborn (Munich, 1933; repr. 1964), 32.
55. Ibid., 71.
56. Erasmus, *De ratione studii* (1512), ed. J.-C. Margolin; Erasmus, *Opera Omnia* (Amsterdam, 1971), 1, pt. 2, 139–140.
57. Ibid., 140–142.
58. Ibid., 142. Note that Erasmus treats this as an exemplary exposition to be imitated by others.
59. Kühlmann, *Gelehrtenrepublik und Fürstenstaat;* P. Burke, "Tacitism," *Tacitus*, ed. T. A. Dorey (New York, 1969), 149–171; J. H. M. Salmon, "Cicero and Tacitus in Sixteenth-Century France," *American Historical Review* 85 (1980), 307–331; M. Stolleis, *Arcana imperii und Ratio status* (Göttingen, 1980); H. Wansink, *Politieke wetenschappen aan de Leidse universiteit, 1575– +/– 1650* (Utrecht, 1981).
60. J. Lipsius, *Ad Annales Corn. Taciti liber commentarius sive notae* (Antwerp, 1581), ep. ded., ed. P. Schrijvers before his article "Justus Lipsius: Grandeur en misère van het pragmatisme," *Rijksuniversiteit te Leiden: Voordrachten, Faculteitendag 1980* (Leiden, 1981), 43–44: "Nec utiles omnes [scil. historiae] nobis pari gradu; ea, ut censeo, maxime, in qua similitudo et imago plurima temporum nostrorum ... Cuius generis si ulla est fuitque, inter Graecos aut Latinos: eam esse Cornelii Taciti Historiam adfirmate apud vos dico, Ordines illustres. Non adfert ille vobis speciosa bella aut triumphos, quorum finis sola voluptas legentis sit; non seditiones aut conciones Tribunicias, agrarias frumentariasve leges; quae nihil ad saecli huius usum: reges ecce vobis et monarchas, et velut theatrum hodiernae vitae. Video alibi Principem in leges et iura, subditosque in Principem insurgentes. Invenio artes machinasque opprimendae, et infelicem impetum recipiendae libertatis. Lego iterum eversos prostratosque tyrannos, et infidam semper potentiam cum nimia est. Nec absunt etiam reciperatae libertatis mala, confusio, aemulatioque inter pares, avaritia, rapinae, et ex publico non in publicum quaesitae opes. Utilem magnumque scriptorem, deus bone! et quem in manibus eorum esse expediat, in quorum manu gubernaculum et reip. clavus."
61. Lipsius to J. Woverius, 3 November 1603, in Lipsius, *Epistolarum selectarum centuria quarta miscellanea postuma* (Antwerp, 1611), ep. 84, 70: "Ego ad sapientiam primus vel solus mei aevi Musas converti: ego e Philologia Philosophiam feci."
62. Oestreich, *Neostoicism; The Medical Renaissance of the Sixteenth Century*, ed. A. Wear et al. (Cambridge, 1985).
63. Seneca, *Epistulae morales* 108.24 on Virgil *Georgics* 3.284. Cf.

Lipsius's enthusiastic *argumentum* for this letter in his edition of Seneca's *Opera* (Antwerp, 1605), 633: "Ostendit deinde, vario fine atque animo ad auctores legendos veniri: nos autem debere illo Philosophi. Pareamus: et legite atque audite vos o Philologi." True, Lipsius did more justice to the alien and historical features of Roman Stoicism than some of his contemporaries; see H. Ettinghausen, *Francisco de Quevedo and the Neostoic Movement* (Oxford, 1972). But the fundamental eclecticism of his approach emerges from G. Abel, *Stoizismus und frühe Neuzeit* (Berlin and New York, 1978), ch. 4.

64. A. Grafton, "Rhetoric, Philology and Egyptomania in the 1570s...," *Journal of the Warburg and Courtauld Institutes* 42 (1979), 193–194.
65. Cf. H. Dreitzel, "Die Entwicklung der Historie zur Wissenschaft," *Zeitschrift für Historische Forschung* 3 (1981), 257–284.
66. Montaigne, *Essais* 3.13; *The Complete Essays of Montaigne,* tr. D. M. Frame (Stanford, 1965), 819. Cf. I. Maclean, "The Place of Interpretation: Montaigne and Humanist Jurists on Words, Intention and Meaning," *Neo-Latin and the Vernacular in Renaissance France,* ed. G. Castor and T. Cave (Oxford, 1984), 252–272.
67. H. H. Gray, "Renaissance Humanism: The Pursuit of Eloquence," *Journal of the History of Ideas* 24 (1963), repr. in *Renaissance Essays,* ed. P. O. Kristeller and P. P. Wiener (New York, 1968), 199–216.
68. Erasmus, *Methodus* (1516), *Ausgewählte Werke,* ed. Holborn and Holborn, 157; J. M. Weiss, "Ecclesiastes and Erasmus: The Mirror and the Image," *Archiv für Reformationsgeschichte* 65 (1974), 101–104.
69. Gray, "Renaissance Humanism," 213–214; V. de Caprio, "Retorica e ideologia nella *Declamatio* di Lorenzo Valla sulla donazione di Costantino," *Paragone* 29, no. 338 (1978), 36–56; G. W. Most, "Rhetorik und Hermeneutik: zur Konstitution der Neuzeitlichkeit," *Antike und Abendland* 30 (1984), 62–79.
70. A. Grafton, *Joseph Scaliger: A Study in the History of Classical Scholarship,* 1: *Textual Criticism and Exegesis* (Oxford, 1983), chs. 1, 4.
71. E. Garin, *L'educazione umanistica in Italia* (Bari, 1953), 7; Murrin, *Veil of Allegory,* 53.
72. Petrarch, *Secretum; Opere,* ed. G. Ponte (Milan, 1968), 504, 522; cf. Kahn, "Figure of the Reader."
73. J. Bodin, *Colloquium Heptaplomeres,* ed. L. Noack (Schwerin, 1857; repr. Hildesheim, 1970), 76.
74. F. A. Wolf, *Prolegomena to Homer* (1795), tr. A. Grafton, G. Most,

and J. E. G. Zetzel (Princeton, 1985), 210; see the introduction, ibid., and Chapter 9 below for further citations.
75. A. Grafton, "Polyhistor into *Philolog:* Notes on the Transformation of German Classical Scholarship, 1780–1850," *History of Universities* 3 (1983 [1984]), 159–192.
76. B. Elman, *From Philosophy to Philology* (Cambridge, Mass., 1984).
77. D. S. Nivison, "Protest against Conventions and Conventions of Protest," *Confucianism and Chinese Civilization,* ed. A. F. Wright (Stanford, 1975), 227–251.
78. Yuan Mei, quoted by Elman, *From Philosophy to Philology,* 200.
79. Kuei Chuang to Ku Yen-wu, 1668; quoted ibid., 30.
80. T. C. Bartlett, "Ku Yen-wu's thought in the Mid-Seventeenth-Century Context" (Ph.D. diss., Princeton, 1985), nicely reveals the civic humanist side of Ch'ing scholarship.
81. Quoted by D. S. Nivison in *The Life and Thought of Chang Hsüeh-ch'eng (1738–1801)* (Stanford, 1966), 176–177; cf. I. Maclean, "Montaigne, Cardano: The Reading of Subtlety/The Subtlety of Reading," *French Studies* 37 (1983), 143–156.
82. Nivison, *Chang Hsüeh-ch'eng,* 178.
83. Cf. U. Muhlack, "Klassische Philologie zwischen Humanismus und Neuhumanismus," *Wissenschaften im Zeitalter der Aufklärung,* ed. R. Vierhaus (Göttingen, 1985), 93–119.

2. The Scholarship of Poliziano and Its Context

1. C. Bellièvre, *Souvenirs de voyages en Italie et en Orient, notes historiques, pièces de vers,* ed. C. Perrat (Geneva, 1956), 4–5. The discussion of "explicit" may refer to Lorenzo Valla's *Elegantiae* 3.41. On the divisions of the Vatican Library see J. Bignami Odier and J. Ruysschaert, *La Bibliothèque Vaticane de Sixte IV à Pie XI* (Vatican City, 1973), 22.
2. Pierpont Morgan Library shelf-mark E 22 B. On Bellièvre see C. Perrat, "Les humanistes amateurs de papyrus," *Bibliothèque de l'Ecole des Chartes* 109 (1951), 173–192.
3. T. Diplovatacius, *De claris iuris consultis,* ed. H. Kantorowicz and F. Schulz (Berlin and Leipzig, 1919).
4. Ibid., 33, 335–336.
5. Ibid., 135, 336; see in general M. Ascheri, *Saggi sul Diplovatazio* (Milan, 1971), 110–116.
6. See in general R. Sabbadini, *Il metodo degli umanisti* (Florence, 1922), 42–45; K. Krautter, *Philologische Methode und humanistische Existenz*

(Munich, 1971); M. T. Casella, "Il metodo degli commentatori umanistici esemplato sul Beroaldo," *Studi Medievali,* ser. 3, 16 (1975), 627–701.
7. V. Zabughin, *Vergilio nel Rinascimento italiano da Dante a Torquato Tasso* (Bologna, 1921), 1:188, 192; *Scholia in P. Ovidi Nasonis Ibin,* ed. A. La Penna (Florence, 1959), xxxix–xl.
8. H. Caplan, *Of Eloquence: Studies in Ancient and Mediaeval Rhetoric,* ed. A. King and H. North (Ithaca and London, 1970), 247–270.
9. For the classicism of the friars see B. Smalley, *English Friars and Antiquity in the Early Fourteenth Century* (Oxford, 1960); J. B. Friedman, *Orpheus in the Middle Ages* (Cambridge, Mass., 1970), ch. 4. For the reception of late medieval scholarship in the Renaissance see M. Beller, *Philemon und Baucis in der europäischen Literatur* (Heidelberg, 1967), 48–49; M. Bonicatti, *Studi sull' Umanesimo, secoli XIV–XVI* (Florence, 1969), 255–260; J. Engels, "Les commentaires d'Ovide au XVIᵉ siècle," *Vivarium* 12 (1974), 3–13.
10. A good general account is provided by R. R. Bolgar, *The Classical Heritage and Its Beneficiaries from the Carolingian Age to the End of the Renaissance* (Cambridge, 1954; repr. New York, 1964); a brisk review of ancient and medieval traditions of commentary can be found in B. Sandkühler, *Die frühen Dantekommentare und ihr Verhältnis zur mittelalterlichen Kommentartradition* (Munich, 1967), 13–24.
11. The practical difficulties emerge from a subscription printed by A. J. Dunston, "A Student's Notes of Lectures by Giulio Pomponio Leto," *Antichthon* 1 (1967), 90: "iuuat haec collegisse Iulio Pomponio praeceptore. o dii immortales quid si mihi notarii manu adderetis ut Iulianas partes quas in lectione retractat assequi possem. saltem et hoc praeberetis me scilicet scabie atrocissima liberar(i). nonne haec uobis magni uidetur?" As many commentaries derive, in at least one recension, from notes taken under these conditions, their textual histories can be complex and bizarre. For two careful case studies see E. L. Bassett, J. Delz, and A. J. Dunston, "Silius Italicus," *Catalogus translationum et commentariorum,* ed. F. E. Cranz et al. (Washington, D.C., 1960–), 3:373–387.
12. M. Filetico to Giovanni Colonna, quoted by G. Mercati, "Tre dettati universitari dell' umanista Martino Filetico sopra Persio, Giovenale, ed Orazio," *Classical and Mediaeval Studies in Honor of Edward Kennard Rand,* ed. L. W. Jones (New York, 1938), 228n46, and by C. Dionisotti, " 'Lavinia venit litora': Polemica virgiliana di M. Filetico," *Italia Medioevale e Umanistica* 1 (1958), 307.

13. For a sample of Poliziano's more elementary lectures, see his *La commedia antica e l'Andria di Terenzio: appunti inediti,* ed. R. Lattanzi Roselli (Florence, 1973); for a more advanced course see his *Commento inedito all' epistola ovidiana di Saffo a Faone,* ed. E. Lazzeri (Florence, 1971); see also the important review of both editions by S. Rizzo in *Annali della Scuola Normale Superiore di Pisa* (classe di lettere e filosofia), ser. 3, 4 (1974), 1707–1711.
14. See *Scholia in Ovidi Ibin,* ed. La Penna, xlvii and n6; E. Barbaro, *Castigationes Plinianae et in Pomponium Melam,* ed. G. Pozzi (Padua, 1973–1979), 1:cxlix–cl.
15. See K. Haebler, *The Study of Incunabula,* tr. L. E. Osborne (New York, 1933), 91.
16. D. Calderini, "Epilogus et *prosphōnēsis* de observationibus," quoted by C. Dionisotti, "Calderini, Poliziano e altri," *Italia Medioevale e Umanistica* 11 (1968), 167. This and the following paragraphs follow Professor Dionisotti's article; see also Barbaro, *Castigationes Plinianae,* ed. Pozzi, 1:cxii–clxviii, and P. Cortesi, *De hominibus doctis dialogus,* ed. M. T. Graziosi (Rome, 1973), 54.
17. D. Calderini, "Ex tertio libro observationum," in *Lampas,* ed. J. Gruter, 1 (Frankfurt, 1602), 316: "*Harpastrum* quid sit longa commentatione Graecis auctoribus explicamus in Epistolam quarti libri Sylvarum, multaque de eo docuimus, quae adhuc (ut opinor) incognita fuerunt: ea siquis volet legere, illic requiret"; ibid., 314: "Repetamus particulam commentariorum nostrorum in Iuvenalem . . ."
18. M. A. Sabellico, *Annotationes in Plinium,* ep. ded.; ibid., 124–125.
19. F. Beroaldo, Sr., *Annotationes centum,* GKW 4113 (Bologna, 1488), fol. a ii verso: "Sane has annotationes nullo servato rerum ordine confecimus, utpote tumultuario sermone dictantes, et perinde ut cuiuslibet loci veniebat in mentem, ut quilibet liber sumebatur in manus, ita indistincte atque promiscue excerpentes annotantesque. Fetus hic plane precox fuit, utpote intra menstruum tempus et conceptus et editus." Beroaldo not only describes his own experience but echoes Gellius, *Noctes Atticae,* praefatio x.
20. L. Ruberto, "Studi sul Poliziano filologo," *Rivista di filologia e di istruzione classica* 12 (1884), 235–237.
21. Poliziano, *Miscellanea,* praefatio; *Opera* (Basel, 1553), 216: "Enimvero ne putent homines male feriati, nos ista quaequae sunt, de fece hausisse, neque grammaticorum transiluisse lineas, Pliniano statim exemplo [cf. *NH,* praefatio, xxi] nomina praetexuimus auctorum, sed honestorum veterumque duntaxat, unde ius ista sumunt, et a quibus versuram fecimus, nec autem, quos alii tantum citaverint,

ipsorum opera temporibus interciderint, sed quorum nosmet ipsi thesauros tractavimus, quorum sumus per literas peregrinati, quanquam et vetustas codicum et nomismatum fides et in aes aut in marmore incisae antiquitates, quae tu nobis Laurenti suppeditasti, plurimum etiam praeter librorum varietatem nostris commentationibus suffragantur."

22. Ibid., cap. 75, 285: "Enarravit Domitius libellum Nasonis in Ibin, praefatus ex Apollodoro se, Lycophrone, Pausania, Strabone, Apollonio, aliisque Graecis etiamque Latinis accepta scribere. Multa in eo commentario vana ridiculaque confingit et comminiscitur ex tempore commodoque suo, quibus fidem facit, aut se frontem penitus amisisse, aut tam magnum sibi fuisse intervallum inter frontem et linguam (sicut ait quidam) ut frons comprimere linguam non potuerit."

23. The text is thus in the Venice 1474 edition of Ovid (Hain 12138), fol. 409 verso.

24. D. Calderini, *Commentarioli in Ibyn Ovidii*, Hain 4242 (Rome, 1474), fol. 25 recto on *Ibis* 569: "Agenor lapsu equi inserta ori manu extinctus est."

25. Poliziano used *Od.* 4.285ff. and Tryphiodorus *Iliou Halōsis* 476ff.; see *Publi Ovidi Nasonis Ibis*, ed. A. La Penna (Florence, 1957), 153–154 *ad loc.*, and *Scholia in Ovidi Ibin*, ed. La Penna, xlviii.

26. Calderini, "Ex tertio libro observationum," *Lampas,* ed. Gruter, 1:316–317: "Equidem *Scauri caussam* actam fuisse apud Ciceronem legi apud Valerium Max. et Pedianum: *Orationem* vero a Cicerone pro illo habitam, unde se haec accipere Laurentius profitetur, legi nusquam, neque exstare arbitror, vereorque ne Grammaticum aliquem ignobilem secutus haec verba recitaverit potius quam legerit usquam apud Ciceronem." Valla had identified a citation in Quintilian *Inst. Or.* 1.5.8 as coming from the *pro Scauro* in his commentary; see e.g. the Venice, 1494 edition (Proctor 4865), fol. c iiii recto.

27. Calderini, "Ex tertio libro observationum," *Lampas,* ed. Gruter, 1:317: "Mendosa est dictio [*medico* for *melico*] apud Plinium [*NH* 7. 24], ut ego quidem arbitror, et heri observavi, cum de Simonide nonnulla apud Graecum scriptorem [*Suda* s.v. Simonides] legi . . ."

28. W. G. Rutherford, *Scholia Aristophanica,* 3: *A Chapter in the History of Annotation* (London, 1905), 387.

29. For an interesting defense of the tradition of double explication see F. Beroaldo, Sr., Commentary on Propertius, in the Venice 1491 edition of Tibullus, Catullus, and Propertius (Proctor 5029), sig. I recto: "Nam, ut inquit divus Hieronymus [*contra Rufinum* 1. 16],

commentatoris officium est multorum sententias exponere, ut prudens lector, cum diversas explanationes legerit, iudicet quid verius sit: et quasi verus trapezita, adulterinae monetae pecuniam reprobet, et probam sinceramque recipiat." Jerome in his turn had defended the practice as the standard one in classical education in his day, citing among other examples that of his teacher Donatus.

30. R. P. Oliver, " 'New Fragments' of Latin Authors in Perotti's *Cornucopiae*," *Transactions of the American Philological Association* 78 (1947), 390–393, 405–406, 411, 412–424.

31. J. Dunston, "Studies in Domizio Calderini," *Italia Medioevale e Umanistica* 11 (1968), 144–149; S. Timpanaro, "Noterelle su Domizio Calderini e Pietro Giordani," *Tra Latino e Volgare: Per Carlo Dionisotti,* ed. G. Bernardoni Trezzini et al. (Padua, 1974), 2:709–712.

32. Dunston in *Antichthon,* 1:91–92.

33. Dunston, in *Italia Medioevale e Umanistica* 11 (1968):134–137, correcting and amplifying R. Sabbadini, *Classici e umanisti da codici Ambrosiani* (Florence, 1933), 59–62; G. Brugnoli, "La 'praefatio in Suetonium' del Poliziano," *Giornale Italiano di Filologia* 10 (1957), 211–220; A. Perosa, "Due lettere di Domizio Calderini," *Rinascimento,* ser. 2, 13 (1973), 6, 13–15.

34. Poliziano, *Opera,* 304: "Neque autem istos Eupolidis poetae versus ex ipsius statim fontibus hausimus, ut cuius opera aetate interciderint, sed eorum partem ex interprete quopiam Aristidae rhetoris accuratissimo, partem ex epistola Plinii Iunioris accepimus."

35. Ibid., 263; cf. Seneca *Epistulae morales* 88.6.

36. Petrarch claimed to follow those "quibus vel similitudo vel autoritas maior ut eis potissimum stetur impetrat"; *Prose,* ed. G. Martellotti et al. (Milan and Naples, 1955), 220. Cf. his statement in his letter "Posteritati," where he says that when he has encountered conflicting historical sources he has followed "quo me vel veri similitudo rerum vel scribentium traxit auctoritas"; ibid., 6.

37. A. Von Martin, *Coluccio Salutati's Traktat 'Vom Tyrannen'* (Berlin and Leipzig, 1913), 77–98; B. L. Ullman, *The Humanism of Coluccio Salutati* (Padua, 1963), 95–99; A. D. Momigliano, "Polybius' Reappearance in Western Europe," *Polybe,* Entretiens sur l'Antiquité Classique, 20 (Vandoeuvres, 1974), 356–357.

38. D. M. Robathan, "Flavio Biondo's *Roma Instaurata,*" *Mediaevalia et Humanistica,* new ser., 1 (1970), 204.

39. In general, the humanists preferred to attribute error to medieval scribes rather than classical authors; e.g. Salutati, who uncovered a conflict between Valerius Maximus and Livy, and concluded: "et

potius credant textum Valerii fuisse corruptum, quam eum in tam supinum errorem, qui in tante scientie virum cadere non debuit, incidisse"; *Il trattato 'De tyranno' e lettere scelte,* ed. F. Ercole (Bologna, 1942), 15. Naturally there were exceptions—e.g. Valla's famous demonstration that Livy had committed an error in genealogy, for which see H. J. Erasmus, *The Origins of Rome in Historiography from Petrarch to Perizonius* (Assen, 1962), 28–29.

40. Poliziano, *Opera,* 259: "Literas igitur Cadmi Phoenicis munus, et papyrum Niloticam, et atramentum scriptorium, et calamum librarium literatoris eius germanum instrumentum videtur mihi Ausonius sub haec involucra complicasse. Cadmus enim literas primus in Graeciam attulit e Phoenice."

41. Ibid.: "Omitto quod et Suidas, aut Zopyrion [who, Poliziano thought, was the source of the *Suda* here] potius *phoinikeia* vocatas literas ait. Omitto Plinium caeterosque permultos, qui dicant eas a Cadmo in Graeciam allatas. Nam cum diversi quae legerant apud Herodotum passim meminerint, satis ipsi fecisse videmur, quod ista suae reddimus autoritati. Nec enim tam numeranda, sicuti putamus, veterum testimonia sunt, quam ponderanda." The final phrase inverts Pliny, Jr., *Ep.* 2.12.5: "Numerantur enim sententiae, non ponderantur..."

42. Beroaldo, *Annotationes centum,* sig. hi verso: "Litteras Cadmi filias vocat, quoniam Cadmus literarum inventor fuisse perhibetur, qui e Phoenice in Graeciam sexdecim numero attulisse fertur a Plinio in vii Na. hi. Vnde antiqui Greci litteras Phoenicias cognominaverunt, ut autor est Herodotus in Terpsicore. Ait idem scriptor se vidisse in templo Apollinis litteras Cadmeas in tripodibus quibusdam incisas, magna ex parte Ionicis litteris consimiles. Quin etiam Cornelius Tacitus Cadmum autorem litterarum fuisse autumat, rudibus adhuc Graecorum populis. Non me preterit alia ab aliis scriptoribus de origine litterarum commemorari, quibus in presentia supersedebimus cum ad rem nostram non pertineant."

43. Naturally there were any number of individual instances where Poliziano's predecessors and contemporaries surpassed him, especially in conjectural emendation. These cases do not, so far as I know, challenge the general rule.

44. E. J. Kenney, *The Classical Text,* 5–6; G. Funaioli, "Lineamenti d'una storia della filologia attraverso i secoli," *Studi di letteratura antica* 1 (Bologna, 1946): 284.

45. Poliziano, *Opera,* 271: "Nam cum ipsa quoque mendosissima plerisque sint locis, vestigia tamen adhuc servant haud obscura verae

indagandae lectionis, quae de novis codicibus ab improbis librariis prorsus obliterantur."

46. Ibid., 282: "Pudet referre quam manifestum, sed nondum tamen a quoquam (quod sciam) nisi nobis indicibus animadversum, mendum Vergilianis codicibus inoleverit libro Aeneid. octavo: Quod fieri ferro, liquidove potestur electro. Caeterum in volumine illo, quod est in intima Vaticana bibliotheca, mire vetustum et grandibus characteribus perscriptum, non potestur offendas, sed potest, usitatius verbum."

47. Ibid., 307: "Locus apud Suetonium in Claudio ita perperam legitur in plerisque voluminibus: Si aut ornatum, aut pegma, vel quid tale aliud parum cessisset: cum veri integrique sic habeant codices: Si automaton vel pegma. Inspice vel Bononiae librum ex divi Dominici, vel item alterum Florentiae ex divi Marci bibliotheca, quam gens Medica publicavit, veterem utrunque. Sed et utroque vetustiorem, quem nunc ipsi domesticum possidemus . . . ubique hanc nimirum posteriorem scripturam invenies."

48. Rizzo, *Lessico filologico*, 147–164; for the quality of Poliziano's collations see R. Ribuoli, *La collazione polizianea del Codice Bembino di Terenzio* (Rome, 1981); for the work of his rival B. Fonzio see V. Fera, "Il primo testo critico di Valerio Flacco," *Giornale italiano di filologia*, new ser. 10 (1979), 230–254.

49. Beroaldus, *Annotationes centum*, sig. [b vi recto]: "Conditam pixidem vocat medicamentis refertam: quae dictio a condio, non a condo, deducitur. Ideoque pronuntiandum est media syllaba producta et ita versus legendus: Turgida nec prodest condita pixide Lide; ubi condita septimus est casus, et cum pixide copulatur. Ego nuperrime versum istum ita scriptum legi in vetusto codice, et olim Angelus Pollicianus, Latine Graeceque doctissimus, mihi retulit se ita locum istum in sincerae fidei libro scriptum animadvertisse."

50. Poliziano, *Opera*, 263: ". . . cacoethes legendum suspicamur, ut sit quod apud Graecos *to kakoēthes*. Eo namque verbo frequentissimo usitatissimoque mala consuetudo significatur. Quod item in vetusto codice Langobardis exarato literis [Vat. lat. 3286, s. xi; Rizzo, *Lessico filologico*, 124–125] reperimus, cuius mihi potestatem legendi fecit Franciscus Gaddius Florentinus summi magistratus a secretis, prudens humanusque vir, nec literis incultus. Sed et ille versus ita in eodem: Turgida nec prodest condita pyxide Lyde."

51. S. Timpanaro, *La genesi del metodo del Lachmann*, repr. of 2d ed. (Padua, 1985), 4–6; G. Kirner, "Contributo alla critica del testo delle *Epistolae ad Familiares* di Cicerone (l. IX–XVI)," *Studi italiani di filologia classica* 9 (1901), 400–406.

52. Poliziano, *Opera*, 246–247: "Nactus sum Ciceronis epistolarum familiarium volumen antiquissimum, de quo etiam supra dixi, tum ex eo ipso alterum desciptum, sicuti quidam putant, Francisci Petrarchae manu. Descriptum autem ex ipso liquet multis argumentis, quae nunc omiserim: sed hic posterior, quem dixi, codex ita est ab indiligente bibliopola conglutinatus, uti una transposita paginarum decuria, contra quam notata [ed. nota] sit numeris deprehendatur. Est autem liber in publica gentis Medicae bibliotheca. De hoc itaque uno quantum coniiciam, cuncti plane quotquot extant [ed. extent] adhuc epistolarum earundem codices, ceu de fonte capiteque manarunt, inque omnibus praeposterus et perversus lectionis ordo, qui mihi nunc loco restituendus quasique instaurandus."
53. Poliziano, *Opera*, 260: "quibusdam etiam, saltem in praefatione, velut ab autore plane et a cogitante atque generante potius, quam a librario et exceptore inductis, expunctis ac superscriptis . . ."
54. See V. Fera, *Una ignota Expositio Suetoni del Poliziano* (Messina, 1983), 224.
55. Poliziano, *Opera*, 261: "igitur in Pandectis his, non iam Pisanis, ut quondam, sed Florentinis, in quibus pura sunt verba: nec ut in caeteris plena maculis et scabie, diffisum [ed. diffissum] reperio, non diffusum."
56. See H. E. Troje, *Graeca leguntur* (Cologne and Vienna, 1971), 21–22.
57. Poliziano, *Opera*, 287: "caeterum in Pandectis istis Florentinis, quas etiam archetypas opinamur, negatio prorsus est nulla. Quo fit, ut interpres legum Florentinus Accursius, mendosum et ipse nactus codicem, pene dixerim miserabiliter se torqueat."
58. M. P. Gilmore, "The Renaissance Conception of the Lessons of History," *Facets of the Renaissance*, ed. W. H. Werkmeister, 2d ed. (New York, Evanston, and London, 1963), 92–95; D. R. Kelley, *Foundations of Modern Historical Scholarship* (New York and London, 1970), 39–43; Beroaldo, *Annotationes centum*, sig. [h iii recto]: "Sexcenta sunt id genus apud iurisconsultos ab Acursio perperam enarrata."
59. A. Poliziano, *Miscellaneorum centuria secunda*, ed V. Branca and M. Pastore Stocchi (Florence, 1972), 4:5.
60. Ibid., 7.
61. R. Sabbadini, *Storia e critica di testi latini*, 2d ed. (Padua, 1971), 77–108.
62. G. Pasquali, *Storia della tradizione e critica del testo*, 2d ed. (Florence, 1952), 61–63.

63. Giovanni Lamola to Guarino of Verona, 31 May 1428, in Sabbadini, *Storia e critica*, 106; see also Rizzo, *Lessico filologico*, 175–177.
64. Giorgio Merula, preface to his edition of Plautus (Venice, 1472); quoted by Rizzo, *Lessico filologico*, 314.
65. Sabbadini, *Storia e critica*, 241–257; C. Questa, *Per la storia del testo di Plauto nell'umanesimo*, 1: *La "recensio" di Poggio Bracciolini* (Rome, 1968), 7–21; Rizzo, *Lessico filologico*, 314–315.
66. L. D. Reynolds and N. G. Wilson, *Scribes and Scholars*, 2d ed. (Oxford, 1974), 128.
67. Poliziano, *Opera*, 228: "Nos de graeco instrumento, quasi de cella proma, non despicabilis nec abrogandae fidei proferemus autoritates, quibus et lectio praestruatur incolumis et interpretamenti nubilum discutiatur."
68. Cf. *Michaelis Marulli Carmina*, ed. A. Perosa (Zürich, 1951), 59, 185.
69. Poliziano, *Opera*, 282: "In elegia eadem Catulli ex Callimacho, Oarion legitur, pro eo quod sit Orion. Quam quoniam integram adhuc inviolatamque dictionem nonnulli temere attentare iam incipiunt, contra hanc sinistram imperitorum audaciam standum mihi est omni (quod aiunt) pede, vel Callimachi eiusdem autoritate, qui sic in hymno ipso in Dianam, etiam nunc extante, ait . . . Sed et Nicander Theriacon libro consimiliter . . . et Pindarus in Isthmiis . . . Et alibi . . . Quare putat Eustathius quinto in Odysseam commentario . . . appellatum. Non igitur Aorion sed Oarion vera lectio."
70. R. Sabbadini, *Il metodo degli umanisti* (Florence, 1922), 58; H. Baron, *From Petrarch to Leonardo Bruni* (Chicago and London, 1968), 203–207, 212–213.
71. Calderini, "Ex tertio libro observationum," in *Lampas*, ed. Gruter, 1:315: "Receptae in usum nostrum fuerunt nonnullae Graecorum dictiones, et in Latinarum seriem [ed.: scite] deductae non elegantiae magis caussa aut festivitatis, quam necessitatis, quod Latina eadem non essent . . ."
72. Ibid., 315, 317.
73. Ibid., 252 (misnumbered 268); 235.
74. Poliziano, *Opera*, 246.
75. Ibid., 282.
76. Ibid., 248–249; *The Tragedies of Ennius: The Fragments*, ed. H. D. Jocelyn (Cambridge, 1967), 118–119, 347.
77. Poliziano, *Opera*, 247–248: "vertit hos nimirum quam potuit ad unguem poeta ingeniosissimus, et sunt tamen in graeco nonnulla, quae noster parum enarrate. Quin si veris concedendum, transmarinam illam nescio quam Venerem ne attigit quidem noster."

78. Poliziano, *Opera*, 248: "Quod vitium linguae potius minus lascivientis, quam parum copiosae."
79. Ibid., 229–230: "... ea ... quae de iusto atque iniusto, deque sui cuique notitia Socrates inibi cum Alcibiade agit, delibasse ex eo ... Persius intelligatur"; "Consimiliter quod ait ibidem, Tecum habita: nonne dialogi eiusdem pervidisse videtur voluntatem? Siquidem (quod Proclus enarrator affirmat) nihil hic aliud Plato, quam literam Delphicam respexit, monentem se quisque ut norit." Cf. Proclus Diadochus, *Commentary on the First Alcibiades of Plato*, ed. L. G. Westerink (Amsterdam, 1954), 6, 11; 19, 11–15. For Landino's use of the same information see R. Cardini, *La critica del Landino* (Florence, 1973), 173n.
80. Poliziano, *Opera*, 230: "Sic item, Dinomaches ego sum, ductum ex eo quod apud Platonem sic est, *o phile pai kleiniou kai deinomachēs* [Persius 4.20; *Alc.* 1 105D]."
81. B. Bischoff, "Living with the Satirists," *Classical Influences on European Culture, A. D. 500–1500*, ed. R. R. Bolgar (Cambridge, 1971), 83–94.
82. F. Jacoby, "Zur Entstehung der römischen Elegie," *Rheinisches Museum für Philologie*, new ser. 60 (1905), 38 and n1.
83. See esp. Macrobius *Sat.* 5.2–22 and P. Courcelle, *Late Latin Writers and Their Greek Sources*, tr. H. E. Wedeck (Cambridge, Mass., 1969), 13–26.
84. R. Sabbadini, *Le scoperte dei codici latini e greci ne' secoli XIV e XV: Nuove ricerche* (Florence, 1914), 9 and n40. De Bury knew only the preface and bks. 1–7.
85. R. de Bury, *Philobiblon*, ed. and tr. E. C. Thomas and M. Maclagan (Oxford, 1960), 110–111, perhaps drawing on Macrobius *Sat.* 5.17.
86. See, e.g., Battista Guarini's assessment of Manuel Chrysoloras, in R. Sabbadini, *La scuola e gli studi di Guarino Guarini Veronese* (Catania, 1896), 219.
87. D. Calderini, *Elucubratio in quaedam Propertii loca quae difficiliora videantur*, Proctor 6949 (Brescia, 1476), sig. [c 6 recto], on 1.20: "Theocritum in primis hoc loco imitatur et transfert de fabula Hylae."
88. Ibid., on 3.21.24: "Muros longos intelligit: qui Graece [space left for Greek] dicebantur. Ii ex urbe in Pyreum usque excurrebant. Historia est apud Thucydidem [1.107.1, 1.108.3] diligentissime perscripta."
89. Ibid., sig. [c 5 recto] on 1.13.21: "Sed dissentit a Strabone Propertius: nam Enipeum amatum a puella scribit Strabo [8.3.32; 9.5.6] esse in Pisana regione, cum tamen alter sit in Thessalia."
90. D. O. Ross, Jr., *Backgrounds to Roman Elegy* (Cambridge, Mass., 1975).

91. Calderini, *Elucubratio*, sigs. [d 5 verso–d 6 recto] on 4.1.64: "Se Romanum Callimachum appellat. Nam Callimachum poetam Graecum versibus Latinis explicat."
92. Ibid., sig c 2 verso on 1.2.1: "Vita. Ex Callimacho quom praecipue imitatur verbum deductum est. Nam et ille blandiens graeca voce amicam [space left for Greek] appellat."
93. Tibullus, Catullus, Propertius, ed. Beroaldo, sig. [s vi verso] on 4.1.64: "Appellat se Romanum Callimachum: quoniam qualis est apud Graecos Callimachus scriptor elegiarum: talis est apud Romanos Propertius: qui in primis Callimachum aemulatus est, et illum in scribendo habet archetypon"; sig. i ii recto on 1.2.1: "Imitatio est Callimachi: a quo [space left for Greek] id est vita amica appellatur."
94. Poliziano, *Opera*, 289: "Iam illud quoque miror, cur et Domitius et alii quidam post illum quocunque momento, quacunque occasione scribere audeant, hoc aut illud imitatione Callimachi dictum fuisse a Propertio, cum praeter hymnos pauculos nihil prorsus extet ad nos poetae istius, nec autem plane quicquam quod amoris argumenta contineat."
95. Ibid., 288–295; cf. *Callimachus*, ed. R. Pfeiffer, 2 (Oxford, 1953), lxvii.
96. Poliziano, *Opera*, 261.
97. H. Baron, *The Crisis of the Early Italian Renaissance* (Princeton, 1966); see also his articles collected as *In Search of Florentine Civic Humanism* (Princeton, 1988).
98. Poliziano, *Opera*, 315, quoting Cyprian *Ep.* 74.9; cf. G. Ladner, *The Idea of Reform* (New York, 1967), 136–139.
99. Cf. C. Dionisotti, *Geografia e storia della letteratura italiana* (Turin, 1967), 154–155, and A. Field, *The Origins of the Platonic Academy of Florence* (Princeton, 1989).
100. E. Bolisani, "Vergilius o Virgilius? L'opinione di un dotto umanista," *Atti dell' Istituto Veneto di Scienze, Lettere ed Arti*, cl. di scienze morali e lettere, 117 (1958–1959), 131–141.
101. M.Baxandall and E. H. Gombrich, "Beroaldus on Francia," *Journal of the Warburg and Courtauld Institutes* 25 (1962), 113–115; but cf. E. Garin, "Note in margine all'opera di Filippo Beroaldo il Vecchio," *Tra Latino e Volgare*, 2:439–441.
102. Krautter, *Philologische Methode*, 129–133.
103. P. Crinito, *De honesta disciplina*, ed. C. Angeleri (Rome, 1955). Cf. M. Santoro, "La polemica Poliziano-Merula," *Giornale italiano di filologia* 5 (1952), 212–233; Dionisotti, "Calderini, Poliziano e altri," 183–185; L. Perotto Sali, "L'opuscolo inedito di Giorgio Merula

contro i *Miscellanea* di Angelo Poliziano," *Interpres* 1 (1978), 146–183; V. Fera, "Polemiche filologiche intorno allo Svetonio di Beroaldo," *The Uses of Greek and Latin: Historical Essays,* ed. A. C. Dionisotti et al. (London, 1988), 71–87.
104. A. Grafton, *Joseph Scaliger: A Study in the History of Classical Scholarship, 1: Textual Criticism and Exegesis* (Oxford, 1983), chs. 2–3.
105. See esp. the work of Cesarini Martinelli referred to in Chapter 1 above, n42.
106. A. Perosa, *Mostra del Poliziano* (Florence, 1955), 13–92.
107. A. M. Adorisio and A. C. Cassio, "Un nuovo incunabolo postillato da Angelo Poliziano," *Italia Medioevale e Umanistica* 16 (1973), 278–281, 284, 286.
108. For the case of P. Vettori see Grafton, *Scaliger,* ch. 2, and Perosa, *Mostra del Poliziano,* 27–28.
109. Poliziano, *Opera,* 3 (*Ep.* 1.2); see N. Rubinstein, "Il Poliziano e la questione delle origini di Firenze," *Il Poliziano e il suo tempo* (Florence, 1957), 101–110.
110. Poliziano, *Opera,* 10–11 (*Ep.* 1.16).
111. Troje, *Graeca leguntur,* 18–20.
112. Poliziano, *Opera,* 13–14 (*Ep.* 1.20). Cf. A. Perosa, "Febris: A Poetic Myth Created by Poliziano," *Journal of the Warburg and Courtauld Institutes* 9 (1946), 74–95, for Poliziano's *arte allusiva.*
113. For a classic case study see E. Garin, " '*Endelecheia*' e '*Entelecheia*' nelle discussioni umanistiche," *Atene e Roma,* ser. 3, 5 (1937), 177–187.

3. Traditions of Invention and Inventions of Tradition in Renaissance Italy: Annius of Viterbo

1. *Scaligerana* (Cologne, 1695), 123.
2. For Berosus see *FrGrHist* 680 F 1; modern translation and commentary by S. M. Burstein (1978). For the general context see S. K. Eddy, *The King Is Dead* (Lincoln, Neb., 1961).
3. J. Selden, *The Reverse . . . of the English Janus,* tr. R. Westcot, quoted by A. L. Owen, *The Famous Druids* (Oxford, 1962), 36.
4. Leiden University Library MS Scal. 10, fol. 2 recto, quoting Helladius from Photius, *Bibliotheca,* cod. 279 and Tatian *Ad Graecos* 36 (= Eus. *Praep. ev.* 10.11.8 = *FrGrHist* 680 T 2).
5. *Thesaurus temporum,* 2d ed. (Amsterdam, 1658), *Notae in Graeca Eusebii,* 407–408.
6. Bodleian Library MS Casaubon 32, fol. 52 verso: "Multa ex Beroso

ipsis verbis recitantur, quae non memini reperire neque apud Eusebium, neque apud Iosephum. Eiusmodi est pericopa de Babylonia et eius mira ubertate. Sed in primis mira natura animalis cuiusdam *oanne,* ex multis composito monstro: cuius vox hominem sonat." For the date and significance of these notes—like those in Leiden MS Scal. 10, not discussed in the recent Teubner ed. of Syncellus (Leipzig, 1984) by A. A. Mosshammer—see A. Grafton, "Protestant versus Prophet: Isaac Casaubon on Hermes Trismegistus," *Journal of the Warburg and Courtauld Institutes* 46 (1983), 93.

7. Saliger, *Notae in Graeca Eusebii,* 408: "Haec quanquam in dubium merito revocari possunt, propter prodigiosa vetustatis et longissimi temporis curricula, tam Chaldaica Berosi, quam Aegyptiaca Manethonis, tamen non solum retinenda sunt, sed etiam in precio habenda propter reverentiam vetustatis, tum etiam, quia medii temporis vera cum illis fabulosis continuantur." For Scaliger's late, polemical euhemerism see Chapter 1 above.

8. Petrarch, *Seniles* 15.5; *Opera* (Basel, 1554), II, 1055–1058; tr. and discussed by P. Burke, *The Renaissance Sense of the Past* (New York, 1970), 50–4; for a text and discussion of Valla's work see W. Setz, *Lorenzo Vallas Schrift gegen die Konstantinische Schenkung* (Tübingen, 1975). See also the standard—and splendid—work of W. Speyer, *Die literarische Fälschung im heidnischen und christlichen Altertum* (Munich, 1971), 99–102.

9. See e.g. Burke, *Sense of the Past;* M. P. Gilmore, *Humanists and Jurists* (Cambridge, Mass., 1963); J. Franklin, *Jean Bodin and the Sixteenth-Century Revolution in the Methodology of Law and History* (New York and London, 1963); D. R. Kelley, *Foundations of Modern Historical Scholarship* (New York and London, 1970).

10. K. O. Müller, *Kleine deutsche Schriften,* I (Breslau, 1847), 445–452 (first published 1837).

11. J. Levine, *Doctor Woodward's Shield* (Berkeley, 1977).

12. Grafton, "Protestant versus Prophet"; G. Parry, "Puritanism, Science, and Capitalism: William Harrison and the Rejection of Hermes Trismegistus," *History of Science* 22 (1984), 245–270.

13. W. Goez, "Die Anfänge der historischen Methoden-Reflexion im italienischen Humanismus," *Geschichte in der Gegenwart: Festschrift für Kurt Kluxen,* ed. E. Heinen and H. J. Schoeps (Paderborn, 1972), 3–21; "Die Anfänge der historischen Methoden-Reflexion in der italienischen Renaissance und ihre Aufnahme in der Geschichtsschreibung des deutschen Humanismus," *Archiv für Kulturgeschichte* 56 (1974), 25–48. See also the more recent interpretation

of W. Stephens, Jr.: "The Etruscans and the Ancient Theology in Annius of Viterbo," *Umanesimo a Roma nel Quattrocento,* ed. P. Brezzi et al. (New York and Rome, 1984), 309–322, and *"De historia gigantum:* Theological Anthropology before Rabelais," *Traditio* 40 (1984), esp. 70–89.
14. A good description of the Rome 1498 ed. can be found in the British Museum catalogue of incunabula, IV, 118–119; I use the texts in the first edition but identify them, for simplicity's sake, by the page numbers of the well-edited and -indexed ed. of Antwerp 1552.
15. Annius, *Commentaria,* 59.
16. Ibid., 80 (the magic took the form of an illusion).
17. Ibid., ep. ded.
18. Ibid., 76–77.
19. For Josephus see F. Blatt, *The Latin Josephus,* I (Aarhus and Copenhagen, 1958), 13–15.
20. M. Luther, *Supputatio annorum mundi,* ed. J. Cohrs; *Werke* (WA), 53 (1920), 33–34, 36, 26–27.
21. I. Lucidus Samotheus, *Opusculum de emendationibus temporum,* 2d ed. (Venice, 1546), II.3, 19 recto.
22. G. Postel, *De Etruriae regionis originibus, institutis, religione et moribus,* ed. G. Cipriani (Rome, 1986), ch. 43, 173.
23. See in general E. Iversen, *The Myth of Egypt and Its Hieroglyphs in European Tradition* (Copenhagen, 1961); D. C. Allen, *Mysteriously Meant* (Baltimore, 1970), ch. 5; R. Wittkower, "Hieroglyphics in the Early Renaissance," *Allegory and the Migration of Symbols* (London, 1977), 113–128.
24. For Viterbo see esp. R. Weiss, "An Unknown Epigraphic Tract by Annius of Viterbo," *Italian Studies Presented to E. R. Vincent* (Cambridge, 1962), 101–120; for Spain, R. B. Tate, "Mitologia en la historiografía española de la edad media e del renacimento," *Ensayos sobre la historiografía peninsular del siglo xv,* tr. J. Diaz (Madrid, 1970), 13–32; for France, R. E. Asher, "Myth, Legend and History in Renaissance France," *Studi francesi* 39 (1969), 409–419; for England, Owen, *The Famous Druids,* and T. D. Kendrick, *British Antiquity* (London, 1950); for Germany, F. Borchardt, *Germany Antiquity in Renaissance Myth* (Baltimore, 1970); for Italy, E. Cochrane, *Historians and Historiography in the Italian Renaissance* (Chicago and London, 1981), 432–435, and G. Cipriani, *Il mito etrusco nel rinascimento fiorentino* (Florence, 1980). Stephens, in "The Etruscans" and *"De historia gigantum,"* emphasizes the transformation that Annius's ideas

underwent in the course of being vulgarized and rewritten in the sixteenth century.
25. Princeton University Library Ex. 2613.1510, with this note on the title page: "Roberti Nicolsoni Londinensis liber Parrhisiis: 1510."
26. See Fumagalli's fine case study, "Un falso tardo-quattrocentesco: Lo pseudo-Catone di Annio da Viterbo," *Vestigia: Studi in onore di Giuseppe Billanovich*, ed. R. Avesani et al. (Rome, 1984), I, 337–360.
27. Annius, *Viterbiae historiae epitoma*, ed. with useful commentary by G. Baffioni in *Annio da Viterbo: Documenti e ricerche*, I (Rome, 1981), 130–131. Baffioni indicates Annius's sources but does not exhaustively interpret his manipulation of them.
28. Annius, *Commentaria*, 432 (Fabius Pictor on the Sabines); 577 (Sempronius on the date and horoscope of the foundation of Rome).
29. The standard account of Annius's life remains R. Weiss, "Traccia per una biografia di Annio da Viterbo," *Italia Medioevale e Umanistica* 5 (1962), 425–441.
30. B. Guenée, *Histoire et culture historique dans l'Occident médiéval* (Paris, 1980), 133–140; B. Smalley, *English Friars and Antiquity in the Early Fourteenth Century* (Oxford, 1960), 233–235, 360–361 (Lathbury, quoting bk. 3, ch. 1 of the Athenian historian "Verosus"). As Smalley points out, the forger took off from the elder Pliny's reference to the Athenians' respect for Berosus's skill in astronomy (*Natural History* 7.123); chronology, like astronomy, was not this forger's long suit.
31. Annius, *Commentaria*, praef.
32. E. Fumagalli, "Aneddoti della vita di Annio da Viterbo O.P. . . . ," *Archivum Fratrum Praedicatorum*, 50 (1980), 167–199; 52 (1982), 197–218; V. Meneghin, *Bernardino da Feltre e i Monti di Pietà* (Vicenza, 1974), 545–550; *Giovanni Rucellai ed il suo Zibaldone*, I: *"Il Zibaldone Quaresimale,"* ed. A. Perosa (London, 1960), 164–170, 179–180.
33. C. Vasoli, *I miti e gli astri* (Naples, 1977), ch. 1; Annius, *Commentaria*, 48–52, in and on Berosus.
34. Annius, *Commentaria*, 15.
35. Ibid., 493, on Cato.
36. Ibid., 238, on Metasthenes. Cf. in general E. N. Tigerstedt, "Ioannes Annius and *Graecia Mendax*," *Classical, Mediaeval and Renaissance Studies in Honor of Berthold Louis Ullman*, ed. C. Henderson, Jr. (Rome, 1964), 2:293–310.
37. Annius, *Commentaria*, 453.
38. Borchardt, *German Antiquity*, 89–91.

39. Diodorus Siculus 1.20.1, 1.27.3–6; Annius, *Commentaria*, 12 (Xenophon on Ninus's monument).
40. See Allen, *Mysteriously Meant;* O. Gruppe, *Geschichte der klassischen Mythologie und Religionsgeschichte* (Leipzig, 1921), 29–31; J. Seznec, *The Survival of the Pagan Gods*, tr. B. F. Sessions (New York, 1953), ch. 1.
41. Annius, *Commentaria*, 188.
42. Beatus Rhenanus, *Rerum Germanicarum libri tres*, 2d ed. (Basel, 1551), 191.
43. Annius, *Commentaria*, 518–519, on Cato; amusingly pulverized by G. Barreiros, *Censura in quendam auctorem qui sub falsa inscriptione Berosi Chaldaei circunfertur* (Rome, 1565), 71.
44. Owen, *The Famous Druids*, 37, using a printed text of the translation (at 5.31.2). The very early manuscript of Poggio's version in Princeton University Library, Garrett 105, already offers the text in the form that I cite (fol. 144 recto); according to A. C. de la Mare, whose opinion was kindly conveyed to me by J. Preston, the manuscript was prepared in the circle of Tortelli and contains indexing notes by Poggio himself.
45. Tatian, *Oratio ad Graecos*, 31; Clement of Alexandria, *Stromateis* 1.117; Eusebius in Syncellus 211 M.
46. Jerome-Eusebius *Chronicle* 108 Fotheringham: "licet Archilocus ... supputet"; the same error is in the Armenian version. The origin of the eight Homers—though not their Olympic victories and achievements in magic and medicine, painting and sculpture—can easily be explained. Jerome gives seven possible dates for Homer; euhemerism turns these into eight Homers who lived at different times.
47. E. F. Rice, Jr., *St. Jerome in the Renaissance* (Baltimore and London, 1985); M. Baxandall, *The Limewood Sculptors of Renaissance Germany* (New Haven and London, 1980).
48. Annius, *Commentaria*, 246.
49. For Geoffrey see G. Gordon, "The Trojans in Britain," *The Discipline of Letters* (Oxford, 1946), 35–58; for Alemannus see Annius, *Commentaria*, 125, on Berosus.
50. Annius, *Commentaria*, 239.
51. Ibid., 460 (Myrsilus); 75–76 (Berosus); 281 (Philo).
52. Ibid., 244 (on Metasthenes).
53. Guenée, *Histoire et culture*, 133–140.
54. Stephens, "The Etruscans" and *"De historia gigantum."* For Josephus see J. R. Bartlett, *Jews in the Hellenistic World: Josephus, Aristeas, the Sibylline Oracles, Eupolemus* (Cambridge, 1985), 86–89; the transla-

tion of *Contra Apionem* 1.8–10 is to be found ibid., 171–176, with useful commentary. The importance of Josephus for Annius had previously received due attention from A. Biondi in the introduction to his translation of M. Cano, *L'autorità della storia profana* (Turin, 1973), xxxviii.

55. Annius, *Commentaria*, 240 (on Metasthenes).
56. O. A. Danielsson, "Annius von Viterbo über die Gründungsgeschichte Roms," *Corolla Archaeologica* (Lund, 1932), 1–16.
57. Rhenanus, *Rerum Germanicarum Libri tres*, 39; P. Crinito, *De honesta disciplina*, ed. C. Angeleri (Rome, 1955), 460; for Vives's comment on Augustine, *City of God* 18, see the interesting transcript made by C. Peutinger in his copy of Annius, published by P. Joachimsen, *Geschichtsauffassung und Geschichtschreibung in Deutschland unter dem Einfluss des Humanismus*, 1 (Leipzig, 1910; repr. Aalen, 1968), 271n24.
58. K. Arnold, *Johannes Trithemius (1462–1516)* (Würzburg, 1971), 167–179; J. Trithemius, "Chronologia Mystica," *Opera historica*, 1 (Frankfurt, 1601), unpaginated.
59. F. von Bezold, "Zur Entstehungsgeschichte der historischen Methodik" (1914), *Aus Mittelalter und Renaissance* (Munich and Berlin, 1918), 362–383.
60. See the literature cited in n9 above. Most previous critiques—e.g. the powerful one of E. Hassinger, *Empirisch-rationaler Historismus* (Bern, 1978)—have addressed issues different from those to be discussed below.
61. See e.g. W. J. Bouwsma, *Concordia Mundi* (Cambridge, Mass., 1957); H. J. Erasmus, *The Origins of Rome in Historiography from Petrarch to Perizonius* (Assen, 1962).
62. G. Postel, *Le thrésor des prophéties de l'univers*, ed. F. Secret (The Hague, 1969), 67. See also 76, where he describes the Annian Cato as drawing material "des monuments publikes,"and Postel, *De Etruriae regionis originibus*.
63. F. Baudouin, *De institutione historiae universae et eius cum iurisprudentia coniunctione* prolegomenôn *libri duo* (Paris, 1561), 48–49.
64. Postel, *De Etruriae regionis originibus*, 195–199. Cipriani shows that Postel's beliefs were shared by many members of that intellectually advanced institution, the Florentine Academy; one of them, Pier Francesco Giambullari, found what seemed vital corroborative evidence in a then unpublished passage in Athenaeus, for which Postel thanked him fervently. It is a pity that Postel's own full-scale defense of the forgeries does not survive. See ibid., esp. 15–23.

65. Baudouin, *De institutione historiae*, 44.
66. For Caius see V. Nutton, "John Caius and the Eton Galen: Medical Philology in the Renaissance," *Medizinhistorisches Journal* 20 (1985), 227–252; for the Oxford/Cambridge debate, a distant ancestor of the Boat Race, see Kendrick, *British Antiquity*.
67. J. Caius, *De antiquitate Cantabrigiensis Academiae libri duo* (London, 1568), 21–25; Caius's etymology of "giant" is an ancient one.
68. See respectively J. Sleidanus, *De quatuor monarchiis libri tres* (Leiden, 1669), 11; Franklin, *Bodin and the Sixteenth-Century Revolution*, 124–125; Savile in Bodleian Library MS Savile 29 (his "Prooemium mathematicum" of 1570), fol. 32 recto, where a reference to Berosus's *defloratio* of Chaldean history is underlined and bracketed. An afterthought?
69. M. Cano, *Loci theologici* 11.6; *Opera* (Venice, 1776), 234; Barreiros, *Censura*, 26–30.
70. Ibid., 35–37; V. Borghini, *Discorsi* (Florence, 1584–85), 1:229. Borghini had help from O. Panvinio (2:305).
71. Barreiros, *Censura*, 56–59, where Barreiros's own chronology seems a bit shaky. For an effort to reply see Postel, *De Etruriae regionis originibus*.
72. Cano, *Loci Theologici*, 230–232.
73. J. Funck, *Commentariorum in praecedentem chronologiam libri decem* (Wittenberg, 1601), fol. B iiij recto.
74. Ibid., fol. [B v verso].
75. Ibid., fol. [A v verso]. Funck says that he has taken his matter "ad verbum fere" from Berosus and the Bible; his genealogy of the descendants of Noah (fols. A iiij verso–[A v recto]) confirms this.
76. J. Bodin, *Methodus*, ch. 10; *Oeuvres philosophiques*, ed. P. Mesnard (Paris, 1951), 254–257.
77. Ibid., ch. 4, 126, praising Metasthenes, Polybius, and Ammianus Marcellinus for their use of *acta publica* and their objectivity when discussing other nations.
78. Ibid., ch. 8, 240. For a modern view of Ctesias, whose ancient critics Bodin knew, see R. Drews, *The Greek Accounts of Near Eastern History* (Cambridge, Mass., 1973), 103–116 (109: "all the details were invented").
79. J. Goropius Becanus, *Origines Antwerpianae* (Antwerp, 1569), 344–345.
80. Ibid., 357–362.
81. Ibid., 362; Goropius goes on to quote a fragment from bk. 43 of Theopompus's *Philippica*, in Latin; this dates Homer 500 years after

the fall of Troy (Clement of Alexandria *Stromateis* 1.117.8 = *FrGrHist* 115 F 205)—that is, to the time of Archilochus.
82. Ibid., ep. ded. See in general A. Borst, *Der Turmbau von Babel* (Stuttgart, 1957–1963), 3.1, 1215–1219.
83. N. Swerdlow, "Pseudodoxia Copernicana: or, Enquiries into very many received tenents and commonly presumed truths, mostly concerning spheres," *Archives internationales pour l'histoire des sciences* 26 (1976), 108–158.
84. S. Petri, *Apologia . . . pro antiquitate et origine Frisiorum* (Franeker, 1603), 15–17, summarizes the Frisian *Urgeschichte*.
85. U. Emmius, *De origine atque antiquitatibus Frisiorum*, in his *Rerum Frisicarum historia* (Leiden, 1616), 7ff.
86. Petri, *Apologia*, 40–41.
87. See further E. H. Walterbolk, "Zeventiende-eeuwers in de Republiek over de grondslagen van het geschiedverhaal: Mondelinge of schriftelijke overlevering," *Bijdragen voor de Geschiedenis der Nederlanden* 12 (1957), 26–44; "Reacties op het historisch pyrrhonisme," ibid., 15 (1960), 81–102; cf. in general S. Schama, *The Embarrassment of Riches* (New York, 1987), ch. 2.
88. See further the classical article by C. Mitchell, "Archaeology and Romance in Renaissance Italy," *Italian Renaissance Studies*, ed. E. F. Jacob (London, 1960), 455–483.

4. Scaliger's Chronology: Philology, Astronomy, World History

1. K. Hartfelder, *Melanchthoniana paedagogica* (Leipzig, 1892), 182.
2. Ptolemy, *Mathematicae constructionis liber primus*, ed. E. Reinhold (Wittenberg, 1549), sig. A 3 recto: "Necessaria est anni descriptio. Quales enim tenebrae retro essent, si nulla fuissent temporum discrimina? qualis in praesenti vita confusio esset, si ignota series esset annorum."
3. P. Haguelon, *Calendarium trilingue* (Paris, 1557), fol. 5 verso: "Intellectis enim quae de anno, mensibus, diebusque sunt cognoscenda, longe facilius intelliges ea in quibus iurisconsulti Hippocratem consulunt. Facilius intelliges ex Aristotelis mente, quo tempore Salpa, Sargus, Torpedo et Squatina pariant. Facilius intelliges locum (ut reliquos taceam) quo dicit Aristoteles coturnices migrare Boedromione, grues autem Maemacterione."
4. Tatian *Oratio ad Graecos* 31; quoted on the title page of Scaliger's *De emendatione temporum* (Paris, 1583) and by Edo Hildericus in his translation of Geminus (Altdorf, 1590), sig. a iiii.

5. Scaliger to S. Calvisius, 3 October 1605; *Epistolae,* ed. D. Heinsius (Leiden, 1627), 611: "Nullae Francofurtenses nundinae sine Chronologorum proventu."
6. Scaliger to Calvisius, 21 May 1607; Göttingen, Niedersächsische Staats- und Universitätsbibliothek, MS Philos. 103, pt. 2, p. 26: "De regibus Israelis, quis sanus promiserit se aliquid dicere, quod certum esse jurare possit?"
7. *De die natali* 16.3–4; quoted by L. G. Giraldi, *De annis et mensibus . . . dissertatio, Opera omnia* (Leiden, 1696), 2: col. 741. Cf. Augustine, *Confessions* 11; Ovid, *Metamorphoses* 15.
8. E. Panofsky, *Studies in Iconology* (New York, 1962), 69–93; F. Saxl, "Veritas filia temporis," *Philosophy and History: Essays Presented to Ernst Cassirer* (Oxford, 1936), 197–222; R. Wittkower, *Allegory and the Migration of Symbols* (London, 1977), 97–106; H. Erskine-Hill, *The Augustan Idea in English Literature* (London, 1983), 267–290.
9. Censorinus *De die natali* 21.1–5; often cited by Scaliger, e.g. in *Thesaurus temporum,* 2d ed. (Amsterdam, 1658), *Animadversiones in chronologica Eusebii,* 4.
10. I. Curio, *Chronologicarum rerum lib. II* (Basel, 1557), 8: "Agnum lupo facilius concordaveris, quam de aetate mundi omnes inter se chronographos."
11. M. Pattison, *Essays* (London, 1909), 1:131; J. Bernays, *Joseph Justus Scaliger* (Berlin, 1855), 49.
12. A. Grafton, *Joseph Scaliger: A Study in the History of Classical Scholarship,* 1: *Textual Criticism and Exegesis* (Oxford, 1983), 128.
13. Censorinus, *Liber de die natali,* ed. E. Vinet (Poitiers, 1568); Bodleian Library D 4.6 Linc., owned by G. J. Vossius, J. Rutgersius, and N. Heinsius after Scaliger.
14. Scaliger, *Lettres françaises inédites,* ed. P. Tamizey de Larroque (Agen and Paris, 1879), 20–21, 25–26, 29–30, 31, 41, 83, 91. The MS is now lost or unidentified; see A. Grafton, "Joseph Scaliger's Collation of the *Codex Pithoei* of Censorinus," *Bodleian Library Record* (Spring 1985) for details.
15. *Heptachordon* (for *azōnion* in Vinet's text and *taxopaon* in Pithou's manuscript) at 13.5.
16. The sheet is now glued to a blank leaf at the end of the book. One note from it was published by G. J. Vossius in his *Etymologicon linguae Latinae* (Amsterdam, 1662), s.v. *Aprilis;* he refers to the sheet as a "scheda . . . inedita."
17. Grafton, *Scaliger,* 1:116–117.
18. "Quomodo mensis Aprilis dici potuerit ab aperiendo, equidem non

video. Primum cum annus ab initio X tantum mensium fuerit, necesse est semper menses vagos fuisse, neque stationes temporum fixas tenuisse. Ita Aprilis semel tantum incidebat in tempus vernum, et post XII annos Romuleos, hoc est X lunares, redibat in orbem. Deinde Aprilis non potest descendere a verbo. Nam nomina quae huius terminationis sunt, si descendunt a verbo, penultimam corripiunt, ut probabilis, utibilis vel utilis, facibilis vel facilis; si autem a nomine, eandem producunt. Equus, equilis; ovis, ovilis; caper, caprilis; aper, aprilis. Praeter pauca: humilis, similis, etc. Ergo Aprilis ab apro, hoc est porco. Vt Athenis *elaphēboliōn*, hoc est cervilis, sic Aprilis *kaprēboliōn*, vel *kaproboliōn*."

19. Vinet, *Annotationes*, in his Censorinus, sig. BB iii verso on 18.8: "VT METONTICVS, QVEM METON. *Metōn* hic, astrologus, Socrati coaevus, genitivum casum habet *Metonos* apud Suidam, Plutarchum, et Ptolemaeum in magnae syntaxeos libro tertio, et alios: unde Metonicus potius fiat, quam Metonticus..."

20. Haguelon, *Calendarium*, fols. 34 verso–35 recto:
    ALPH. Calendae *apo tou kalō*, ut tradunt Macrobius et Polydorus, dicuntur. Theodorus vero Gaza scripsit Plutarchum asseruisse Calendas dici a praepositione clam.
    CYAN. Ha ha he.
    ALPH. Quid rides? mene habes ludibrio?
    CYAN. Non profecto, sed cum abs te audio Latinum a Graeco hujus vocis Calendae etymologiam, contra vero Graecum ejusdem nominis notationem a Latino mutuari voluisse, ob oculos quorundam stultitiam posuisti: ob quam vel Agelastus Crassus, vel etiam Heraclitus lachrymis confectus risu emoreretur.

21. G. Castellani, "Un traité inédit en grec de Cyriaque d'Ancône," *Revue des études grecques* 9 (1896), 225–230, at 229.

22. That is, if year 1 of a series begins in Romulean and Julian March, its Romulean April will be roughly the Julian April; but in year 2 April will fall in Julian February, and so on.

23. "Sextus annus Romuleus est intercalaris, ut congruat cum lunari. Eo no[mine] lustrum institutum. Et sex anni Romulei consummant annos v lunare[s] sicut in Graecia [MS: Graeci] Olympico certamine intercalabatur. Nam eo nomine insti[tutum] fuerat."

24. J. Gines de Sepulveda, *De correctione anni mensiumque Romanorum... commentatio* (Paris, 1547), fol. 4 verso: "Hanc igitur naturalem commodissimamque rationem Romani ut sequerentur, iam inde ab urbe condita sibi proposuerant, sed ignoratione motuum caelestium non statim assequi potuerunt, quod petebant. Nam Romulus decem dun-

taxat mensium annum constituit, Albanos, unde Romani orti erant, imitatus . . ."

25. Giraldi, *Opera*, 2: cols. 758–759; O. Panvinio, *Fastorum libri V* (Heidelberg, 1588), 25–26; Censorinus, ed. Vinet, sig. DD iii recto.
26. G. Postel, *Cosmographicae disciplinae compendium* (Basel, 1561), 60–61: "Vnde fingens se in Martis honorem velle anni ponere principium, abstulit a capite anni Ianuarium, in quo Iani memoria (vel teste in vetustis Problematis Plutarcho) coniuncta cum primo solis ad nos redeuntis signi gradu destrueretur. Et sic nebulo tyrannorum maximus, sui novi et in decimestri spacio impossibilis anni constitutione, evertit funditus leges temporum: ita ut nunc ob illius erratum, capita signorum a capitibus mensium, quae una per Iani institutionem semper incedebant, distent 21 diebus." Posel uses his sources—notably Plutarch *Numa* 19—with great independence. See H. J. Erasmus, *The Origins of Rome in Historiography from Petrarch to Perizonius* (Assen, 1962), 40–46.
27. J. Goropius Becanus, *Origines Antwerpianae* (Antwerp, 1569), 428: "Cuius [= Iani] statuam hac de re ita dedicarunt memoriae posteritatis, ut numerus dierum totius anni in digitorum positu et flexu manus eius signaretur . . . Mirandum igitur tam crasse quosdam hallucinatos esse, ut decem duntaxat menses tempore Romuli fuisse scribant. Quamvis Romulus (ut volunt) barbaro et militari ingenio esset, utpote proles Martis inimici mortalis omni rationi rectae et disciplinis omnibus humanis, non tamen id probe inde colligi potest, non alios eo tempore fuisse, qui meliora scirent, et annum rectius computarent."
28. See Ovid, *Fasti*, ed. J. G. Frazer (London, 1929), 2: 8–29.
29. Scaliger, *De emendatione*, 117: "Quia igitur exploso illo decimestri anno ad omnia et per omnia inutilissimo, praeterea exibilata sententia, quae Romulum ex Pastore et Rustico Metonem et Calippum constituebat, ad veras rationes confugiendum est . . ."
30. "Si Hecatombaeonis initium a luna prima quae solstitio quam proxima erat incip[eret], quomodo octavo quoque anno intercalaretur non video. Hoc enim fieri non potera[t] cum ad eam rem cyclo enneadecaterico uterentur, qui intercalationi locu[m] non sinit, cum per annos embolimos hoc expediatur. Sed octavo quoque ann[o] 30 menses accedere animadverterant."
31. Plato *Laws* 767C; previously quoted by (e.g.) M. Beroaldus, *Chronicum, Scripturae Sacrae autoritate constitutum* (Geneva, 1575), 59. For Scaliger's acquaintance with the text see Grafton, *Scaliger,* 1:105–106, 276n32.

32. J. Lalamant, *Exterarum fere omnium et praecipuarum gentium anni ratio et cum Romano collatio* (Geneva, 1571), 111: "Verum enim intelligere licet Attici anni principium, seu si vis primum diem Hecatombaeonis primi illorum mensis, non stato semper die, sed ne mense quidem occurrere, sed pro aliis atque aliis octennii annis anticipare et in anteriora progredi per XI. quotannis dierum spatium."
33. Lalamant, *Collatio,* 103; Beroaldus, *Chronicum,* 51, 59.
34. H. Savile, "Prooemium mathematicum," Bodleian Library, MS Savile 29, fol. 34 recto: "vidit enim aequinoctia solstitiaque accurate fere ad sua principia retrahi posse assumto annorum solarium novendecim spacio, quo 235 menses synodici absolvuntur: hac ratione ut praeter eum diem, quem alioqui olympiadi cuique ex veteri instituto usitatum erat, ad annos undeviginti lunares 7 menses attexerentur: quorum sex triginta singuli constarent diebus, septimus viginti novem definiretur. Hanc periodum Graeci vocarunt *enneadecaetērida* annumque magnum Metonis."
35. Scaliger, *De emendatione,* 55–57, using Censorinus 18.8 to show that the true Metonic cycle contained 6,940 days; Geminus 8 was not accessible to Scaliger until after 1590.
36. E.g. Paul of Middelburg, *Paulina* (Fossombrone, 1513), sig. d iiii recto.
37. A. Manuzio, Jr., *De quaesitis per epistolam libri iii* (Venice, 1576), 82–83: "sed de lunari [sc. anno] Metonem intellexisse, ideo credendum, quia Graeci omnes, praeter Arcades et Acarnanes, annum diebus cccliv computabant: Metonem autem in Graecia natum, Athenis Theophrastus, Diodorus, Censorinus et Tzetzes, Spartae Aelianus dixit, nemo apud Arcades... aut apud Acarnanes..."
38. Censorinus, ed. Vinet, sig. BB iij verso: "Ista autem Metonis decaënneateris... nostris computatoribus cyclus decemnovenalis dictus est et hinc eius numerus aureus."
39. "Quod Dio ait Caesarem ab Aegyptiis intercalandi rationem didicisse, i[...] perquam ridiculum. Accepit enim Dio a Lucano, et quaecunque Lucan[us] de bello civili pronunciavit per licentiam poeticam, id Plutarchus e[t] Dio tanquam *historikōs* dictum acceperunt. Atqui Aegyptii non intercalabant, nisi anno Cynico confecto, quod memoria hominum sem[el] tantum accidit. Graeci quarto quoque anno intercalabant. Ergo ab illis Cae[sarem] didicisse quid vetat?"
40. Censorinus, ed. Vinet, sig. EE ii recto: "... annumque ad Solis cursum accommodavit, quemadmodum ab Aegyptiis Alexandriae acceperat: sicut scribit Dio libro quadragesimo tertio."
41. Manuzio, *De quaesitis,* 48; Giraldi, *Opera,* 2: col. 763. Cf. also Ma-

crobius *Sat.* 1.14.3–4, 1.16.38–40; Pliny *NH* 18.210–212. In many of these texts the "Egyptians" are Hellenistic Greeks in Alexandria.
42. Thus when Scaliger describes the Egyptians as intercalating once in 1,460 years, he means that they add 365 days after 1,460 Egyptian years, since 1,461 Egyptian years = 1,460 Julian; see I. Stadius, *Tabulae Bergenses* (Cologne, 1560), 14, a book Scaliger owned.
43. E. Reinhold, *Prutenicae tabulae* (Tübingen, 1551), fol. 23 recto.
44. Scaliger, *De emendatione*, 13: "Resumamus igitur eos annos, ex quibus tanquam elementis, ad tot tamque diversa genera annorum progressus factus est."
45. Ibid., 156: "Caesar igitur devicto Pompeio postquam Sapientibus Aegypti aliquot dierum operam dedisset, ab iisque anni Canicularis modulum didicisset, ex eo tempore in animo habuit annum emendare..."
46. Scaliger, *Lettres*, 101–102, 104, 106–107, 111.
47. Scaliger to Chrestien, 4 September 1581; Bibliothèque Nationale, MS Dupuy 496, fol. 206 (copy); edited in the original version of this chapter, as appendix I: *Journal of the Warburg and Courtauld Institutes* 48 (1985), 135–136.
48. See O. Neugebauer, *Ethiopic Astronomy and Computus*, SB Wien, Phil.-hist. kl., 347 (Vienna, 1979), 55, 163, 203, 219.
49. Edited in the original appendix I, 136: "Ie n'ai veu les vers qu'aves faict sur le Computus Aethiopicus. Il vous en fauldroit doncques faire [MS faires] plusieurs aultres: Car i'ai des Computus en papier plus qu'en bourse. Car i'ai tous les computus de toutes les nations Chrestiennes, Mahometaines, et qui plus est, i'ai bien avant fouillé en terre, ou i'ai trouvé tous les Computus des anciens Grecs. Souvienne vous ie vous prie, de ce que ie vous dis, que i'ai changé les temps, comme Daniel dit [7.25] d'Antiochus Epiphanien."
50. Leiden University Library MS Or. 225 (Gregory 49).
51. Ibid., rear endpapers (Oriental style): "Ad calcem Evangelii secundum Iohannem librarius dicit librum scriptum fuisse die ultima Iulii (is enim est Thamuz secundum Christianos Syroarabas) feria iii, indictione xii. Iam Graeci cum Syris, Eusebium et lxx secuti, epocham Christi ab mundi [MS anni] conditu ponunt 5500. Noster iste ait librum scriptum fuisse anno mundi 6687. A quibus si detrahas 5500, relinquuntur 1187. Sed tamen in illa aera neque ultima Iulii erit feria tertia [MS sexta], neque indictio duodecima. Apparet igitur viii annis maiorem Epocham mundi eos facere, quam Syros, et Christo nascente annos a mundi conditu 5508 fluxisse. Ita ille liber a Christo nato scriptus fuit anno 1179. Quae aera habet indictionem xii, nu-

merum cycli Solaris xii. Litera fuit Dominicalis G. Sic B ultima Iulii mensis litera fuit feria tertia. Et indictio non a mense Decembri, sed Kal. Ian. incipit."

52. The 15-year indiction cycle is built into the Julian ecclesiastical calendar. To find the indiction of a year in the Christian era, add 3 and divide by 15. The remainder is the indiction (if there is no remainder the indiction is 15). In this case, $(1187 + 3)/15 = 79$ rem 5; $(1179 + 3)/15 = 78$ rem 12, which was sought. One can find the indictions for all years of the Christian era handily tabulated in I. Lucidus Samotheus, *Opusculum de emendationibus temporum* (Venice, 1546).

53. The first seven days of every year in the Julian ecclesiastical calendar are designated by the letters A–G. The letter matching the day on which the first Sunday of the year falls is the year's Dominical letter. Leap years have two, one designating Sundays before and one Sundays after the intercalated day, which changes the order of weekdays. To find the position in the 28-year cycle of Dominical letters of a year in the Christian era, add 9 and divide by 28. The remainder is the year's place in the cycle; if there is no remainder the place is 28. One then takes this number to a table of the 28 Dominical letter combinations to find the year's letter. In this case, $(1187 + 9) / 28 = 42$ rem 20, which has letter D; $(1179 + 9) / 28 = 42$ rem 12, which has letter G, which was sought. One can also look up the letter(s) for any year in the Christian era in Lucidus, *Opusculum*.

54. All days in the Julian year have letters from A to G. If one knows the *littera ferialis* (or *kalendarum*) of a given day and the Dominical letter of its year, one can determine the weekday on which the date falls. Lucidus provides these in the *Opusculum*.

55. Scaliger, *De emendatione*, 249, 367.

56. H. Gelzer, *Sextus Julius Africanus und die byzantinische Chronographie* (Leipzig, 1880–1898), 1:46–51; A. A. Mosshammer, *The Chronicle of Eusebius and Greek Chronographic Tradition* (Lewisburg, 1979), 146–157.

57. Anastasius of Sinai, *Quaestiones et responsiones*, in *Sacrae bibliothecae sanctorum tomus sextus*, ed. M. de la Bigne (Paris 1575), quaest. 92, cols. 237–240.

58. As, for example, A. Pontacus rightly argued in his "Apparatus," *Chronica trium illustrium auctorum* (Bordeaux, 1602), 14–15.

59. Scaliger knew the Byzantine era perfectly well by the time he completed the *De emendatione;* see his treatment of the same document there, 249.

60. The Greek indiction begins on 1 September, the imperial on 24 September, the pontifical on 25 December or 1 January; none of these would have different results in this case, as is obvious.
61. Leiden MS Or. 225, rear endpapers (Oriental style): "Temporibus Bedae annus secundum Graecos a conditu mundi 6276. Annus Christi temporibus Bedae 703. Differentia, anni a conditu mundi ad Christum 5573. Sed haec corrupta. Infra autem manifesto ait Beda a Christo ad sua tempora fluxisse annos 778. Aufer a 6276. Subsident, 5498. Quae aera minor est 2 annis, quam Graecorum communis 5500."
62. The "Canones annalium, lunarium ac decennovenalium circulorum," most accessible in *Patrologia latina* 90, cols. 877ff. See C. W. Jones, *Bedae pseudepigrapha* (Ithaca, N.Y., 1939), 82–83.
63. Bede, *Opuscula cumplura de temporum ratione*, ed. Ioh. Noviomagus (Cologne, 1537), fol. 100 recto: "Si nosse vis quotus sit annus secundum Graecos duntaxat, multiplica xv, per ccccxviii, fiunt vimcclxx. Adde indictiones anni cuius volueris, quae presentis anni ii esse comprobantur, et fiunt vicclxxvi."
64. See the *Vita Bedae ex Iohanne Trithemio,* ibid., fols [A iij recto - A iiij recto], which gives A.D. 732.
65. Scaliger, *De emendatione*, 316: "Si incipit [Tisri] quarta feria, decima erit sexta feria, in qua celebratur solenne Kipphurim . . . Atqui duo Sabbata continua incommoda erant in tractu Hierosolymitano, propter cibos, quos coctos biduum continuum asservare in regionibus calidis praesertim tempore anni ferventissimo periculosum erat. Item corpora mortuorum asservari non possunt citra putoris et corruptionis periculum."
66. S. Münster, *Kalendarium hebraicum* (Basel, 1527), 128–129: "Sin neomenia observaretur feria 4 aut 6 caderet festum propitiationis in feriam sextam aut feriam primam, et tunc cogerentur duo continuare sabbata, quum festum illud expiationis non minus sacrum sit sabbato: quod esset eis grave propter olera (ut aiunt) et mortuos. Oporteret enim cocta olera servare in tertium diem, et similiter defunctorum cadavera, quod in calidis regionibus citra periculum fieri non posset." Casaubon entered cross-references in his working copy of the *Kalendarium* (British Library 481.c.2) which show that he had seen the connection between Münster's arguments and those advanced by Scaliger in the second edition of the *De emendatione* (Leiden, 1598), 604.
67. Scaliger, *De emendatione* (1583), fol. a v recto.

68. See G. Levi della Vida, *Documenti intorno alle relazioni delle chiese orientali con la S. Sede durante il pontificato di Gregorio XIII*, Studi e testi 143 (Vatican City, 1948), 22–25.
69. Scaliger, *De emendatione*, 226–227.
70. T. Campanella, "De libris propriis et recta ratione studendi syntagma," H. *Grotii et aliorum dissertationes de studiis instituendis* (Amsterdam, 1645), 406: "Historiam Temporum Scaligeri admirantur Germani, eamque sequuntur etiam Plurimi nostrates, inter quos recensetur Vechettus Florentinus, qui scripsit de anno Sacro tam subtiliter quam imprudenter: voluit enim per eclipses atque Cyclos Lunares in Historiis antiquiorum extantes numerum annorum corrigere..."
71. See M. Sondheim, *Thomas Murner als Astrolog* (Strasbourg, 1938); J. North, "Astrology and the Fortunes of Churches," *Centaurus* 24 (1980), 181–211; K. M. Woody, "Dante and the Doctrine of the Great Conjunctions," *Dante Studies* 95 (1977), 119–134.
72. Eusebius/Jerome, *Chronicon* (Venice, 1483), sig. [b 6 recto]: "... lucidissima veritate secundum Alphonsium regem Castellae astronomorum celebratissimum ab Adam usque ad Cathaclismum anni 2402.d.315. differentia 161.d.50 ad superius notatum numerum, videlicet 2242."
73. This was pointed out by a seventeenth-century reader of the Bodleian copy (K 3.20 Auct.): "Alphonsi regis Chronologia haec aliter apud alios se habet ubi aetas 1a annorum est 3882 et 167 dierum"—though he has copied a printer's error, 167, for the correct figure of 267 days. The Alphonsine date for the Flood is in fact the Indian era of the Kaliyuga, 17 February 3102 B.C., often connected with the Flood by Islamic sources—and quite impossible to reconcile with the Old Testament.
74. In this respect the technical chronology of the later Middle Ages resembles the efforts made in the same period to impose quantification on physics, pharmacy, and other previously qualitative disciplines.
75. Copernicus, *De revolutionibus orbium coelestium libri sex* (Nürnberg, 1543), 3.11; see N. Swerdlow and O. Neugebauer, *Mathematical Astronomy in Copernicus's De revolutionibus* (Heidelberg, 1984), 1:183–188.
76. T. Bibliander, *De ratione temporum... liber unus* (Basel, 1551), 121: "[after quoting Censorinus 21] Hactenus Censorinus ex Varrone. Tenemus igitur clavem, qua historiae sacrae et exterae seriem pandamus."

77. Reinhold, *Prutenicae tabulae,* fol. 21 verso: "Sed quod ad historiam adtinet, primum Mardocempadus, cuius ut Babylonici regis annos in trib. Lunae deliquiis numerat Ptolemaeus, eumque Nabonassaro 26. annis posteriorem facit, alius esse non potest, quam qui tum a Metasthene et aliis scriptoribus, tum vero in sacris literis Merodach nominatur . . ."
78. J. Funck, *Chronologia* (Basel, 1554), 2d pagination, 17: "non potui melius invenire consilium, quam cuius me de hac re ambigentem . . . Osiander participem fecit: nempe ut tempora ad certissimas Ptolemaei astrologi aeras . . . accommodarem: initio sumpto a magna Nabonassaris aera, quae a cunctis Ptolemaei sectatoribus, cum pro certissima, tum pro antiquissima omnium habetur."
79. Curio, *Chronologicarum rerum lib.,* 31: "Ptolemaeus noster Salmanassarem hunc . . . appellat Aegyptio sono Nabonassarem, et a regni ipsius auspicio, quasdam motuum coeli supputationes deduxit . . ."
80. J. Bodin, *Methodus ad facilem historiarum cognitionem* (Paris, 1566), ch. 8, 388–389.
81. Scaliger, *De emendatione,* 276.
82. P. Apianus, *Astronomicum Caesareum* (Ingolstadt, 1540), Enunctiatum 29, sig. I iii verso: "Quod quidem tam grande malum [i.e. the inaccuracy of historians] sola ecleipsium cognitio emendare et in melius vertere potest. Per ecleipses enim omnia certos in annos reduci possunt, Christum praecedentes non minus quam sequentes."
83. Apianus knew that Eusebius set the defeat of Darius in A.M. 4871 and the birth of Christ in A.M. 5199; $5199 - 4871 = 328$. Hence Guagamela and the accompanying eclipse took place in 328 B.C. But Plutarch dated the eclipse to Boedromion (*Alex.* 31). Apianus, going against the standard authority, Theodore Gaza, put Boedromion next to Julian June. The only eclipse that fell in June near 328 B.C. was Oppolzer 1360, 28 June 326 B.C., which Apianus identified as the Gaugamela eclipse. The correct eclipse and date were given by P. Crusius, *Liber de epochis seu aeris temporum et imperiorum* (Basel, 1578), 19.
84. Scaliger, *Lettres,* 113, 115, 117.
85. Scaliger, *Commentarius et castigationes,* in *M. Manilii Astronomicôn libri quinque* (Paris, 1579), 243 on 4.776: "L. Tarrutius Mahematicus conceptum ait anno I. Olymp. II, hora tertia, xxiii die mensis *Choiac*. Is est December, quando fuit Solis *pantelēs ekleipsis*. Sed *tēn emphanē genesin,* quam Ptolemaeus vocat *proodon, ekptōsin, kai ektropēn,* xxi die Thot, Sole Oriente. Is est mensis September."
86. Crusius, *De epochis,* 62: "Haec cum ita se habeant, suspicor in iis

locis Plutarchum aliorum vestigia sequutum esse, qui tempora et motus a posteriore, ut vocant, conati sunt investigare. Verum exactiore Mathematicarum artium cognitione destituti, certitudinem attingere non potuerunt."

87. A. Grafton and N. Swerdlow, "Technical Chronology and Astrological History in Varro, Censorinus and Others," *Classical Quarterly*, new ser., 35 (1985).
88. Scaliger, *De emendatione*, 213. The ecliptic syzygy (not visible as an eclipse at Rome) in question really took place on 24 April 750 B.C., as one of a pair of glancing eclipses that fell a month apart (the other on 25 March 750 B.C.). These pairs are fairly common.
89. See the third ed. of the *De emendatione* (Geneva, 1629), 396.
90. G. Postel, *De originibus* (Basel, 1553), 72.
91. W. Lazius, *De aliquot gentium migrationibus . . . libri xii* (Basel, 1572), 23, 20: "Hic deest dies et mensis. Est autem monumentum antiquum. nam post latam legem fluxissent tantum dlx anni."
92. See Chapter 3 above.
93. Scaliger, *De emendatione*, 202: "Porro, ut Simplicius scribit, Callisthenes retulit, anno, quo capta fuit Babylon ab Alexandro, Chaldaeos non amplius quam mille nongentorum et trium annorum memoriam repetivisse, cum eos ipse de antiquitate et originibus rerum Chaldaicarum percontaretur. Babylon capta est anno periodi Iulianae 4383, qui annus praecedit primum Calippicae periodi. De quibus annis deducto tempore antiquitatis et originum Chaldaicarum, nempe annorum 1903, remanent anni periodi Iulianae 2480. quorum et diluvii differentia est 60 duntaxat. Igitur in tanta obscuritate rerum non multum abludit epilogismus Chaldaicus a Mosaico. Sane verus iste est, quem proposuimus."
94. Simplicius on *De caelo* 2.12, 293a4, as printed (in a Greek back-translation from Moerbeke's Latin) in the Aldine edition (Venice, 1526), fol. 123 recto. The modern edition by Heiberg (*CAG* 7, Berlin, 1894) based on the Greek MSS gives a much larger figure, 31,000 years, for the duration of Babylonian astronomical records (and reads *tērēseis*, not *paratērēseis*).
95. He did know the text, as it is clear from his collections of the fragments of the pre-Socratics in Leiden University Library MS Scal. 25, fols. 102ff. and Bodleian Library 8° C 238 Art., [32], [47] (this is Scaliger's working copy of his *Poesis philosophica* [Geneva, 1573]).
96. P. d'Abano, "De differentiis et epicyclis," ed. G. Federici Vescovini, *Medioevo* 11 (1985), 198; G. Pico della Mirandola, *Disputationes ad-*

versus astrologiam divinatricem 11.2; *Opera omnia* (Basel, 1572), 714: "est in re nostra Porphirii testimonium evidentissimum, qui quidem quas Calisthenes ex Babylone, temporibus Aristotelis et Alexandri, Chaldaeorum de astris annotationes in Graeciam misit, Mille ait et nongentorum trium annorum observationem duntaxat continuisse, quos annos ab Alexandro retro si numeres in Abraham Patriarchae vel paulo superiora tempora desinere eos comperies."

97. Bodin, *Methodus*, ch. 8, 385; P. Ramus, *Prooemium mathematicum* (Paris, 1567), 8–9; I. Th. Freigius, *Mosaicus* (Basel, 1583), 116–117.
98. Bodleian Library, MS Savile 29, fols. 29 verso–30 recto: "Illud tamen praetereundum non est, quod ipse quoque Ptolemaeus 70 3i libri capite videtur innuere, solum ab initio regni Nabonassari, id est quadringentorum ante Alexandrum annorum, observationes suo tempore extitisse . . ."
99. Scaliger, *Epistolae*, 364; cf. A. Borst, *Der Turmbau von Babel* (Stuttgart, 1957–1963), 1215–1219.
100. Goropius, *Origines*, 434–435: "Hic cum Callisthenem carissimum sibi discipulum apud Alexandrum haberet, rogavit eum, ut diligentissime a Babylonicae antiquitatis peritis inquireret astrorum observationes et annorum seriem. Qui libenter admodum hac in re praeceptori suo obtemperans, omnes observationes ipsorum in Graeciam transmisit. Has autem Porphyrius dicit mille nongentorum et trium annorum fuisse, nec plurium omnino, de qua re Simplicium licet videre, qui in secundum librum Aristotelis de caelo scribens doctissime et copiosissime veterum de caelestium orbium numero opiniones recenset. Cum ergo Alexander Babyloniam occuparit, cum tria millia sexcenti triginta sex anni ab orbe condito numerarentur, necesse erit fateri observationes ipsorum ad mundi annum millesimum septingentesimum et tricesimum primum ascendere, quo tempore Nembrod potuerit esse natus annos paulo plus minus quadraginta quinque: si Chus statim a diluvio genitus Nimbrodum tricesimo aetatis suae anno procrearit . . . Habemus ergo inter Chaldaeos et eos, qui e Sacris Bibliis computant, insignia consensus documenta . . . Quocirca annos illos, quos Septuaginta Interpretes inter Ninum et diluvium Hebraico calculo adiecerunt, audacter reiiciemus."
101. G. B. Vico, *Scienza Nuova giusta l'edizione del 1744*, ed. F. Nicolini (Bari, 1911–1916), 1:72–75.
102. A. Bucholzer, *Isagoge chronologica* (n.p., 1580); *Chronologia* (n.p., 1594), e.g. 575, 585; Eusebius, *Ecclesiastica historia* (Basel, 1570),

Prooemium, repr. in I. I. Grynaeus, *Epistolarum selectarum (quae sunt ad pietatem veram incentivum) libri duo*, ed. A. Scultetus (Offenbach, 1612), 44–84.

103. Scaliger, *De emendatione,* 252: "Vide quid fit ubi autoritas antevertit veritati. Quicunque enim illa legunt, non dubitant vera esse, quia ab Eusebio profecta." Cf. *HE* 2.17 and Grynaeus's comments.

104. Scaliger to Beza, 7 December 1584 (Julian); Leiden University Library MS Perizonianus Q 5, fols. 11 recto–14 recto; published with commentary in the original version of this chapter: *Journal of the Warburg and Courtauld Institutes* 48 (1985), 141–143.

105. Bibliothèque Nationale Rés. G. 141, 252: "Haec maxime voluimus tractare propter incunabula [MS incunabila] Christianismi. Neque enim sunt aliena huic loco."

106. Scaliger, *De emendatione,* 136.

107. Scaliger, *Epistolae,* 90; *Lettres,* 123.

108. Scaliger, *Lettres,* 123–124.

109. Scaliger's notion that the lengths of the months in his year match the duration of the sun's stay in the appropriate zodiacal signs had been discussed by Zarlino, *De vera anni forma* (Venice, 1580), 7. Scaliger presumably first derived the idea from the calendar of Dionysius discussed by Ptolemy in the *Almagest.* For his (undistinguished) contribution to the debate see the still standard work of F. Kaltenbrunner, *Die Polemik über die Gregorianische Kalenderreform* (Vienna, 1877), 70–76.

110. See the letter from Panvinio to Mercator in Mercator's *Chronologia* (Cologne, 1569).

111. G. Génébrard, *Chronographiae libri quatuor* (Paris, 1580), sig. * 5 recto; 69: "Quicquid sit, apud eum nomina et saecula regum Babylonicorum et Assyriorum videntur ficta aut corrupta, ut proinde ex eo nihil certo statui possit, nedum chronologia, ut isti faciunt, describi, nisi forte imaginaria."

112. Crusius, *De epochis,* 89–90.

113. M. Pattison, *Isaac Casaubon, 1559–1614* (London, 1875), 370 n7.

114. T. E. Mommsen, *Medieval and Renaissance Studies,* ed. E. F. Rice, Jr. (Ithaca, N.Y., 1959), chs. 12–14; B. Guenée, *Histoire et Culture historique dans l'Occident médiéval* (Paris, 1980), 148–154.

115. See M. Reeves, *The Influence of Prophecy in the Later Middle Ages* (Oxford, 1969).

116. I adapt the phrase from Guenée's "fièvre computistique": *Histoire,* 152–153.

117. Bucholzer, *Isagoge,* 45 recto, 52 verso–53 recto.

118. Goropius, *Origines*, 436–438.
119. A. Bouché-Leclercq, *L'astrologie grecque* (Paris, 1899); M. Screech, "The Magi and the Star (Matthew 2)," *Histoire de l'exégèse au XVIe siècle*, ed. O. Fatio and P. Fraenkel (Geneva, 1978), 385–409.
120. North, "Astrology"; M. Haeusler, *Das Ende der Geschichte in der mittelalterlichen Weltchronistik* (Cologne and Vienna, 1980), ch. 10.
121. P. d'Ailly, *Vigintiloquium de concordia astronomice veritatis cum theologia*, 19–20 (Venice, 1490), sigs. b 4 recto ff.
122. Ibid. *Elucidarium*, 16–17; sigs f2 recto–f3 recto; 29, sig. [f8 verso].
123. Tycho Brahe, *Opera omnia*, ed. I. L. E. Dreyer, 6 (Copenhagen, 1919), 160.
124. See G. J. Rheticus, *Narratio prima*, ed. and tr. H. Hugonnard-Roche et al. (Warsaw, 1982), 47–48, 155–156; N. Swerdlow, "Long-Period Motions of the Earth in *De revolutionibus*," *Centaurus* 24 (1980), 239–241. For Osiander's reaction see the fragment of a letter published by M. List in *Studia Copernicana* 16 (1978), 455–456.
125. G. Bock, *Thomas Campanella* (Tübingen, 1974), ch. 4.
126. Scaliger, *De emendatione*, 429; Casaubon has underlined part of the passage and written "No" beside it in his copy, British Library 582.1.9.
127. Ibid., 414: "DE ANNO SABBATICO Abiectis septenariis de annis mundi, relinquitur annus Hebdomadis currentis. Et sane mirum est, occulto quodam Dei opt. max. consilio in annis mundi eam observationem anni Sabbatici latere."
128. From the text cited in n104 above.
129. C. Vivanti, *Lotta politica e pace religiosa in Francia fra Cinque e Seicento* (Turin, 1963).
130. Later polemics on the calendar and other matters fully eradicated Scaliger's original, subtle ecumenism.

5. Protestant versus Prophet: Isaac Casaubon on Hermes Trismegistus

1. See in general F. Yates, *Giordano Bruno and the Hermetic Tradition* (London, 1964); E. Garin, *Ermetismo del Rinascimento* (Rome, 1988).
2. The most important accounts are E. Garin, *La cultura filosofica del Rinascimento italiano* (Florence, 1961), 143–154; Yates, *Bruno*; Garin, *Ermetismo*; and M. Pope, *The Story of Archaeological Decipherment* (London, 1975), 34–35.
3. The best general account is still M. Pattison, *Isaac Casaubon, 1559–1614*, 2d ed. (Oxford, 1892); other literature is listed by C. B.

Schmitt, "Theophrastus," *Catalogus translationum et commentariorum,* ed. F. E. Crantz et al. (Washington, D.C., 1960—), 2:262–263.
4. *Secunda Scaligerana,* s.v. CASAUBON; *Scaligerana,* ed. P. Desmaizeaux (Amsterdam, 1740), 2:259. For more modern judgments see Aeschylus, *Agamemnon,* ed. E. Fraenkel (Oxford, 1950), 1:36–38, 62–78; J. Glucker, "Casaubon's Aristotle," *Classica et Medievalia* 25 (1964), 274–296.
5. I. Casaubon, *Ephemerides,* ed. J. Russell (Oxford, 1850), 1:47.
6. Ibid., 7.
7. C. Vivanti, *Lotta politica e pace religiosa in Francia fra Cinque e Seicento* (Turin, 1963), pt. 2, ch. 4.
8. Pattison, *Casaubon,* 441, 104.
9. Casaubon's notes on the New Testament appeared in 1587; ibid., 475.
10. *B. Gregorii Nysseni ad Eustathiam, Ambrosiam et Basilissam epistola,* ed. I. Casaubon (Paris, 1606), 105–109.
11. Ibid., sig. i recto-verso, 91; *Gregorii Nysseni Epistulae,* ed. G. Pasquali (Berlin, 1925), xxxvii–xli, lxxiv.
12. *Historiae Augustae scriptores,* ed. I. Casaubon (Paris, 1603), *Emendationes ac Notae,* 3–4: "Satis enim mirum videtur nobis quod de Aelio Spartiano, Aelio Lampridio et Iulio Capitolino vulgati libri suggerunt: omnes hos tres cum sub Diocletiano cepissent florere, ad Constantini tempora durasse; omnes Imperatorum omnium vitas tempore eodem scribere aggressos (et quidem stylo ita parum dissimili, ut discrimen vix ullum liceat notare) pariter in opere instituto progressos, pariter desiisse"; cf. 177 and Prolegomena, sig. e ii verso.
13. Casaubon's notes are in his copy of *Poetae Graeci principes heroici carminis* (Geneva, 1566), now in Cambridge University Library, Adv.a.3.3.; pt. 1,726: "hic certe non fuit caecus"; 729: "Hom. hac voce nunquam sic est usus: nec Hom. est hoc carmen; nec ullius optimi poetae, nedum Hom." Estienne had already remarked in his printed notes that the poem's diction was not Homeric, and Casaubon had underlined the comment (pt. 2, 488).
14. Casaubon, *Ephemerides,* 2:928.
15. In the notes that he took on the chronicle of George Syncellus, now Bodleian Library MS Casaubon 32, fol. 53 verso, Casaubon had copied out a text ascribed to the Egyptian priest Manetho about the translation of hieroglyphic books written by Thoth, the first Hermes, into Greek after the Flood (Syncellus 40–41 M). Casaubon commented: "No. quomodo potest verum esse scriptum istud Trism. haec nugae." He certainly took these notes before leaving France

for England in 1610, and the comment thus seems to be his first explicit critical reference to the *Hermetica* and their claims.

16. *Mercurii Trismegisti Poemander, seu de potestate ac sapientia divina. Aesculapii definitiones ad Ammonem regem*, ed. A. Turnèbe (Paris, 1554); British Library 491.d.14. This edition is hereafter referred to as "*CH*, ed. Turnèbe"; since Casaubon's copy bears another owner's name and the date 1604, it was evidently a late acquisition.

17. Pattison, *Casaubon*, 428. For more recent studies of Casaubon's habits as a reader see A. Momigliano, "Un appunto di I. Casaubon dalle 'Variae' di Cassiodoro," *Tra Latino e Volgare: Per Carlo Dionisotti*, ed G. Bernardoni Trezzini et al. (Padua, 1974), 2:615–617; T. A. Birrell, "The Reconstruction of the Library of Isaac Casaubon," *Hellinga Festschrift* (Amsterdam, 1980), 59–68.

18. Here and elsewhere divisions of the text refer to the modern edition of the *Corpus Hermeticum*, ed. A. D. Nock, tr. A.-J. Festugière (Paris, 1945), 1–2. This edition is hereafter referred to as *CH*, ed. Nock.

19. *CH*, ed. Turnèbe, 20: "Ad baptismum alludit."

20. *CH*, ed. Turnèbe, 5: "No. si hoc verum est et hic scripsit ante Mosem per hunc non per Mosem Deus mysteria revelavit."

21. Yates, *Bruno;* D. P. Walker, *The Ancient Theology* (London, 1972), introduction.

22. Ibid., ch. 2; J. Harrie, "Duplessis-Mornay, Foix-Candale and the Hermetic Religion of the World," *Renaissance Quarterly* 31 (1978), 499–514.

23. *T. Lucretii Cari De rerum natura libri sex* (Lyons, 1576); Leiden University Library 755 H 9, 110: "stultissime Lucreti. putasne eandem esse eorum vermium [in 3.719ff.] et hominum animam?"

24. Ibid., 80: "No. nugas Epicur. quib. mundus est creatus a natura, temere, incassum, frustra. Quanto melius philosophus, *hē phusis matēn ouden poiei*."

25. I. Casaubon, *De rebus sacris et ecclesiasticis exercitationes xvi* (Geneva, 1654), 67: "quod apud Platonem, Aristotelem, Theophrastum, et tot alios e Paganis, curiosissimos omnium disciplinarum scrutatores, eorum quae hodie apud Mercurium istum aut Sibyllas miramur nullum penitus extat vestigium.'

26. Ibid., 66: "Ante omnia movet illud me; quod Verbo Dei contrarium videtur, existimare, tam profunda Mysteria Gentibus fuisse clarius proposita, quam illi Populo quem Deus Opt. Max. ut peculiariter suum dilexit . . ."

27. Ibid.: "Nam alioqui scimus quam exiles scintillae veritatis priscis temporibus Populo Iudaico illuxerint."

28. Ibid., 67: "Haec sacrae Scripturae testimonia et his similia, quî stare possunt, si verum est, pleraque et ea quidem praecipua doctrinae Christianae mysteria, etiam ante Mosem Gentibus fuisse proposita? Nam Mercurium Trismegistum ante Mosem vixisse, siquidem certa sunt quae de eo traduntur, nullum est dubium."
29. See also Pattison, *Casaubon*, 335.
30. *CH*, ed. Turnèbe, 6: "Joh. 12.46."
31. Ibid., 21: Casaubon underlines *philēsai ou dunasai* and *prōton to sōma* and writes "Ex Johanne." This is a better parallel than the last; see *CH*, ed. Nock, ad loc.
32. *CH*, ed. Turnèbe, 10, on the phrase *hoi sunodeusantes tēi planēi kai sugkoinōnēsantes tēi agnoiai*, in 11.12: "Saepe in N.T."; 56, on *katargētheis*: "Vox N.T."
33. *CH* 12.22; cf. 2.4–6.
34. *CH*, ed. Turnèbe, 12: "Nota locum Orig. in Celsum p. 329. Eleganter Dionys. Areop. p. 240 *huperkeitai tōn ousiōn* etc. et p. 243. *huperousiotēs* Dion. Areop. p. 239." Casaubon refers to *Dionysii Areopagitae opera quae extant* (Paris, 1561–1562), 1:240, 243, 239.
35. *CH*, ed. Turnèbe, 11: "No. ter vox *hagios*. vide caussam apud Damascenum pag. 520." For the text at issue see *Exercitationes*, 76.
36. Bodleian Library, MS Casaubon 52, fol. 99 recto: ". . . quia ea fuit simplicitas primorum temporum, ut titulos in quib. posteriora saecula lascivierunt (vel teste Plinio in praefatione) ignorarent, cuius rei argumentum est, quod libri orientalibus linguis scripti tales titulos etiam nunc ut plurimum nesciunt."
37. Casaubon, *Exercitationes*, 79: ". . . stylus huius libri alienissimus est a sermone illo quo Graeci Hermetis aequales sunt usi. Nam illa vetus lingua multa habuit vocabula, multas phrases, imo totius elocutionis ideam ab illa diversissimam qua posteriores Graeci sunt usi. Hic nullum penitus vestigium antiquitatis, nullus *chnous,* nullus *pinos tēs archaiotētos* qualem praestantissimi veterum Criticorum etiam in Platone observant, nedum in Hippocrate, Herodoto et aliis antiquioribus. Contra, multa hic vocabula, quae ne vetustior quidem Hellenismus agnoscat eo qui vigebat circa nativitatem Domini."
38. Ibid., 76: "Amat enim iste planus, simul cum doctrina sacra, verba quoque furari Sacrae Scripturae."
39. See esp. ibid., 72, on the word *homousios*.
40. *CH*, ed. Turnèbe, 19: "Plato in Tim. p. 477."
41. *CH*, ed. Turnèbe, 34: "Nihil perit rerum creatarum"; "Mundus *deuteros theos*."
42. Casaubon, *Exercitationes*, 77–79: "Vnicum adhuc proferam Dogma,

non Platonis quidem aut Platonicorum proprium, sed multorum Graecorum Sapientum commune, quod ita Aegyptius iste scilicet Mercurius tractat, ut purum putum ipsum fuisse philosophum, Graecorum disciplinis, non autem veteris Aegyptiorum sapientiae mysteriis imbutum vel ex eo possit intelligi. Fuit igitur opinio vetustissimorum Graecorum, nihil e rebus creatis perire sed mutari dumtaxat. Mortem, inane esse nomen sine re: nam revera mutationem esse, quam vulgus dicit mortem: quippe nihil creatum interire, sed formam aliam assumere. Hippocrates: . . . *apollutai men oun ouden hapantōn chrēmatōn, oude ginetai, hoti mē prosthen ēn. xummisgomena de kai dialuomena alloioutai* . . . Ad hoc igitur Graecorum Dogma saepe respicit fictus iste Hermes Aegyptius . . . Initio octavi sermonis rationem cur nihil pereat reddit hanc: quod mundus sit animal aeternum, Deus quidam secundus: omnia autem quae sunt in mundo partes esse illius: maxime autem hominem. Eandem rationem repetit in xii. ubi addit: *hē dialusis ou thanatos estin alla kramatos dialusis. dialuetai de ouch hina apolētai, all' hina nea genētai.* Sed totus liber hic esset transscribendus, si singula vellem persequi doctrinae capita, quae iste falsus Mercurius e Graecis philosophis in rem suam vertit: nam si excipias quae derivata sunt e fonte Scripturarum, omnia illius inde sunt."

43. *CH*, ed. Turnèbe, 97: "Phidias." Cf. Casaubon, *Exercitationes*, 79.
44. *CH*, ed. Turnèbe, 98: "Historia extat apud Strabonem, Clem. etc." Cf. *Exercitationes*, 79.
45. Reading *hē tōn aiguptiōn onomatōn dunamis*, as in Turnèbe's text.
46. *CH*, ed. Turnèbe, 90: "Lingua Aeg. Vide Jamblichum p. 150. 151." Casaubon refers to Iamblichus. *De mysteriis Aegyptiorum, Chaldaeorum, Assyriorum* (Lyons, 1549).
47. *CH*, ed. Turnèbe, 90: "No. *husteron,* Nae hic homo suaviludius est, qui haec scribat."
48. Casaubon, *Exercitationes*, 79: "Omnia *gnēsiōs* Graeca, et Hellenismo eius quam designavi aetatis sua constat ratio . . . in nono dicit, *eikotōs keklētai kosmos. kosmei gar ta panta.* an . . . *kosmos* et *kosmein* voces sunt antiqui sermonis Aegyptiaci?"
49. Ibid., 70: "Postremo, quum omnes scientiae atque disciplinae ab inventione Mercurii Trismegisti Aegyptii crederentur profectae, certatim omnes fere scriptores cunctarum scientiarum aut gloriam illius celebrarunt aut de eius nomine suis scriptis gratiam et famam quaesiverunt. Quid igitur miramur, primis illis seculis nascentis Christianismi, quum tanta licentia libri falsis titulis inscripti quotidie fingerentur, extitisse aliquem sacris nostrae pietatis leviter imbutum,

qui hoc ipsum in scientia Theologiae et rebus fidei sibi putarit tentandum?"

50. F. Purnell, Jr., "Francesco Patrizi and the Critics of Hermes Trismegistus," *Journal of Medieval and Renaissance Studies* 6 (1976), 155–178.
51. In *CH*, ed. Turnèbe, the section of the Latin version where Phidias is mentioned is set off by itself on 119–123, with a printed heading: "De animo ab affectione corporis impedito. Aesculapii esse non videtur."
52. Purnell, "Patrizi," 159–164.
53. *CH*, ed. Turnèbe, title page: "De hoc auctore quaedam Beroaldus in Chro. p. 23."
54. M. Beroaldus, *Chronicum, Scripturae Sacrae autoritate constitutum* (Geneva, 1575), 23 (Casaubon's working copy is British Library C.79.3.12[1]): "Quod vero aiunt Mercurium istum fuisse antiquiorem Pharaone, sicuti prodidit Suidas, falsum esse arguit ipsum Paemandri, quod ad illum refertur, scriptum. Nam Sibyllarum meminit quae multis post Pharaonem seculis extiterunt. Item Aesculapius is, ad quem scribit iste Mercurius, Phidiae meminit, qui Periclis aetate vixit. Ita librum illum Mercurii Trismegisti subdititium esse apparet, in quo multa dicuntur, quae nec ratione nec autoritate fulciantur idonea, qui Ilerdam mittendus sit cum suo autore." On the appearance of the Sibyl in Ficino's version of the *Corpus* see F. Purnell, Jr., "Hermes and the Sibyl: A Note on Ficino's *Pimander*," *Renaissance Quarterly* 30 (1977), 305–310; M. J. B. Allen, "The Sibyl in Ficino's Oaktree," *Modern Language Notes* 95 (1980), 205–210.
55. See Purnell, "Patrizi."
56. *Isaaci Casauboni notae*, 2d ed., in *Diog. Laert. De vitis, dogm. et apophth. clarorum philosophorum, libri x* (Geneva, 1593), 9: "Postremo, vel ipse stylus me in eam impellit sententiam, ut putem verissimam esse inscriptionem quam Michael Sophianus in scriptis codicibus reperit, *Mousaiou tou Grammatikou*. Nam siquis conferat (quod nos fecimus aliquando diligenter) id poema cum aliorum recentiorum poetarum Graecorum, maxime Nonni, qui Dionysiaca conscripsit, carminibus: non dubitabit hanc sententiam esse verissimam."
57. A. Grafton, "Rhetoric, Philology and Egyptomania in the 1570s ...," *Journal of the Warburg and Courtauld Institutes* 42 (1979), 183n11.
58. Sixtus Senensis, *Bibliotheca sancta* (Venice, 1566), 76.
59. H. Rosweyde, *Lex talionis XII. Tabularum Cardinali Baronio ab Isaaco Casaubono dicta* (Antwerp, 1614), praef., sig. [*8 recto].
60. J. C. Bulenger, *Diatribae ad Isaaci Casauboni exercitationes adversus*

illustrissimum Cardinalem Baronium (Lyons, 1617), 81: "Iudicium Illustriss. Baronii de Mercurio Trismegisto multi ex veteribus et recentioribus probant. Tuum tamen de eodem non improbo. Recentiorem illum auctorem fuisse oportet, qui Phidiae et Eunomii Locri Citharoedi meminerit, et trophaea, per quercum vaticinia, prytanea commemoret. De Sibyllis non persuades."

61. Rosweyde, *Lex talionis,* 177–193; Bulenger, *Diatribae,* 81–84; A. Eudaemon-Joannes, *Refutatio exercitationum Isaaci Casauboni* (Cologne, 1617), 36–47; cf. J. Cappell, *Vindiciae pro Isaaco Casaubono* (Frankfurt, 1619), cols. 19–25.

62. G. J. Vossius, *De theologia gentili et physiologia Christiana,* 1.10 (Frankfurt, 1668), 1:75; T. Stanley, *Historia philosophiae,* 4.4 (Leipzig, 1711), 287; J. Marsham, *Chronicus canon Aegyptiacus, Ebraicus, Graecus et disquisitiones* (London, 1672), 234–235; H. Witsius, *Aegyptiaca,* 2.5.6 (Amsterdam, 1683), 94–95.

63. H. Conring, *De Hermetica Aegyptiorum vetere et Paracelsicorum nova medicina liber unus* (Helmstedt, 1648), 46: "Ex his omnibus manifestum est id quod diximus, ea quae uno Poemandri nomine princeps edidit Marsilius Ficinus et vertit Latine, variae notae esse et variorum scriptorum; et quidem alia videri a Christianis, alia a Platonicis supposita, et Hermetis nomine conficta, paucissima redolere Aegyptiacam veterem doctrinam. Etsi enim Plato ipse fortassis theologiam suam debeat Aegyptiis ... tamen propria multa Aegyptus in sacris habuit quae Platoni placuisse non videntur. Nec tamen vel illa pauca, quae diximus prae aliis Aegyptiaca esse, aut ipsius antiquissimi Mercurii aut omnino veteris alicuius sunt scriptoris. Rectissime enim Casaubonus de omnibus illius libri opusculis: ... ."

64. I. Vossius, *De Sibyllinis aliisque quae Christi natalem praecessere oraculis* (Oxford, 1680), 39–40: "sed vero ita omnino credimus, omnes istos quos diximus libros, vergentibus ad finem Danielis hebdomadis, a Judaeis toto orbe dispersis fuisse compositos, Deo impellente ipsorum mentes ad significandum gentibus Christi adventum. Infinita itaque illi edidere volumina partim sub Patriarcharum et Prophetarum suorum nominibus, quales fuere libri qui olim lecti fuere sub nominibus Adami, Enochi, Abrahami, Moysis, Eliae, Esaiae et Jeremiae; partim vero sub nominibus illorum, quorum magna apud gentiles esset existimatio, veluti Hystaspis, Mercurii Trismegisti, Zoroastris, Sibyllarum, Orphei, Phocyllidae et complurium aliorum."

65. Walker, *Ancient Theology,* 241.

66. Casaubon, *Notae,* 16–17, 7.

67. J. Brucker, *Institutiones historiae philosophicae* (Leipzig, 1756), 73:

"... quamquam vix negari posse videtur, nonnullas iis inesse veteris doctrinae Aegyptiacae reliquias, quae tamen a spuriis discerni non possunt, nec ad *gnēsiotēta* librorum istorum quicquam faciunt."

68. A. Kircher, *Prodromus Coptus sive Aegyptiacus* (Rome, 1636), 172: "Erat autem tunc, quantum quidem coniectura assequi possum, ea inter Aegyptiorum vernaculam et Graecam distinctio, quae modo inter Italicam, Hispanicam aut etiam Gallicam et Latinam; adeo ut nec Graecis Aegyptiam, nec Aegyptiis Graecam linguam addiscere esset difficile."

69. Ibid., 174: "Quid enim aliud Osiris, nisi *hosios hiros*, Sacrosanctus? Quid aliud Isis nisi prudens naturae progressus, *apo tou iesthai*, hoc est ab eundo deducto nomine?"

70. Ibid., 185.

71. Ibid., 264: "Quid aliud per alatum globum nisi circulus ille, cuius centrum ubique, circumferentia nusquam, ut Trismegistus loquitur, denotatur; videlicet intellectus ille abstractus, supramundanus, mens prima, Pater inquam coelestis...?"

72. Quoted by E. Iversen, *Obelisks in Exile*, 1: *The Obelisks of Rome* (Copenhagen, 1968), 92.

73. Brucker, *Institutiones*, 73: "Quod ita evenisse testatur IAMBLICHVS: Solitos fuisse sacerdotes Aegypti sui ingenii foetus Mercurio inscribere, et sapientiam suam illi tribuere: quod et fastum et fraudem horum hominum prodit, et quo pacto tot libri Hermetici Aegyptus parturiverit, manifestat: verisimile autem serius id, et potissimum invalescente in Aegypto philosophia Pythagorico-Platonica, sub Ptolemaeis accidisse. Cum enim prisca Mercurii primi monumenta theogonias adumbrarent, facile fuit, amissa clave, cuilibet sui ingenii commenta sub hac larva vendere et imprimis in rationes physicas convertere. Et huius quoque generis sunt libri Hermetici, inter quos Paemander et Asclepius eminent, qui hodie feruntur, qui pleni sunt philosophiae Alexandrinae somniis..."

74. *Epistolae Ho-Elianae*, 4.43; ed. J. Jacobs (London, 1890), 632, 629. The learned historian of the Druids Elias Schedius also believed; see his *De diis Germanis... syngrammata quatuor* (Amsterdam, 1648), 109–110.

75. P. E. Jablonski, *Pantheon Aegyptiorum* (Frankfurt, 1750–1752), Prolegomena, xcvii–xcviii.

76. Ibid., 1:43–44.

77. Yates, *Bruno*, 401, and cf. Chapter 6 below.

78. I. Casaubon, *De satyrica Graecorum poesi et Romanorum satira libri duo*, 1.2; ed. J. J. Rambach (Halle, 1774), 27: "ORPHEVS, cuius hodie

extant hymni, poeta et ipse procul dubio perantiquus, etsi non est, quod facile probari queat, theologus ille vetustissimus, de quo tam multa Graeci . . ."
79. Bodleian Library, MS Casaubon 60, fol. 247 recto: "Eum libellum etsi non putamus eius quam prae se fert antiquitatis: credimus tamen admodum antiquum, et plane eiusmodi, quem legisse non poeniteat studiosum aut Gr. linguae aut antiquitatis." By the time he finished taking these notes, Casaubon had changed his mind for the worse, deciding that "antiquiss. est auctor" and that "ille ipse est qui rebus gestis interfuit, quae hic memorantur" (ibid., fol. 253 recto). In the *Exercitationes* Casaubon cites Aristeas as an old and authoritative source for Jewish customs in general and the actions of the Seventy Translators in particular: "Narrat Aristeas vetustissimus scriptor Septuaginta Seniores, qui vertendis libris sacris Alexandriae operam posuerunt, solitos prius quam ad lectionem se accingerent, manus lavare . . ." (536).
80. Casaubon, *Exercitationes,* 66: "Nam equidem ingenue profiteor, omnia huius generis sive Oracula sive Enuntiata, quo apertiora sunt, eo fieri mihi suspectiora."

6. The Strange Deaths of Hermes and the Sibyls

1. See e.g. P. Burke, *The Renaissance Sense of the Past* (New York, 1969), 55–59. W. Setz synthesizes earlier research on and provides a new text of *Lorenzo Vallas Schrift gegen die Konstantinische Schenkung* (Tübingen, 1975); for an interesting recent analysis see V. de Caprio, "Retorica e ideologia nella *Declamatio* di Lorenzo Valla sulla donazione di Costantino," *Paragone* 29, no. 338 (1978), 36–56. For Valla's attack on Dionysius see his *Collatio Novi Testamenti,* ed. A. Perosa (Florence, 1970), 167–168, on Acts 7.22–23.
2. For the *Ad Herennium* see A. Grafton and L. Jardine, *From Humanism to the Humanities* (London, 1987), 26–27; for Aristotle's *Theology,* J. Kraye, "The Pseudo-Aristotelian *Theology* in Sixteenth- and Seventeenth-Century Europe," in *Pseudo-Aristotle in the Middle Ages: The Theology and Other Texts,* ed. J. Kraye et al. (London, 1986), 265–286; for Seneca, L. Panizza, "Biography in Italy from the Middle Ages to the Renaissance: Seneca, Pagan or Christian?," *Nouvelles de la République des Lettres* 2 (1984), 47–98.
3. See the important recent synthesis of P. G. Schmidt, "Kritische Philologie und pseudoantike Literatur," *Die Antike-Rezeption in den*

Wissenschaften während der Renaissance, ed. A. Buck et al. (Weinheim, 1983), 117–128.
4. See the classic study by P. Lehmann, *Pseudo-Antike Literatur des Mittelalters* (Leipzig and Berlin, 1927; repr. Darmstadt, 1964) and the informative recent survey by G. Constable, "Forgery and Plagiarism in the Middle Ages," *Archiv für Diplomatik, Schriftgeschichte, Siegel- und Wappenkunde* 29 (1983), 1–41.
5. See in general E. N. Tigerstedt, "Ioannes Annius and *Graecia Mendax*," *Classical, Mediaeval and Renaissance Studies in Honor of Berthold Louis Ullman,* ed. C. Henderson, Jr. (Rome, 1964), 2:293–310.
6. For the supplement to Curtius see P. G. Schmidt, *Supplemente lateinischer Prosa in der Neuzeit* (Göttingen, 1964), 16–17 and inserted Nachtrag; for Horapollo see Scaliger's *Opus novum de emendatione temporum* (Paris, 1583), 134 (3rd ed. [Geneva, 1629], 196) and *Secunda Scaligerana* s.v. ORUS: "ORUS est bon"; ed. P. Desmaizeaux (Amsterdam, 1740), 480; for Hermes see *Prima Scaligerana* s.v. PHILO: "PHILO Judaeus, mirabilis Auctor est et lectione dignissimus. Mirabilior et Antiquissimus *Pimander* Mercurii Trismegisti" (ibid., 136). Scaliger also collaborated on an edition of the Hermetic Corpus in 1574, though his part in it has not been clearly defined.
7. G. Fowden's *The Egyptian Hermes* (Cambridge, 1986) surveys (and largely supersedes) the extensive earlier literature on the Hermetic Corpus.
8. F. A. Yates, *Giordano Bruno and the Hermetic Tradition* (London, 1964).
9. See Chapter 5 above.
10. See e.g. Burke, *Sense of the Past,* 62–63.
11. Schmidt, "Kritische Philologie," 119–120; W. Speyer, *Die literarische Fälschung im heidnischen und christlichen Altertum* (Munich, 1971), 112–128.
12. For Diogenes see Kraye, "The *Theology*"; for the commentators see e.g. Ammonius, *In Aristotelis De interpretatione commentarius,* ed. A. Busse, Commentaria in Aristotelem Graeca, 4.5 (Berlin, 1897), 5–7; Philoponus (olim Ammonius), *In Aristotelis Categorias Commentarium,* ed. A. Busse, Commentaria in Aristotelem Graeca, 13.1 (Berlin, 1898), 7; both cited and discussed by Speyer, *Fälschung,* 125. For a more detailed study see C. W. Müller, "Die neuplatonischen Aristoteleskommentatoren über die Ursachen der Pseudepigraphie," *Rheinisches Museum für Philologie,* N.F. 112 (1969), 120–126, repr. in *Pseudepigraphie in der heidnischen und jüdisch-christlichen Antike,* ed. N. Brox (Darmstadt, 1977), 264–271.

13. For a detailed and critical study see W. D. Smith, *The Hippocratic Tradition* (Ithaca, N.Y., and London, 1979), chs. 2–3. For Galen and Lucian see G. Strohmaier, "Übersehenes zur Biographie Lukians," *Philologus* 120 (1976), 117–122.
14. Galen 16.5 K.
15. E.g. Galen 15.105 K.; see Speyer, *Fälschung,* 112; Smith, *Hippocratic Tradition,* 201; and C. W. Müller, *Die Kurzdialoge der Appendix Platonica* (Munich, 1975), 12–16, who emphasizes the hypothetical and unsatisfactory character of Galen's explanation. But discussion is by no means closed. Despite their divergences *de omni re scibili et quibusdam aliis,* J. E. G. Zetzel and S. Timpanaro converge in taking Galen's testimony as historical; see respectively *"Emendaui ad Tironem,"* *Harvard Studies in Classical Philology* 77 (1973), 227–245, and *Per la storia della filologia virgiliana antica* (Rome, 1986), 34n28.
16. Speyer, *Fälschung,* 120n7, 321; Smith, *Hippocratic Tradition,* 172–175. Ancient forgers also knew how to use scholars' tools and language to give their wares the appearance of authenticity. Thus Aristeas uses Alexandrian critical standards to validate his argument that the Hebrew texts of the Bible are inexact and corrupt (30–31)—see E. Bickerman, *Studies in Jewish and Christian History,* 1 (Leiden, 1976): 228–229.
17. J. Wowerius, *De polymathia tractatio* 16; *Thesaurus Graecarum Antiquitatum,* ed. Jac. Gronovius, 10 (Leiden, 1701): cols. 1067–1074. See also the still more elaborate collection of pagan and Christian evidence in J. Clericus, *Ars Critica,* 4th ed. (Amsterdam, 1712), 2:302–323 (3.2.1).
18. Hieronymus, *Opera,* I: *Opera exegetica,* 5: *Commentariorum in Danielem libri III <IV>*, Corpus Christianorum, series Latina, 75 A (Turnhout, 1964), prologus, 771–775 = Porphyry, *"Gegen die Christen,"* 15 *Bücher: Zeugnisse, Fragmente und Referate,* ed. A. von Harnack, Abhandlungen der Königlich Preussischen Akademie der Wissenschaften, phil.-hist. Kl., 1916, frag. 43, 67–68. For Porphyry see in general J. Geffcken, *The Last Days of Greco-Roman Paganism,* tr. S. MacCormack (Amsterdam, 1978), 56–81.
19. See the new text and translation of this work, with elaborate introduction and commentary, by N. de Lange, *Sources Chrétiennes* 302 (Paris, 1983), 471ff.
20. Africanus, Letter to Origen, 9; ibid., 520–521.
21. Ibid., 4–5; 516–517—a point also made by Porphyry ap. Hieron. *Opera* 1,5:773.
22. Africanus, Letter to Origen, 6; 518–519.

23. Ibid., 3; 514–517.
24. The exact extent of Porphyry's debt to Christian scholars remains problematic; see P. M. Casey, "Porphyry and the Origin of the Book of Daniel," *Journal of Theological Studies*, n.s. 27 (1976), 15–33, and R. L. Wilken, *The Christians as the Romans Saw Them* (New Haven and London, 1984), ch. 6.
25. For the diffusion of Africanus's correspondence with Origen see de Lange's introduction, p. 508. Campanella saw Bodin's critique of modern efforts to use Daniel's four empires as the basic division of world history as a revival of Porphyry's ideas; see his *Articuli Prophetales*, ed. G. Ernst (Florence, 1977), 233. Heinsius took Africanus's letter as part of the inspiration for his own effort to reconstruct the Oriental character of New Testament Greek. See his *Aristarchus Sacer* (Leiden, 1627), prolegomena, fol. *****IV.
26. For a case study see A. Grafton, "Appendix: The Availability of Ancient Sources," *Cambridge History of Renaissance Philosophy*, ed. C. B. Schmitt et al. (Cambridge, 1988).
27. For the context—and the vital but sometimes indirect influence— of Estienne's *Thesaurus* see F. L. Schoell, *Etudes sur l'humanisme continental en Angleterre* (Paris, 1926), 140–159; A. C. Dionisotti, "From Stephanus to Du Cange: Glossary Stories," *Revue d'Histoire des Textes* 14–15 (1984–1985 [1986]), 303–336.
28. See Chapter 5 above.
29. Ibid., n37.
30. *Vita Plotini* 16; *Plotinus*, ed. and tr. A. Armstrong (Cambridge, Mass., and London, 1966), 1:44–45.
31. The whole text of *De mysteriis* tries to present itself as a letter from the Egyptian priest Abammon replying to doubts and questions raised by Porphyry; for the occurrence of "the language of the philosophers" in supposedly Egyptian Hermetica see 8.4. Fowden, *Hermes*, treats the text as the work of Iamblichus; see 131–141.
32. Iamblichus, *De mysteriis* 8.4.
33. Fowden, *Hermes*.
34. *Civitas Dei* 8.22–27, 39; cf. Fowden, *Hermes*, 209–210.
35. Fowden, *Hermes*, ch. 8.
36. Ibid., ch. 7.
37. Ibid., chs. 1–2; for Dictys and "Punic"—not the sort of parallel cited by Fowden—see E. Champlin, "Serenus Sammonicus," *Harvard Studies in Classical Philology* 85 (1981), 189–212.
38. The blurb appears on the verso of a first leaf, the recto of which is blank, in the Treviso 1471 ed. of Ficino's Latin: "Tu quicunque es

qui haec legis, sive grammaticus sive orator seu philosophus aut theologus, scito: Mercurius Trismegistus sum, quem singulari mea doctrina et theologia [ed.: theologica] Aegyptii prius et barbari, mox Christiani antiqui theologi, ingenti stupore attoniti admirati sunt. Quare si me emes et leges, hoc tibi erit commodi, quod parvo aere comparatus summa te legentem voluptate et utilitate afficiam." It is by F. Rolandello; see E. Garin, *Ermetismo del Rinascimento* (Rome, 1988).

39. A Vergicius, ep. ded., *Mercurii Trismegisti Poemander* (Paris, 1554), fol. alpha 2v; *Hermetis Trismegisti Poemander,* edited by G. Parthey (Berlin, 1854), viii.

40. F. Patrizi da Cherso, *Lettere ed opuscoli inediti,* ed. D. Aguzzi Barbagli (Florence, 1975), 175-177.

41. F. de Foix de Candale, ep. lectori, *Mercurii Trismegisti Pimandras utraque lingua restitutus* (Bordeaux, 1574), +iiiiv; *Poemander,* ed. Parthey, xiii: "... hic Mercurii libellus philosophiam Divino permiscet oraculo."

42. *Hermetica,* ed. W. Scott, 4 (Oxford, 1936), 245; cf. Fowden, *Hermes,* 9, and N. G. Wilson, *Scholars of Byzantium* (London and Baltimore, 1983), 158-159.

43. Cf. in general D. P. Walker, *The Ancient Theology* (London, 1972), ch. 3, and *The Prefatory Epistles of Jacques Lefèvre d'Étaples and Related Texts,* ed. E. F. Rice, Jr. (New York and London, 1971), epp. 9, 43, 55, 64, 133 (which reveal some interesting ambivalences).

44. See Chapter 5 above and G. Cozzi, *Paolo Sarpi tra Venezia e l'Europa* (Turin, 1979), 3-133.

45. I. Casaubon, *Ephemerides,* ed. J. Russell (Oxford, 1850), 1:21.

46. Ibid., 129-130.

47. Cozzi, *Sarpi,* 131.

48. Erasmus, *Opera omnia,* ed. J. Clericus (Leiden, 1703-1706), 4, col. 587.

49. Erasmus, *Lingua,* quoted by C. Christ-von Wedel, *Das Nichtwissen bei Erasmus von Rotterdam* (Basel, 1981), 117.

50. Erasmus, *Declarationes ad Censuras Facultatis Theologiae Parisiensis, Opera Omnia,* ed. Clericus, 9:917: "Tum enim piis etiam viris persuasum erat, Deo gratum esse, si tali fuco populus ad aviditatem legendi provocaretur." For Seneca and St. Paul see A. Momigliano, *Contributo alla storia degli studi classici* (Rome, 1955), 28-29; for Vives and Aristeas see R. Pfeiffer, *Geschichte der Klassischen Philologie: Von den Anfängen bis zum Ende des Hellenismus,* 2nd ed. (Munich, 1978), 129n84.

51. Scaliger to Casaubon, 30 October (Julian) 1605, in Scaliger, *Epistolae,* ed. D. Heinsius (Leiden, 1627), 303–304: "Callisthenem illum nunquam vidi. et quia a Pseudogurionide citatur, omnino Latinum fuisse, non Graecum, quem ille vidit, necesse est. Nam Graecismi imperitum eum vincunt scripta eius. Istiusmodi *hupobolimaiōn* scriptorum monstra olim fuerunt, in quibus Dares Phyrgius, Dictys Cretensis, qui hodie Latini exstant. eos etiam Graeci habuerunt, quorum fragmenta in Cedreno exstant. Quid dicas de Epitome Iliados Epica Latina? quam ridicule Pindarum Thebanum proscripserunt? Quid Aristaeas ille, quam antiquus est, ut etiam a Iosepho citatur? quod est *tōn hellēnistōn Ioudaiōn paregcheirēma.* Quid Ecataeus de Iudaeis, quem ab iisdem Hellenistis antiquitus confictum fuisse, manifesto ex Origene colligitur? Quid Pseudosibyllina oracula, quae Christiani Gentibus obiiciebant, quum tamen e Christianorum officina prodiissent, in Gentium autem bibliothecis non reperirentur? Adeo verbum Dei inefficax esse censuerunt, ut regnum Christi sine mendaciis promoveri posse diffiderent. Atque utinam illi primi mentiri coepissent." For Scaliger's attack on Aristeas see J. C. H. Lebram, "Ein Streit um die Hebräische Bibel und die Septuaginta," *Leiden University in the Seventeenth Century: An Exchange of Learning,* ed. T. H. Lunsingh Scheurleer (Leiden, 1975), 21–63; for Phocylides see *The Sentences of Pseudo-Phocylides,* ed. P. W. van der Horst (Leiden, 1978), 4–6. See also H. J. de Jonge, "J. J. Scaliger's *De LXXXV canonibus apostolorum diatribe,*" *Lias* 2 (1975), 115–124, 263. And for more general discussion of Scaliger and Erasmus see A. Flitner, *Erasmus im Urteil seiner Nachwelt* (Tübingen, 1952), 94–105; B. Mansfield, *Phoenix of His Age* (Toronto, 1979), 121–131; and the exhibition catalogue *Erasmus en Leiden* (Leiden, 1986), esp. 30–45.
52. *De divinatione* 2.54.110–112.
53. For Virgil see R. G. M. Nisbet, "Virgil's Fourth Eclogue: Easterners and Westerners," *Bulletin of the Institute of Classical Studies* 25 (1978), 59–78; A. Wallace-Hadrill, "The Golden Age and Sin in Augustan Ideology," *Past and Present* 95 (1982), 19–36. For Constantine see his *Oratio ad sanctorum coetum* 18–19.
54. *Sibyllinorum Oraculorum libro octo,* ed. X. Betuleius (Basel, 1545), 7; *Oracula Sibyllina,* ed. Alexandre (Paris, 1841), vii: "Quae cum caeca gentilitas intelligeret minime, pro insanis fuerunt habitae."
55. Ibid., 7 (vii): "Non inquam illis Enthusiasmus fuit furibundus, id quod vel ex Acrostichis animadvertere est . . ."
56. Ibid., 7 (vii): "Non vendico illis propheticam maiestatem, quam veneramur in Hebraeorum prophetis . . . Sed interim quorundam pro-

phetarum voces clare expressas dissimulare non possumus. An non audimus Zachariam manifeste in his versiculis... An non prorsus idem cum Osea dicit... Quid quod eandem civitatem coelestem describit, cuius *ekphrasis* speciosa in Apocalypsi legitur." On Betuleius see J. A. Parente, Jr., *Religious Drama and the Humanist Tradition* (Leiden, 1987), 91–93.

57. See Geffcken's ed. of the *Oracula* (1902), xii.
58. *Sibyllinorum oraculorum libri viii,* ed. S. Castellio (Basel, 1555); Alexandre xii: "Sed sunt nonnulli, quibus haec oracula nimis aperta videantur: ideoque ficta putent ab aliquo Christiano..."
59. Ibid.: "Nonne quae de Christo gentilibus praedicta sunt, ea clariora esse oportuit, quod Mose et caetera disciplina carebant, quae eis ad Christi lumen quasi praeluceret: ut quod hic deerat, id oraculorum perspicuitate compensaretur?"
60. Ibid., xiii: "haec perspicuitas eo valet, ut vera esse, non falsa appareat."
61. *Monumenta S. Patrum orthodoxographa* (Basel, 1569), fol. a5v: "Atqui multa sunt in his Sibyllinis carminibus de Christo dicta, quae et illustriora sunt quibusdam locis Prophetarum, et historiam magis de exhibito Christo, quam de venturo vaticinium, sapiunt. Vnde colligo non esse vetustissimum hoc scriptum. Sed licet, ut loquitur Scriptura, unicuique abundare suo sensu in hac re."
62. Goertz 4159, 207: "Alius Sibyllae haec vident[ur]." Cf. 151 on 3.809ff.: "Sibylla ipsa se negat Erythraeam"; 238 on 8.216ff.: "Hoc loco Acrostichis emendatior videtur ea, quae infra legitur 265."
63. Ibid., 88 on 2.339ff.: "Deplorat sua peccata et dein horum veniam poscit suppliciter Sibylla"; 220 on 7.151ff.: "Sibylla scelera sua fatetur et detestatur." On 320 the annotator summarizes Castellio's argument that the Sibyl started her career in the Ark and that her reappearances over a life span of several hundred years had given rise to the legend that there were many Sibyls: "Sibylla nurus Noae"; "Caussa, cur plures Sibyllae numeratae sint."
64. I. Temporarius, *Chronologicarum demonstrationum libri tres* (Frankfurt, 1596), 13–19; 2nd ed. (La Rochelle, 1600), 14–21. In the first ed. Temporarius sums up thus: the Oracula "carmina videntur ab homine pio concinnata fuisse, quibus inseruerit quaecunque Sibyllina ex autorum variorum lectione conquirere potuit." His model in using the Sibyl as a historical source is perhaps Josephus *Jewish Antiquities* 1.118.
65. See *Oracula,* ed. Geffcken, xi–xii.
66. *Sibyllina Oracula,* ed. Opsopoeus (Paris, 1599), praef.; Alexandre, xxvii: "Deinde exemplar Sibyllarum ab Aemaro Ranconeto praeside

suo tempore doctissimo alicubi emendatum, quaternionem unum m.s. . . . Nic. Turnebus . . . communicavit."

67. The working copy (of Betuleius's 1545 ed.) and the quaternio form part of a Gennadius Library Sammelband (G B/C 2152); I owe this information to F. R. Walton.

68. Grafton, "Rhetoric, Philology," 182–183.

69. Opsopoeus, praef; Alexandre, xxv: "Esayas indefinite praedixit, Ecce virgo pariet puerum. At Sibylla nominatim, Ecce virgo Maria pariet puerum Iesum in Bethlehem. Quasi vero Prophetae minus instinctu divino agitati futura praedixerint quam Sibyllae . . ."

70. Ibid. (ibid., xxiv–xxv): "Venio nunc ad res ipsas in Sibyllinis expositas, quarum pleraeque adeo plene, copiose, et aperte describuntur, ut vel inde tantum coniicere liceat, illas non praedictas, sed prius visas, auditas, aut lectas, ita perspicue recitatas fuisse."

71. Ibid. (ibid., xxiii): "ita concinne denique et perspicue omnia referuntur, ut historiae potius quam oraculi, animi sedati potius quam furoris, perspicuitatis potius quam obscuritatis indicia de se praebeant" Opsopoeus's balanced phrases reveal his debt to Cicero clearly, despite his practice of elegant variation in the actual wording.

72. *Epistolae Ho-Elianae: The Familiar Letters of James Howell*, ed. J. Jacobs (London, 1890), 629 (ep. 4.43). Even if this letter, like others in the collection, is partly or entirely a fiction, the attitude it expresses remains of interest.

73. Ibid., 630.

74. Cf. Schmidt, "Kritische Philologie," for a different view. I follow E. Bickerman, *Studies in Jewish and Christian History* 3 (Leiden, 1986), 196–197.

75. W. King, *Dialogues of the Dead*, 7: "Chronology," *A Miscellany of the Wits*, ed. K. N. Colville (London, 1920), 61–62.

76. Cf. A. Grafton, "Sleuths and Analysts," *TLS* (8 August 1986), 867–868, and the Introduction above.

7. Humanism and Science in Rudolphine Prague: Kepler in Context

1. J. Kepler, *Gesammelte Werke* (Munich, 1937–), 19:350.
2. R. J. W. Evans, *Rudolf II and His World* (Oxford, 1973), 156.
3. On Goldast see B. Hertenstein, *Joachim von Watt (Vadianus), Bartholomäus Schöbinger, Melchior Goldast* (Berlin and New York, 1975), 115–135.
4. L. P. Smith, *Life and Letters of Sir Henry Wotton* (Oxford, 1907), 2:205–206.

5. Ibid., 2:205n4.
6. Evans, *Rudolf II;* T. D. Kaufmann, *L'école de Prague* (Paris, 1985). See also N. Mout, *Bohemen en de Nederlanden in de zestiende eeuw* (Leiden, 1975).
7. Kepler, *Ges. Werke*, 19:350.
8. F. Seck, "Johannes Kepler als Dichter," *Internationales Kepler-Symposium Weil der Stadt 1971*, ed. F. Krafft et al. (Hildesheim, 1973), 427–450.
9. N. Jardine, *The Birth of History and Philosophy of Science* (Cambridge, 1984).
10. Kepler, *Ges. Werke*, 19:328.
11. F. J. Stopp, *The Emblems of the Altdorf Academy* (London, 1974).
12. See Seck, "Kepler als Dichter," for an exemplary treatment of these aspects of Kepler's writing.
13. Earlier technical treatments of Kepler's work are listed in the bibliography to B. Stephenson's elegant analysis of *Kepler's Physical Astronomy* (New York, 1987), 206–208, which supersedes most of them. Much biographical detail can be found in the late E. Rosen's characteristically learned and cranky *Three Imperial Mathematicians* (New York, 1986).
14. See D. C. Allen, *The Legend of Noah* (Urbana, 1949).
15. A. Grafton, "The World of the Polyhistors: Humanism and Encyclopedism," *Central European History* 18 (1985), 31–47.
16. J. Kepler, *The Six-Cornered Snowflake*, ed. and tr. C. Hardie (Oxford, 1966), 6–7. For Wacker and Bruno see Evans, *Rudolf II*, 232.
17. Kepler, *Ges. Werke*, 13:330.
18. Heraclitus *Problemata Homerica* 53; cf. Plutarch *De audiendis poetis* 19E.
19. Kepler, *Ges. Werke*, 14:45.
20. See in general the excellent survey of T. Bleicher, *Homer in der deutschen Literatur (1450–1740)* (Stuttgart, 1972). For Crusius's own view see *Diarium Martini Crusii 1598–1599*, ed. W. Göz and E. Conrad (Tübingen, 1931), 201–202. He says of his commentary: "posui ubique doctrinas Ethicas, Oeconomicas, Politicas, Physicas, etc. . . . Ex his commentariis demum intelligeretur: quantus sit Poëta Homerus, quanta sapientia."
21. *Anneus Lucanus cum duobus commentis Omniboni et Sulpitii* (Venice, 1505), fols. 23–24.
22. Kepler, *Ges. Werke*, 13:393.
23. Kepler, *Ges. Werke*, 13:132–133. For a brief discussion of Kepler's views on Nigidius and Lucan see F. Boll, *Sphaera* (Leipzig, 1903), 362n1.

24. Kepler, *Ges. Werke*, 13:134–135.
25. Ibid., 13:148.
26. Ibid., 13:158.
27. Ibid., 14:46.
28. M. *Annaei Lucani Belli civilis libri decem*, ed. A. E. Housman (Oxford, 1950), "Astronomical Appendix," 325–337; R. J. Getty, "The Astrology of P. Nigidius Figulus (Lucan I, 649–65)," *Classical Quarterly* 35 (1941), 17–22.
29. Kepler, *Ges. Werke*, 14:45.
30. Ibid., 19:329.
31. Ibid., 13:127.
32. P. Crusius had already pointed out that Xerxes' crossing into Greece in 480 B.C. could not have been accompanied by a solar eclipse; see his *Liber de epochis seu aeris temporum et imperiorum*, ed. I. T. Freigius (Basel, 1578), 57: "Non convenit synchronismus. Nam eclipsis illa [of 481 B.C.] toto anno praecessit expeditionem seu traiectum Xerxis in Europam." This did not stop Joseph Scaliger from redating the crossing to Ol. 74, 3, on the basis of the synchronism between it and a solar eclipse attested by Herodotus. He concluded: "Beavit tamen nos Herodotus, qui characterem apposuit, ex quo illum annum dignosceremus." *Opus novum de emendatione temporum*, (Paris, 1583), 222–223.
33. Scaliger, *De emendatione*, 15, 47.
34. Kepler, *Ges. Werke*, 15:208.
35. Scaliger, *De emendatione*, 15.
36. Kepler, *Ges. Werke*, 15:209.
37. See Kaufmann, *L'école de Prague*, 217.
38. W. Lazius, *De aliquot gentium migrationibus . . . libri xii* (Basel, 1572), 23.
39. M. von Aitzing, *Pentaplus regnorum mundi* (Antwerp, 1579); see G. N. Clark, *War and Society in the Seventeenth Century* (Cambridge, 1958), 134–140.
40. Evans, *Rudolf II*, 280n4. Another example is C. Leovitius's book *De coniunctionibus magnis* (1564); ibid., 221. This was dedicated to Maximilian II.
41. Kepler, *Ges. Werke*, 1:183.
42. Ibid., 1:398.
43. Ibid., 1:330. I borrow the translation of Jardine, *Birth of History and Philosophy of Science*, 277; his discussion of the passage is excellent.
44. Jardine, 279.
45. Kepler, *Ges. Werke*, 1:331.
46. See M. Caspar's comments in Kepler, *Ges. Werke*, 1:445, 462–464.

47. L. Valla, "Oratio in principio sui studii 1455," in M. Baxandall, *Giotto and the Orators* (Oxford, 1971), 177; tr. and discussed ibid., 119–120. Velleius's doctrine is peculiar to him; see K. Heldmann, *Antike Theorien über Entwicklung und Verfall der Redekunst* (Munich, 1982). I cannot account for Valla's knowledge of this text, normally thought to have been discovered in the sixteenth century. See *Texts and Transmission*, ed. L. D. Reynolds (Oxford, 1983), 431–433.
48. G. Cardano, *In Cl. Ptolemaei Pelusiensis IIII de astrorum iudiciis aut, ut vulgo vocant, Quadripartitae constructionis libros Commentaria* (Lyons, 1555), 59–63 on 1.3.25: "Manifestum est ex generalibus constitutionibus coeli principaliter ista pendere. Non inficiamur etiam aliquid in rem facere quae adducuntur a Paterculo."
49. See in general Jardine, *Birth of History and Philosophy of Science*, ch. 8; A. Grafton, *Joseph Scaliger: A Study in the History of Classical Scholarship*, 1: *Textual Criticism and Exegesis* (Oxford, 1983), ch. 7.
50. Savile's *Prooemium mathematicum* is preserved in Oxford, Bodleian Library, MS Savile 29.
51. F. E. Manuel, *Isaac Newton, Historian* (Cambridge, Mass., 1963).
52. Jardine, *Birth of History and Philosophy of Science*, 116–117, 181, 276–277.
53. Kepler, *Ges. Werke*, 10:36–41.
54. G. Pico della Mirandola, *Opera Omnia* (Basel, 1572), 715.
55. Kepler, *Ges. Werke*, 14:285.
56. See for example Croll's genealogy of knowledge, described in O. Hannaway, *The Chemists and the Word* (Baltimore and London, 1975), 18–20, 51. Libavius's critique of the Hermetic tradition (discussed ibid., 78, 98–105) makes an interesting parallel to Kepler.

8. Isaac La Peyrère and the Old Testament

1. F. Berriot, "Hétérodoxie religieuse et utopie politique dans les 'erreurs estranges' de Noël Journet (1582)," *Bulletin de la Société de l'Histoire du Protestantisme Français* 124 (1978), 236–248; R. Peter, "Noël Journet détracteur de l'Ecriture Sainte (1582)," *Croyants et sceptiques au XVIe siècle*, ed. M. Lienhard (Strasbourg, 1981), 147–156.
2. See most recently M. Olender, *Les langues du Paradis* (Paris, 1989), 39–48, 109–111.
3. H.-J. Schoeps, *Philosemitismus im Barock* (Tübingen, 1952), 3–18.
4. G. Patin, *Lettres*, ed. J.-H. Reveillé-Parise (Paris, 1846), 1:296–297, 2:175; M. Mersenne, *Correspondance*, ed. C. de Waard et al. (Paris, 1932–), 12:364; 15:98.
5. G. Gliozzi, *Adamo e il nuovo mondo* (Florence, 1977).

6. For this and the following paragraph see R. H. Popkin, *Isaac La Peyrère, 1596–1676* (Leiden, 1987).
7. I. La Peyrère, *Prae-Adamitae* (n.p., 1655), ch. 7, 29; *Men before Adam* (London, 1656), 18.
8. See Mersenne, *Correspondance*, 16:198–199; and for a sample of Worm's scholarship see his *Fasti Danici* (Copenhagen, 1633).
9. H. L. Brugmans, *Le séjour de Christian Huyghens à Paris et ses relations avec les Milieux scientifiques français* (Paris, 1935), 154, 155, 156, 159.
10. M. Sossus, *De numine historiae liber* (Paris, 1632), 199.
11. Ibid., 214. For the classical and later sources of this dating see J. Bidez and F. Cumont, *Les mages hellénisés* (Paris, 1973), 2:7–62.
12. I. La Peyrère, *Systema theologicum* (n.p., 1655), 3.6, 151–152; *A Theological Systeme* (London, 1655), 168.
13. *M. Manilii Astronomicon liber I*, ed. A. E. Housman, 2d ed. (Cambridge, 1937), xv.
14. C. Saumaise, *De annis climactericis et antiqua astrologia diatribae* (Leiden, 1648), prae.
15. La Peyrère, *Systema*, 3.7, 160; *Systeme*, 177.
16. P. le Prieur, *Animadversiones in librum Praeadamitarum* (n.p., 1666), 121.
17. R. Simon, *Lettres choisies*, 2 (Rotterdam, 1704), 18.
18. Ibid., 27.
19. La Peyrère, *Systema*, 4.1, 185; *Systeme*, 204–205. The English abridges the original: "Hae causae me movent, quare libros quinque illos, non Mosis archetypos, sed excerptos et exscriptos ab alio credam."
20. Peter, "Journet détracteur," 152–153.
21. Simon, *Lettres*, 2:27.
22. Cf. C. Ginzburg, *The Cheese and the Worms*, tr. J. and A. Tedeschi (Baltimore and London, 1980).
23. D. Pastine, "Le origini del poligenismo e Isaac La Peyrère," *Miscellanea Seicento* (Florence, 1971), 7–234.

9. Prolegomena to Friedrich August Wolf

1. L. Hatvany, *Die Wissenschaft des nicht Wissenswerten: Ein Kollegienheft*, 2nd ed. (Munich, 1914), 6. On Hatvany see F. Lilge, *The Abuse of Learning* (New York, 1948), 109; *Mythology and Humanism: The Correspondence of Thomas Mann and Karl Kerényi*, tr. A. Gelley (Ithaca, N.Y., and London, 1975), 112, 171–172.

2. Hatvany, *Wissenschaft,* 14.
3. Ibid., 7.
4. Ibid., 91–106: "Sappho und die sapphische Liebe. (Eine missratene Seminararbeit)." On the interpretations of which Hatvany was making fun, see e.g. H. Rüdiger, *Sappho, Ihr Ruf und Ruhm bei der Nachwelt* (Leipzig, 1933), 102–109 (Welcker); 150–153 (Wilamowitz).
5. See esp. Hatvany, *Wissenschaft,* 15.
6. On Wolf, see in general M. Fuhrmann, "Friedrich August Wolf," *Deutsche Vierteljahrsschrift für Literaturwissenschaft und Geistesgeschichte* 33 (1959), 187–236, with ample references to the earlier literature; the best treatments in English are M. Pattison, "F. A. Wolf," *Essays,* (Oxford, 1889), 1:337–414, and R. Pfeiffer, *History of Classical Scholarship from 1300 to 1800* (Oxford, 1976), 173–177.
7. Cf. C. E. McClelland, "The Aristocracy and University Reform in Eighteenth-Century Germany," *Schooling and Society,* ed. L. Stone (Baltimore and London, 1976), 146–173.
8. Wolf, *Darstellung der Alterthums-Wissenschaft* (1807), in his *Kleine Schriften,* ed. G. Bernhardy (Halle, 1869) (hereafter *Kl. Schr.*), 2:883.
9. For Wolf's list of subsidiary disciplines, see ibid., 894–895; for discussion see esp. A. Böckh, *Encyklopädie und Methodologie der philologischen Wissenschaften,* ed. E. Bratuscheck (Leipzig, 1877), 39–44; G. Pasquali, *Filologia e storia,* 2d ed. (Florence, 1964), 67–73; J. Bolter, "Friedrich August Wolf and the Scientific Study of Antiquity," *Greek, Roman and Byzantine Studies* 21 (1980), 83–99.
10. Wolf, *Darstellung, Kl. Schr.,* 2:825–26.
11. W. Mettler, *Der junge Friedrich Schlegel und die griechische Literatur: Ein Beitrag zum Problem der Historie* (Zürich, 1955); C. Menze, *Wilhelm von Humboldt und Christian Gottlob Heyne* (Ratingen, 1966).
12. See the description and samples from W. von Humboldt's notes, in the Akademie-Ausgabe of Humboldt's *Gesammelte Schriften,* 7, pt. 2 (1907): 550–553; cf. P. B. Stadler, *Wilhelm von Humboldts Bild der Antike* (Zürich and Stuttgart, 1959), 17–25.
13. Humboldt, *Ges. Schr.,* 7, pt. 2:553.
14. Cf. Menze, *Humboldt und Heyne,* 12–13.
15. See e.g. H. Butterfield, *Man on His Past,* repr. (Boston, 1960), ch. 2; P. H. Reill, *The German Enlightenment and the Rise of Historicism* (Berkeley, 1975). On Heyne's self-conscious use of the methods of "statistics," see C. Antoni, *La lotta contro la ragione,* 2nd ed. (Florence, 1968), 153–156.
16. C. G. Heyne, *Opuscula academica,* 1 (Göttingen, 1785), 287.
17. Ibid. 76–134; elaborate discussion in Mettler, *Schlegel,* 46–97.

18. Wolf, *Prolegomena ad Homerum*, 3rd ed., ed. R. Peppmüller (Halle, 1884; repr. Hildesheim, 1963), ch. 41, 145. Wolf's n59 *ad loc.*: "Erudite haec persecutus est Heynius in Diss. de genio saeculi Ptolemaeorum..."
19. J. D. Michaelis to R. Wood, 1770, in R. Wood, *An Essay on the Original Genius of Homer (1769 and 1775)*, ed. B. Fabian (Hildesheim and New York, 1976), bibliographical note, xiii*: "Tribus enim nobis ita videbatur, Heynio, Becmanno et mihi, sic Homerum legi oportere, ut Tu legisti: mallemusque neminem in Germania aliter eum legere" (reference kindly supplied by B. Fabian).
20. Gravina argued that the differences between Virgil and Homer stemmed less from their talents per se than from the different cultural and political worlds for which they wrote. He explicitly treated Homer as a primitive.
21. J. F. Christ, *Noctes Academicae...*, 2 (Halle, 1727), 99.
22. Ibid., 102–111. Christ takes Venus as being about to undergo intercourse from the front and the rear simultaneously, 105.
23. Ibid., 105–106.
24. Ibid., 104. On Christ, see E. Schmidt, *Lessing*, 3rd ed., 1 (Berlin, 1909), 40–48; and A. D. Potts, "Winckelmann's Interpretation of the History of Ancient Art in Its Eighteenth-Century Context," Diss. London 1978 (copy in the Warburg Institute), 1:94–98. For Wolf's own assessment of Christ, see his contribution to Goethe's *Winckelmann und sein Jahrhundert*, repr. in *Goethes Briefe an F. A. Wolf*, ed. M. Bernays (Berlin, 1868), 129.
25. M. Wegner, *Altertumskunde* (Freiburg and Munich, 1951), 85–87.
26. On Holstenius see R. Almagià, *L'opera geografica di Luca Holstenio*, Studi e testi cii (Vatican City, 1942); for the study of the early church see in general S. Bertelli, *Ribelli, libertini e ortodossi nella storiografia barocca* (Florence, 1973), ch. 3 (to be used with caution); G. Wataghin Cantino, "Roma sotterranea: Appunti sulle origini dell'Archeologia cristiana," *Ricerche di storia dell'arte* 10 (1980), 5–14 (with good bibliography); H. Gamrath, *Roma sancta renovata* (Rome, 1987).
27. *Passio sanctarum martyrum Perpetuae et Felicitatis*, ed. L. Holstenius (Rome, 1663), 118.
28. For Cluverius and Scaliger, see e.g. H. J. Erasmus, *The Origins of Rome in Historiography from Petrarch to Perizonius* (Assen, 1962). No adequate history of scholarship in early modern Europe has yet been written. The best introduction is perhaps Wegner, *Altertumskunde*, 75–101; the most incisive analysis is A. D. Momigliano, "Ancient History and the Antiquarian," *Journal of the Warburg and Courtauld*

Institutes 13 (1950), 285–315. Some of the more informative and original monographs are: T. D. Kendrick, *British Antiquity* (London, 1950); J. G. A. Pocock, *The Ancient Constitution and the Feudal Law* (Cambridge, 1957); A. Ellenius, *De arte pingendi: Latin Art Literature in Seventeenth-Century Sweden and Its International Background* (Uppsala and Stockholm, 1960); E. Iversen, *The Myth of Egypt and Its Hieroglyphs in European Tradition* (Copenhagen, 1961); P. Rossi, *Le sterminate antichità: Studi vichiani* (Pisa, 1969); G. Cantelli, *Vico e Bayle: Premesse per un confronto*, Studi vichiani 4 (Naples, 1971); A. Dupront, *L. A. Muratori et la société européenne des pré-Lumières* (Florence, 1976). See now the far-reaching survey of C. Borghero, *La certezza e la storia* (Milan, 1983).

29. Wolf, *Prolegomena*, in *Demosthenis oratio adversus Leptinem cum scholiis veteribus et commentario perpetuo*, ed. Wolf (Halle, 1789), cxxxxv: "Hactenus scribendo ea sum potissimum secutus, quae ad hanc Leptineam caussam partim illustrandam valere, partim disci ex ea queant" (for the second passage quoted see n30 below). On Wolf's hermeneutics see J. Wach, *Das Verstehen*, 1 (Tübingen, 1926), 62–82, and the recent symposium volume on *Philologie und Hermeneutik im 19. Jahrhundert*, ed. H. Flashar et al. (Göttingen, 1979). Wolf's debt to Semler—on whom see H. W. Frei, *The Eclipse of Biblical Narrative: A Study in Eighteenth and Nineteenth Century Hermeneutics* (New Haven and London, 1974), esp. 246–248—would repay further study.

30. Wolf, *Prolegomena in Leptineam*, lxxxv: "Satis diximus de argumento Orationis et statu caussae: itaque videmur nobis eos quoque adiuvasse, qui forte a graecae linguae cognitione imparatiores ad hanc lectionem accesserint. Patet iam, quo in orbe rerum et sententiarum versetur Oratio; in quo multum adiumenti est ad melius intelligendum. Attamen ut ea a nobis, qui legimus, quoad eius fieri possit, eodem modo intelligatur, quo intellecta quondam est ab iis, qui Oratorem coram audierunt; inprimis cognoscenda est ratio, quae Athenis obtinuit, publicorum munerum curandorum, nec non legum ad Populum ferendarum abrogandarumque."

31. See *The Speech of Demosthenes against the Law of Leptines*, ed. J. E. Sandys (Cambridge, 1890), xliii–xlv; cf. Böckh, *Encyklopädie*, 166: "Für das Studium werden die Ausleger die besten sein, welche die richtige Mitte halten, wozu Wolf in seinem Commentar zur Leptinea . . . das erste Muster gegeben hat."

32. Jacques de Tourreil, "Préface historique" to the Philippics, *Oeuvres de Mr. de Tourreil*, quarto ed. (Paris, 1721), 1:266. Tourreil died in

1714, but I have seen only the quarto and duodecimo editions of 1721 which are apparently much revised versions of the earlier editions. On Tourreil's life and the various forms in which his Demosthenes translations appeared, see G. Duhain, *Un traducteur de la fin du XVIIe siècle et du commencement du XVIIIe siècle. Jacques de Tourreil, traducteur de Démosthène (1656–1714)* (Paris, 1910).

33. See e.g. Tourreil's remark on the "three statues sixteen cubits high" mentioned in *De corona*, 91, *Oeuvres de Mr. de Tourreil*, 2:529: "Que sur le Port l'on érige trois Statues de seize coudées chacune. Quelqu'un trouvera peut-être ces Statues d'une grandeur démésurée. Il est vrai que cette hauteur est énorme, par rapport à nos usages. Mais elle n'a rien de surprenant par rapport aux usages des Anciens, qui pour marquer leur reconnoissance envers leurs bienfacteurs, leur élevoient souvent, comme l'on sçait, des Statues Colossales." Cf. Duhain, *Jacques de Tourreil*, 252.

34. For example, his crisp account of the trierarchy, *Oeuvres de Mr. de Tourreil*, 2:535–539.

35. Wolf, *Prolegomena in Leptineam*, lxxxv–vi, n58, on lxxxvi: "Qui, quae virtus est Gallorum paene propria, dilucide, multoque, quam Petitus et alii, aptius ad communem intelligentiam de ea re [scil. the trierarchy] scripsit."

36. Wolf, *Prolegomena in Leptineam*, c.

37. *Oeuvres de Mr. de Tourreil*, 2:535. The point was not new with Tourreil; see S. Petit's standard *Leges Atticae* (Paris, 1635), 3.4, e.g. 271: "Ratio autem Trierarchiarum subinde mutavit..."

38. Wolf, *Praefatio in Leptineam*, xxxxix–l: "Sed ego sic omnino statuo, aliter oratione affici legentem, aliter audientem."

39. "Préface historique," *Oeuvres de Mr. de Tourreil*, 1:266.

40. Both Duhain, *Jacques de Tourreil*, and U. Schindel, *Demosthenes im 18. Jahrhundert* (Munich, 1963), 78–81, slightly overstate the singularity of Tourreil's (highly distinguished) work. For the context see esp. L. Gossman, *Medievalism and the Ideologies of the Enlightenment* (Baltimore, 1968).

41. See e.g. the quite accurate review of Wolf's 1787 *Tetralogia dramatum Graecorum* in the *Humanistisches Magazin* (1788), 271: "Vom Aristophanes wählte er die *Ekklēsiazousai* auch vornehmlich aus dem Grunde, weil diess Lustspiel die Denkungsart, die Sitten und die Staatsverfassung der Athenienser so trefflich schildert." Cf. Wolf's preface, *Kl. Schr.*, 1:287: "Ex Aristophane quam addidi fabulam, in hac praeterea spectavi notationem ingenii, morum reique publicae Atheniensium..."

42. Cf. the comparison Menze draws between Heyne and Humboldt, in his *Humboldt und Heyne*.
43. J. Bernays, *Phokion und seine neuere Beurtheiler* (Berlin, 1881), 1–14, 100–102.
44. See his "Epistola . . . ad auctorem," in *Fragmenta Stesichori Lyrici*, ed. J. A. Suchfort (Göttingen, 1771), xliii–xlvii.
45. Text in the Akademie-Ausgabe of Humboldt's *Ges. Schr.*, 1 (1903), 255–281; cf. Stadler, *Humboldt*, 32–53.
46. Ibid., 53–56.
47. Humboldt to Wolf, 23 January 1793; Humboldt, *Gesammelte Werke*, 5 (Berlin, 1846), 18.
48. Cf. Wolf, *Darstellung, Kl. Schr.*, 2:870 with Humboldt to Wolf, 16 June 1804; Humboldt, *Gesammelte Werke*, 5:266–267.
49. Wolf, *Kl. Schr.*, 2:884–885n.
50. Böckh, *Encyklopädie;* cf. B. Bravo, *Philologie, histoire, philosophie de l'histoire* (Wrocław, Warsaw, and Cracow, 1968).
51. See e.g. R. C. Jebb, *Homer*, 5th ed. (Glasgow, 1894), ch. 5; Broccia, *La questione omerica* (Florence, 1979), 121–125 (violently *con*); M. Murrin, *The Allegorical Epic* (Chicago and London, 1980), 189–196 (judiciously *pro*).
52. R. Volkmann, *Geschichte und Kritik der Wolfschen Prolegomena zu Homer* (Leipzig, 1874), chs. 4–5.
53. R. Wood, *An Essay on the Original Genius and Writings of Homer* (London, 1775), 279; previously quoted by J. L. Myres, *Homer and His Critics* (London, 1958).
54. Wolf learned of Vico from Melchior Cesarotti; see Cesarotti, *Prose edite e inedite*, ed. G. Mazzoni (Bologna, 1882), 393, 399, and Wolf, *Kl. Schr.*, 2: 1157–1166. For Rousseau see the *Prolegomena*, ch. 20, n54.
55. Jebb, *Homer*, 114–115.
56. *HomeriIlias seu potius omnia eius quae extant opera*, ed. O. Giphanius (Strassburg, 1572), 1:15: "Confusionis quoque prioris vestigia quaedam esse reliqua monet Iosephus in Apionem, quod quaedam in Homero reperiantur inter se pugnantia."
57. Ibid.: "Aiunt autem Pisistratum non satis bona fide in his componendis esse versatum: quosdam enim versus de suo admiscuisse." Cf. Giphanius's note on *Il.* 2.557ff.
58. Quoted by Wolf, *Prolegomena*, chs. 39 and 27, Bentley not quite accurately. Cf. J. Bernays, *Gesammelte Abhandlungen*, ed. H. Usener (Berlin, 1885), 2:356–359, and Jebb, *Homer*, 105–106. Both Giphanius and Bentley responded to the brilliant critique of Homer leveled by J. C. Scaliger, *Poetices libri septem* (Lyons, 1561).

59. Wolf, *Kl. Schr.*, 1:237–239.
60. E.g. Broccia, *La questione omerica*, 22–31.
61. E.g. R. Pfeiffer, *History of Classical Scholarship from the Beginnings to the End of the Hellenistic Age* (Oxford, 1968), 214; S. Timpanaro, *La genesi del metodo del Lachmann* (Florence, 1963), 26.
62. Wolf, *Prolegomena*, ch. 47, 186n27.
63. Cf. J. E. G. Zetzel, "*Emendaui ad Tironem:* Some Notes on Scholarship in the Second Century A.D.," *Harvard Studies in Classical Philology* 77 (1973), 225–243; "The Subscriptions in the Manuscripts of Livy and Fronto and the Meaning of *Emendatio*," *Classical Philology* 75 (1980), 56–57.
64. Wolf, *Prolegomena*, ch. 38, 132–133.
65. Ibid., ch. 47, 183.
66. *Albi Tibulli Carmina,* ed. C. G. Heyne, 3d ed. (Leipzig, 1798), xiii–lxiv.
67. Ibid., xxiii, xlvi–xlvii, lvii. On xxx Heyne writes: "Exposita est tanquam in stemmate *prosapia ac stirps lectionis Tibullianae* per tot editiones deducta . . ."
68. See in general G. Pasquali, *Storia della tradizione e critica del testo,* repr. of 2d ed. (Florence, 1971), 4–5; and cf. *Tibulli Carmina,* ed. Heyne, xxxvi–xl, for an interesting description of the Guelpherbytani.
69. Ibid., xv, xxn**.
70. Wolf, *Prolegomena,* ch. 7, 17n2.
71. Wolf, *Kl. Schr.*, 1:587–590.
72. For an interesting parallel cf. D. Wyttenbach's preface to Plutarch's *Moralia,* 1794, repr. in his *Opuscula* (Leiden and Amsterdam, 1821), 1:266–434. Wyttenbach makes clear (300–301) that he would trace the history of his text in antiquity if the evidence allowed him to.
73. *Homeri Ilias,* ed. Giphanius, 1:16: "Hac autem re factum est, ut multi versus a veteribus proferantur, ut Aristotele et aliis, qui in nostris libris hodie non extent: quod illi aliis editionibus, nos Aristarchica utamur."
74. J. R. Wettstein, *Pro Graeca et genuina linguae Graecae pronunciatione . . . orationes apologeticae* (Basel, 1686), 2:155.
75. L. Küster, *Historia critica Homeri* (Frankfurt a. O., 1696), 101–102; cf. Wolf, *Kl. Schr.*, 1:196–197 and, for further information on what Baroque scholarship knew of the history of the Homeric texts, T. Bleicher, *Homer in der deutschen Literatur (1450–1740)* (Stuttgart, 1972), 166–177.
76. J. M. Chladenius, *De praestantia et usu scholiorum Graecorum in poetas*

diatribe secunda (Wittenberg, 1732), 17: "Vti enim non melius de praestantia artificis iudicatur, quam si in sua officina opereque aliquo conficiendo conspiciatur . . ."
77. Ibid.: "Et dabunt veteres exemplum Criticae magis pietati quam integritati Poetarum inservientis, cum nihil in iis tolerarent, nisi quod cum religione vel potius superstitione sua conveniebat."
78. T. Burgess, *Appendix*, in R. Dawes, *Miscellanea critica*, 2d ed. (Oxford, 1781), 416–417.
79. *Theogonia Hesiodea*, ed. Wolf (Halle, 1783), 58. This edition (now much criticized) was greeted with great respect; see Heyne's "ad editorem epistola," ibid., 143–166, and G. Hermann's annotated copy (Cambridge University Library Adv.d.83.21).
80. J. B. G. D'Ansse de Villoison, *Prolegomena*, in *Homeri Ilias* (Venice, 1788), vin; *Addenda*, lvi–lvii; Murrin, *Allegorical Epic*, 257nn53,56.
81. Wolf owed Villoison much: e.g. the suggestion that the second recension attributed in antiquity to Aristarchus was produced not by him but by others after his death, for which cf. Wolf, *Prolegomena*, 184, and Villoison, *Prolegomena*, xxvii; also the comparison between the textual histories of Homer and the Koran (Wolf, 119; Villoison, xxiiin1). Yet Wolf's historical analysis was far more coherent and sophisticated; cf. e.g. his analysis of the texts called *ek tōn poleōn* and Villoison's, 137 and xxvi respectively.
82. Wolf, *Vorlesungen über die Alterthumswissenschaft*, ed. Gürtler/Hoffmann (Leipzig, 1839), 2:167.
83. L. C. Valckenaer, *Dissertatio de praestantissimo codice Leidensi*, in his *Hectoris interitus carmen Homeri sive Iliadis liber xxii cum scholiis vetustis Porphyrii et aliorum* (Leeuwarden, 1747), 133–135.
84. Ibid., 146.
85. Ibid., 12–13. Valckenaer's heavily annotated copy of Hemsterhusius's edition is in Cambridge University Library (Adv.d.72.8).
86. *Aristophanis comoedia Plutus*, ed. T. Hemsterhusius (Harlingen, 1744), xii–xiii.
87. Ibid., 218; see J. G. Gerretzen, "Schola Hemsterhusiana," Diss. Nijmegen 1940.
88. *Euripidis Tragoedia Hippolytus*, ed. Valckenaer (Leiden, 1768), xvi–xx.
89. E. Hulshoff Pol, "Studia Ruhnkeniana," Diss. Leiden 1935, 134–136, 166–167.
90. Wolf's accounts of his motives contradict one another. Cf. Wolf to Schütz, 13 September 1796, in S. Reiter, *Friedrich August Wolf: Ein Leben in Briefen* (Stuttgart, 1935) (hereafter "Reiter"), 1:215, with

W. Peters, *Zur Geschichte der Wolfschen Prolegomena zu Homer,* Beilage zum Progr. des Königlichen Kaiser-Friedrichs-Gymnasiums in Frankfurt a. M. (1890), 40.
91. *Aristophanis Nubes,* ed. J. A. Ernesti (Leipzig, 1753), vi.
92. I infer these facts from Hermann's notes on Reiz's lectures from the winter semester 1787–88, in Hermann's copy of Ernesti's edition; Cambridge University Library Adv.d.83.10.
93. *Opuscula Ruhnkeniana,* ed. T. Kidd (London, 1807), prae., lix; sect. 12.
94. J. P. Siebenkees, "Nachricht von einer merkwürdigen Handschrift der Iliade des Homer, in der venetianischen S. Markusbibliothek," *Bibliothek der alten Litteratur und Kunst* 1 (1786), 70–71.
95. *Graeca scholia scriptoris anonymi in Homeri Iliados lib. I,* ed. A. Bongiovanni (Venice, 1740); cf. his descriptions of both A and B in his *Graeca D. Marci Bibliotheca codicum manuscriptorum per titulos digesta* (Venice, 1740), 243–244. On this and other early publications of Homeric scholia see *Scholia Graeca in Homeri Iliadem (Scholia Vetera),* ed. H. Erbse (Berlin, 1969–), 1:lxvii.
96. C. D. Beck, *De ratione qua Scholiastae . . . adhiberi recte possint* (Leipzig, 1785), viii–ix.
97. *Oratorum Graecorum, quorum princeps est Demosthenes, quae supersunt monumenta ingenii,* 2 (Leipzig, 1770): pt. 2, 4.
98. See e.g. Beck, *De ratione,* ixn21; Siebenkees, "Nachricht," 1:63.
99. Volkmann, *Geschichte,* 40–43.
100. *Bibliothek der alten Litteratur und Kunst,* 5 (1789), 26–55, esp. 41–51.
101. J. A. Fabricius, *Bibliotheca Graeca,* 4th ed., ed. G. C. Harles, 1 (Hamburg and Leipzig, 1790).
102. *Allgemeine Literatur-Zeitung,* 30 January 1796, col. 271. Heyne was enraged by the laudatory tone of the review; Heyne to Herder, 18 February 1796, *Von und an Herder,* ed. H. Düntzer and F. G. von Herder, 2 (Leipzig, 1861), 232.
103. For a discussion of scholia—the Demosthenes scholia by Ulpian—by Wolf see his commentary on the *Leptinea,* 210–211.
104. See in general *The Massoreth Ha-Massoreth of Elias Levita,* ed. C. D. Ginsburg (London, 1867), 40–61; M. Olender, *Les langues du Paradis* (Paris, 1989).
105. See P. Hazard, *The European Mind (1680–1715),* tr. J. Lewis May (New York, 1963), 180–197.
106. Wolf, *Prolegomena,* ch. 4, 11.
107. *Allgemeine Literatur-Zeitung,* 1 February 1791, col. 246.
108. Wolf, *Prolegomena,* ch. 4, 9–10; ch. 42, 153.

109. Ibid., ch. 3, 7.
110. See J. F. J. Arnoldt, *Fr. Aug. Wolf* (Braunschweig, 1861–1862), 2:387–406; Reiter, 1:320, 322, 335, 339, 354, 366 on Griesbach; 1:50, 68, 84, 92, 95, 96, 99–100, 101, 102, and 2:345 on Semler.
111. E. Sehmsdorf, *Die Prophetenauslegung bei J. G. Eichhorn* (Göttingen, 1971).
112. E. S. Shaffer, *"Kubla Khan" and the Fall of Jerusalem* (Cambridge, 1975).
113. Wolf, *Prolegomena*, ch. 15, 46–47 and 47n25; *Vorlesungen*, 1:305.
114. J. G. Eichhorn, *Einleitung ins Alte Testament*, 2d ed. (Leipzig, 1787), 1:14.
115. Ch. 2 of the *Einleitung* bears the title "Geschichte des Textes der Schriften des A.T.," 1:133.
116. Ibid., 1:260–261; here and elsewhere I use the partial translation by G. T. Gollop, privately printed in 1888.
117. Ibid., 1:299–300; for Isaiah 53.9 read 55.11.
118. Quoted by J. Buxtorf, Jr., *Anticritica* (Basel, 1653), 478.
119. *Jacob ben Chajim ibn Adonijah's Introduction to the Rabbinic Bible*, ed. and tr. C. D. Ginsburg (1867), 42–57; *Elias Levita*, ed. Ginsburg, 102–119.
120. L. Cappell, *Critica sacra* 3.2.4 (Halle, 1775–1786), 1:182–183.
121. Buxtorf, *Anticritica*, 478–479.
122. B. Walton, *In Biblia Polyglotta Prolegomena*, ed. F. Wrangham (Cambridge, 1828), 1:474; cf. 480–483.
123. *Vetus Testamentum Hebraicum*, ed. B. Kennicott (Oxford, 1776–1780), 1:16–17; G. B. de Rossi, *Variae lectiones Veteris Testamenti*, 1 (Parma, 1784), lii. See W. McKane, "Benjamin Kennicott: An Eighteenth-Century Researcher," *Journal of Theological Studies*, n.s. 28 (1977), 445–464.
124. Wolf, *Prolegomena*, ch. 47, 186.
125. Ibid., n26.
126. Eichhorn, *Einleitung*, 1:260 (tr. Gollop).
127. Wolf, *Prolegomena*, ch. 47, 186–187.
128. Eichhorn, *Einleitung*, 1:309.
129. See Michaelis's *Vorrede*, in his *Deutsche Uebersetzung des Alten Testaments*, 1 *(welcher das Buch Hiobs enthält)* (Göttingen and Gotha, 1769), sigs. b 2 verso–b 3 recto; Sehmsdorf, *Prophetenauslegung*, 121–124.
130. Michaelis, *Vorrede*, sig. b verso.
131. *Tibulli Carmina*, ed. Heyne, xlvi–xlvii; Sehmsdorf, *Prophetenauslegung*, 125–128.
132. Eichhorn, "Ueber die Quellen, aus denen die verschiedenen Erzäh-

lungen von der Entstehung der alexandrinischen Uebersetzung geflossen sind," *Repertorium für Biblische und Morgenländische Litteratur* 1 (1777), 266–280.
133. Wolf, *Prolegomena*, ch. 33, 112–114 and n12.
134. H. K. A. Eichstädt, *Oratio de Io. Godofr. Eichhornio*, in his *Opuscula oratoria*, 2d ed. (Jena, 1850), 607; cf. 634–635 n13.
135. See *Crabb Robinson in Germany 1800–1805*, ed. E. J. Morley (London, 1929), 161–162.
136. Wolf, *Prolegomena*, 10.
137. Wolf, *Prolegomena*, 2.2; available in the Berlin 1876 ed. of the *Prolegomena*, 178–179.
138. Wolf, *Vorlesungen*, 1:311.
139. Wolf, *Kl. Schr.*, 1:252.
140. B. M. Metzger, *The Text of the New Testament*, 2d ed. (Oxford, 1968), 119–121; Timpanaro, *Lachmann*, 22; *J. J. Griesbach: Synoptic and Text-Critical Studies 1776–1976*, ed. B. Orchard and T. R. W. Longstaff (Cambridge, 1978).
141. Cf. Pasquali, *Storia*, 3–12; Timpanaro, *Lachmann*.
142. *Allgemeine Literatur-Zeitung*, 16 June 1795 = *Ges. Schr.*, 1: 370–376.
143. Wolf, *Prolegomena*, ch. 2, 4; for precedents see Timpanaro, *Lachmann*, ch. 2.
144. Wolf, *Prolegomena*, ch. 27, 87–88. For a precedent see Michaelis's preface to his edition of R. Lowth, *De sacra poesi Hebraeorum* (Göttingen, 1758), sig. a 2 verso: "Poetam de poesi Hebraeorum dicere statim intelligent lectores, ingenio praestantem, quod elegantissimis Graecorum Latinorumque literis excoluerat. Rarum hoc et exoptatum munus Hebraicis carminibus contigit: aliter enim poetas vates tractabit, aliter merus grammaticus."
145. Wolf, *Prolegomena*, ch. 22, 71; for a precedent see Wood, *Essay*, 259–260: "But the oral traditions of a learned and enlightened age will greatly mislead us, if from them we form our judgement on those of a period, when History had no other resource. What we observed at Palmyra puts this matter to a much fairer trial; nor can we, in this age of Dictionaries, and other technical aids to memory, judge, what her use and powers were, at a time, when all a man could know, was what he could remember."
146. B. G. Niebuhr, "Die Sikeler in der Odysee," *Rheinisches Museum* 1 (1827), 257; K. A. Varnhagen von Ense, *Tagebücher* (Leipzig, 1861), 1:106; 2:68.
147. Quoted by C. Diehl, *Americans and German Scholarship, 1770–1870* (New Haven and London, 1978), 71; cf. E. R. Dodds, *Missing Persons* (Oxford, 1977), 27.

148. H. Aarsleff, *The Study of Language in England, 1780–1860* (Princeton, 1967), 134, 154–159.
149. See e.g. K. Lehrs, "Einleitung zu Homer," *Kleine Schriften*, ed. A. Ludwich (Königsberg, 1902), 21–25.
150. Lehrs, *De Aristarchi studiis Homericis* (Königsberg, 1833), 36–38.
151. G. Parthey, *Das Alexandrinische Museum* (Berlin, 1838), 111–135; F. Ritschl, *Die Alexandrinischen Bibliotheken unter den ersten Ptolemäern und die Sammlung der Homerischen Gedichte durch Pisistratus* (Breslau, 1838).
152. A. Gräfenhan, *Geschichte der klassischen Philologie im Alterthum* (Bonn, 1843–1850).
153. See e.g. S. Lieberman, *Hellenism in Jewish Palestine* (New York, 1950), 28–37; B. Gerhardsson, *Memory and Manuscript* (Lund and Copenhagen, 1961).
154. Peters, *Zur Geschichte*, 36.

ACKNOWLEDGMENTS

THE ESSAYS collected here were first written for the following publications: Chapter 1: *Renaissance Quarterly* 38 (1985), 615–649. Chapter 2: *Journal of the Warburg and Courtauld Institutes* (hereafter *JWCI*) 40 (1977), 150–188. Chapter 3: *The Transmission of Culture in Early Modern Europe*, ed. A. Grafton and A. Blair (Philadelphia: University of Pennsylvania Press, 1990), 8–38. Chapter 4: *JWCI* 48 (1985), 100–143. Chapter 5: *JWCI* 46 (1983), 78–93. Chapter 6: *The Uses of Greek and Latin: Historical Essays*, ed. A. C. Dionisotti, A. Grafton, and J. Kraye (London: The Warburg Institute, 1988), 155–170. Chapter 7: *Literary Culture in the Holy Roman Empire*, ed. J. A. Parente, Jr., et al. (Chapel Hill: University of North Carolina Press, 1991). Chapter 8: *Times Literary Supplement*, 12–18 February 1988, 151–152. Chapter 9: *JWCI* 44 (1981), 101–129. My thanks to the editors of *Renaissance Quarterly*, the *Journal of the Warburg and Courtauld Institutes*, and the *Times Literary Supplement* for permission to reprint articles. All of them have been edited in minor ways; but I have made no effort to bring them fully up to date.

Warm thanks also to the History Department of Princeton University, which provided funds to cover the cost of keyboarding some of the articles; to Julie Peterson, Faye Angelozzi, and Peggy Reilly, who put the texts on floppy disks with miraculous speed and accuracy; to John O'Malley and Katharine Park for their comments on the original plan for this work; to Richard Kroll for guidance into Bentley's world; and to Joseph Levine, David Quint, Nancy Siraisi, Noel Swerdlow, and Lindsay Waters for their criticism of an earlier draft of the introduction, which has not appeared before. My greatest debt is to J. B. Trapp, who long ago initiated me into the mysteries of the Warburg Institute and whose firm editorial hand chastened and enriched the original versions of several of these chapters. This book is dedicated to him, as an inadequate token of gratitude and affection.

INDEX

Accursius, 61
Achenwall, G., 218
Aeschylus, 27–28, 36, 55
Aitzing, Michael von, 193–195
Alanus, 49
Alcibiades I, 68
Alciato, Andrea, 33
Alexander, 129, 131, 134–136
Alfonso of Castile, 129, 200
Allen, Don Cameron, 31
Allen, Michael, 31
Anastasius of Sinai, 124
Annius, Joannes, 29, 76–103, 137–139
Antiochus Epiphanes, 121
Apianus, Petrus, 131
Apion, 92
Apollodorus, 37
Apollonius Rhodius, 67, 230
Appian, 118
Apuleius, 66, 169
Aratus, 200
Archilochus, 80, 88–89, 100
Aristarchus, 89, 224, 226, 228, 234, 237, 238
(pseudo-) Aristeas, 16, 160–161, 172, 297n79, 299n16
Aristonicus, 231

Aristophanes, 230, 231
Aristophanes of Byzantium, 224, 234, 237
Aristotle, 105, 135–136, 149, 162, 164
Asconius Pedianus, Q., 53–54
Augustine, 93, 168, 172–173
Ausonius, 56–57, 66, 108

Bacon, Francis, 1–3, 30–31, 32, 34, 179–180
Bancroft, George, 242
Baptista Mantuanus, 73
Baronio, Cesare, 145–161
Barreiros, Gasper, 96–97, 134
Basil of Caesarea, 146–147
Batrachomyomachia, 148
Baudouin, François, 95–96
Bayle, Pierre, 204
Beck, C. D., 231–232
Beckmann, Johann, 219
Bede, 125–126
Bellièvre, Claude, 47–48
ben Chajim, Jacob, 236–237
Bengel, J. A., 241
Bentley, Richard, 12–21, 177, 225
Bernard, Edward, 15–16, 249n22
Bernays, Jacob, 107, 143

Beroaldo, Filippo, 37, 51, 57, 58–59, 61, 71, 73–74, 108, 261–262n29
Beroaldus, Matthaeus, 116, 137, 154–155
Berosus, 76–78, 80, 82, 85, 86, 87, 88, 90–91, 92, 95–100, 101–103, 134, 210
Bersuire, Pierre, 26
Bertram, Corneille, 138
Betuleius, Xystus, 173
Beza, Theodore, 138
Bezold, Friedrich von, 94
Bibliander, Theodore, 130
Billanovich, Giuseppe, 27
Biondo, Flavio, 56
Birago, Lampugnino, 92–93
Bochart, Samuel, 17
Bodin, Jean, 5, 7–8, 28–29, 43, 78, 98–99, 131, 135–136, 142, 300n25
Böckh, August, 221, 223
Bolgar, Robert, 3
Boulliau, Ismaël, 2–3
Bongiovanni, A., 231
Bracciolini, Poggio, 63, 88
Brahe, Tycho, 142, 184, 199–200
Branca, Vittore, 9
Brown, Sibylla, 176
Brucker, Jacob, 158–160
Brugnoli, G., 9
Bruni, Leonardo, 55–56, 81
Bruno, Giordano, 185
Bucholzer, Abraham, 137–138, 140–141, 143
Bulenger, Julius Caesar, 156
Burgess, Thomas, 228–229
Bury, Richard de, 69–70
Buxtorf, J., 237
Buxtorf, J., Jr., 234, 237

Caesar, 118–119
Caius, John, 96
Caius, Thomas, 96
Calderini, Domizio, 9, 35, 50–51, 52–53, 55, 66, 70–71
Callimachus, 65–67, 70–71
Callippus, 115, 135
Callisthenes, 134–137
Calvin, John, 147, 160
Campanella, Tommaso, 129, 142, 166, 300n25
Cano, Melchior, 78, 96–97, 134
Canter, Willem, 175
Cappell, Jacques, 156
Cappell, Louis, 237
Cardano, Girolamo, 198
Cardini, Roberto, 32–33
Casaubon, Isaac, 16, 21–22, 36, 77, 78, 79, 145–177, 225
Casaubon, Meric, 3, 13
Casella, Maria Teresa, 9
Cassius Dio, 118
Castellio, Sebastian, 173–174
Cato, 80, 83
Catullus, 65–67, 214–215
Caylus, A. C. P. de Tubières, 220
Censorinus, 105–106, 108–119, 137
Cesarini Martinelli, Lucia, 9, 34–35
Chassanion, Jean, 211–212
Chladenius, J. M., 228
Chrestien, Florent, 121, 126, 131
Christ, J. F., 219–220
Chrysippus, 34
Chytraeus, David, 96
Cicero, 23–25, 34, 52–53, 59–60, 61–63, 67, 141, 162, 172–173, 175, 209–210
Clement of Alexandria, 18, 100
Cluverius, Philipp, 220
Code (of Justinian), 33
Conring, Hermann, 157, 158
Constantine, 172–173
Conway, Anne, 17–21
Copernicus, Nicolaus, 97, 129–131, 161, 181, 183, 200
Coppini, Donatella, 28
Corpus hermeticum, 18, 145–177
Crinito, Pietro, 74, 93
Croll, Morris, 32
Crusius, Martin, 178, 186–187
Crusius, Paulus, 131–133, 139, 285n83, 306n32
Ctesias, 91, 98
Cudworth, Ralph, 17–20, 158–159

Index

Curio, Iacobus, 106, 130
Curtius Rufus, Q., 163
Cyprian, 73
Cyriac of Ancona, 110

d'Ailly, Pierre, 129, 141–142
D'Amico, John, 32
Daniel, 121, 137, 140, 158, 165–166, 195
Danielsson, O. A., 92
Dares, 78, 99
Darius, 129
de Foix de Candale, François, 169
Demosthenes, 220–222, 232
de Nolhac, Pierre, 27
de Rossi, G. B., 237
Descartes, René, 1–3, 19
de Thou, J. A., 144
Dictys, 78, 99, 168
Digest (of Justinian), 60–61, 74
Diodorus Siculus, 87, 88, 119, 207, 209–210
Diogenes Laertius, 151, 164, 225
Dionisotti, Carlo, 9
Dionisotti, Carlotta, 28
Dionysius Areopagites, 150–151, 162, 167, 172
Dionysius of Halicarnassus, 87, 92–93, 99
Diplovatacius, Thomas, 48
Domitian, 35
Donation of Constantine, 162
Dorat, Jean, 36, 174–175
Dousa, Janus, 103
Dunston, John, 9, 28, 55
Dupuy, Claude, 121, 123, 131, 139, 144, 174

Eichhorn, J. G., 235–241
Eichstädt, H. K. A., 239
Elman, Benjamin, 45
Emmius, Ubbo, 102–103
Ennius, 55, 67
Erasmus, Desiderius, 7–8, 27, 37–38, 41–42, 171–172
Eratosthenes, 89, 92–93
Ernesti, J. A., 231

Estienne, Henri, 28, 30, 78, 148, 155, 166–167, 290n13
Eudaemon-Joannes, Andreas, 156
Eudoxus, 118, 119, 135
Euhemerus, 37
Eunomus, 152, 156
Eupolis, 55
Euripides, 67, 230
Eusebius, 89, 98, 100, 101–102, 122, 124, 129, 132, 137–138, 140, 143
Eustathius, 66, 228, 229, 232
Evans, Robert, 180

Fabius Pictor, Q., 80, 81, 83–84
Fabricius, J. A., 232
Fera, Vincenzo, 9
Ficino, Marsilio, 31, 149, 157, 163, 168–169
Filetico, Martino, 50
Firmicus Maternus, 141, 190
Fowden, Garth, 168
Freigius, Ioannes, 135
Fumagalli, Eduardo, 83
Fumaroli, Marc, 32
Funck, Johann, 97–99, 130

Gale, Theophilus, 17–18
Galen, 164–165
Galilei, Galileo, 2, 179, 184
Gamaliel, 234
Garin, Eugenio, 9, 32–33, 42
Gatterer, Johann Christoph, 218
Gaza, Theodore, 110
Geffcken, Johannes, 173
Gellius, Aulus, 62, 69, 108
Génébrard, Gilbert, 139, 154, 170
Geoffrey of Monmouth, 90
George Syncellus, 77, 102, 269–270n6
Giambullari, Pier Francesco, 274n64
Giphanius, Obertus, 225, 228
Giraldi, L. G., 112
Goez, Werner, 80, 84, 100
Goldast, Melchior, 178–181
Goropius Becanus, Joannes, 99–101, 112–113, 136–137, 141, 154
Gräfenhan, A., 243

Gravina, Gian Vincenzo, 219
Gregory of Nyssa, 147
Griesbach, J. J., 235, 240, 241
Gronovius, J., 3
Grotius, Hugo, 148, 206
Grynaeus, Jacob, 137, 174
Guarini, Guarino, 49, 63, 66
Guenée, Bernard, 28, 84, 91
Guicciardini, Francesco, 26
Guidetti, Lorenzo, 23–26

Haguelon, Pierre, 105, 109–110
Harles, G. C., 232
Hassinger, Erich, 31
Hatvany, Ludwig, 214–215
Hebenstreit, J. B., 201
(pseudo-) Hecataeus, 16, 18
Heinsius, Daniel, 166
Heldmann, K., 306–307n47
Hemsterhusius, T., 230
Heraclitus, 186–187
Hercules, 37
Herennius Philo, 16, 79
Hermes Trismegistus, 78, 79
Herodotus, 56–57, 98, 100, 151, 192
Hervet, Gentian, 124
Herwart von Hohenburg, Johann, 187–190, 202
Hesiod, 78, 167, 229
Hesychius, 13
Heyne, Christian Gottlob, 10, 31, 215, 217–219, 221, 227, 231, 232, 235, 238–239, 241, 316n102
Hipparchus, 198–202
Hippocrates, 151–152, 164–165, 167
Holstenius, Lucas, 220
Homer, 31, 36, 37, 53, 55, 69, 78, 88–89, 148, 167, 186–187, 217–218, 223–243
Horapollo, 163
Howell, James, 160, 176
Humboldt, Wilhelm von, 10, 222–223, 241
Huyghens, Christiaan, 208

Iamblichus, 153, 155, 159, 160–161, 168

Ibn Ezra, 211
Ignatius Na matallah, 128
Isidore of Seville, 53, 88

Janus, 95, 112–113
Jardine, Nicholas, 181
Jerome, 37, 89, 100, 124, 140, 165, 233, 261–262n29
Joachim of Fiore, 140
Job, 171
John of Damascus, 167
Josephus, 81, 91–92, 97, 100, 101–102, 209, 225, 303n64
Journet, Noël, 204, 211–212
Julius Africanus, Sextus, 124, 165–166
Justinian, 60
Justin Martyr, 15
Juvenal, 59, 71

Kaufmann, Thomas, 180
Kenney, E. J., 28, 57
Kennicott, Benjamin, 234, 235, 237
Kepler, Johannes, 2, 178–203
King, William, 177
Kircher, Athanasius, 159, 184
Krautter, Konrad, 9
Kraye, Jill, 28
Kühlmann, Wilhelm, 32, 39
Küster, Ludolf, 228

Lachmann, Karl, 241
Lactantius, 168
Lalamant, Jean, 116, 126
Lamola, Giovanni, 62–63
Landino, Cristoforo, 23–25, 33, 68
Le Peyrère, Isaac, 22, 204–213
Lavinius, Petrus, 26
Lazius, Wolfgang, 133–134
Le Clerc, Jean, 37, 255n50
Lefèvre d'Etaples, Jacques, 149
Lehrs, Karl, 226, 242
Lemmi, Charles, 31
Le Prieur, Philippe, 210
Leto, Giulio Pomponio, 49, 55, 74
Levine, Alice, 31
Levine, Joseph, 79

Levita, Elias, 236–237
Lilly, William, 177
Livy, 97, 103, 111
Lipsius, Justus, 26, 39–40
Lowry, Martin, 27
Lowth, R., 318n144
Lucan, 35, 118, 187–190
Lucian, 164
Lucidus Samotheus, Ioannes, 82, 125
Lucretius, 149
Luther, Martin, 81–82, 86–87, 98

Machiavelli, Niccolò, 26
Macrobius, 69, 108, 110, 112, 116
Maestlin, Michael, 182–183, 186–187, 191–192
Maffei, Scipione, 219
Maïer, Ida, 31
Maimonides, 127–128, 170–171
Malalas, John, 13–14
Mandrou, Robert, 3
Manetho, 76–77, 80, 99, 101–103, 134
Manilius, 107–108, 132–133
Manuzio, Aldo, 27–28, 108–109
Manuzio, Aldo, Jr., 117–118
Marius Rusticus, 55
Mars, 112–113
Marsham, John, 157
Martial, 50, 55
Massari, Buonaccorso, 23–26
Medici, Lorenzo de', 60, 72–73
Melanchthon, Philipp, 104
Mencke, J. B., 184
Menze, C., 217
Mercator, Gerardus, 131
Mersenne, Marin, 205
Merula, Giorgio, 63–64
Metasthenes, 29, 80, 90–91, 92, 97–98, 99, 138
Meton, 109, 115, 116, 200
Mettler, W., 217
Michaelis, J. D., 219, 235, 238, 318n144
Modesti, Jacopo, 48
Momigliano, Arnaldo, 6–7

Montaigne, Michel de, 40–41
More, Henry, 17–21
Morhof, D. G., 184
Moses, 82
Moss, Ann, 31
Müller, Karl Otfried, 79
Münster, Sebastian, 127–128
Muhlack, Ulrich, 31
Mund-Dopchie, Monique, 27–28
Muret, Marc-Antoine, 39
Murrin, Michael, 31, 42
Musaeus, 30, 78, 147, 154–155, 167
Myrsilus, 87

Nabonassar, 98, 129–131, 135, 202, 210
Nabuchodonosor, 98
Nero, 60
Newton, Isaac, 14, 20–21, 32, 199
Nicander, 66
Nicolson, Robert, 83
Niebuhr, B. G., 114, 242
Nigidius Figulus, 187–190
Nimrod, 136–137
Noah, 81, 83, 95
Nonnus, 155, 167
Nostitz, Hans von, 180–181, 202
Noviomagus, Ioh., 125–126
Numa, 112
Nutton, V., 248n9

Ognibene da Lonigo, 188
Opsopoeus, Johannes, 174–176
Origen, 143, 165–166, 167
Orphica, 15, 30, 147, 161, 167
Orosius, 140
Osiander, Andreas, 130, 142
Ovid, 26, 31, 37, 52–53, 67–68, 111

Panizza, Letizia, 28
Panvinio, Onofrio, 112, 139
Pareus, David, 36–37, 77–78
Parthey, G., 242–243
Pascal, Blaise, 2
Pastore Stocchi, Manlio, 9

Patin, Gui, 205
Patisson, Mamert, 121, 138
Patrizi, Francesco, 157, 169
Pattison, Mark, 107, 143, 147
Perosa, Alessandro, 9
Perotti, Niccolò, 50, 54
Perpetua, 220
Persius, 68
Petavius, Dionysius, 193
Petrarch, Francesco, 7, 27, 36, 42–43, 55, 59–60, 78, 163
Petri, Suffridus, 102–103
Phalaris, 177
Phidias, 152, 154, 156, 164
Philo, 137–138, 163
(pseudo-) Philo, 99
Phocion, 222
Phocylides, 172
Pico della Mirandola, G., 135, 202, 210
Pindar, 66
Pithou, François, 174
Plato, 31, 38, 78, 115–116, 119, 151–152, 157, 159, 163, 169, 210, 214
Plautus, 55, 63–64
Pliny, 51, 52, 56–57, 66, 112
Plotinus, 167–168
Plutarch, 66, 83–84, 132, 187, 192–193
Poliziano, Angelo, 6–9, 27, 28, 31, 33, 34–36, 42, 47–75
Polybius, 28
Popkin, Richard, 207
Porphyry, 135–136, 165–166, 167–168, 229
Portus, Franciscus, 146
Postel, Guillaume, 82, 93, 94–95, 112–113, 133
Priapeia, 66
Proclus, 68
Propertius, 28, 70–71, 80
Psellus, Michael, 169
Ptolemy, 98, 117, 129–131, 132–133, 135, 137–138, 198–202, 210, 288n109
Purnell, Frederick, 154

Quint, David, 31
Quintilian, 33, 53, 68

Rabelais, François, 26, 251n11
Ramus, Petrus, 135
Ranconet, Aimar, 174–175
Rashi, 127
Regiomontanus, Johannes, 200
Reinhold, Erasmus, 104, 119, 130, 200
Reiske, J. J., 232
Reiz, F. W., 231
Rhenanus, Beatus, 88, 93
Rheticus, G. J., 142
Ribuoli, Roberto, 9
Ritschl, F., 243
Rizzo, Silvia, 27, 58
Rolevinck, Werner, 129
Romulus, 95, 110–115, 132, 141
Röslin, Helisaeus, 191
Rossi, Paolo, 32
Rosweyde, Heribert, 156
Rothmann, Christopher, 142
Rothstein, Max, 28
Ruhnken, David, 230, 231

Sabbadini, Remigio, 27
Sabellico, M. A., 51
Salmanassar, 98, 129–131
Salutati, Coluccio, 36, 55–56, 262–263n39
Santritter, J. L., 129
Sappho, 214
Saumaise, Claude, 209–210
Savile, Henry, 117, 135, 199
Scaliger, Joseph, 5, 9–10, 13, 16, 26, 33, 36–37, 40, 42, 76–78, 101–103, 104–144, 146, 163, 172, 191–193, 209–210, 251n12, 254n38, 255n49, 306n32
Schlegel, F., 242
Schreckenfuchs, Erasmus Oswald, 126
Schütz, C. G., 232–233
Scriptores Historiae Augustae, 33, 95–96, 148
Seck, Friedrich, 181
Selden, John, 77

Semler, J. S., 235
Sempronius, 84, 92
Senacherim, 229
Seneca, 28, 40, 162, 170, 172
Sepulveda, Juan Gines de, 111–112
Servius, 49–50, 69
Servius Tullius, 111
Seznec, Jean, 31, 32
Sibyls, 148, 158, 172–175
Sicherl, Martin, 27
Siebenkees, J. P., 231
Simon, Richard, 206, 210, 211, 212, 234, 237
Simonides, 54
Simplicius, 135–136
Sixtus Senensis, 155
Sleidanus, Joannes, 96
Smalley, Beryl, 84, 272n30
Solinus, 92
Solon, 158, 192
Sophianus, Michael, 155
Sophocles, 14–16, 18, 156
Sossus, Gulielmus, 209
Spinoza, B., 206
Stanley, Thomas, 157
Statius, 33, 34–36, 55
Stephens, Walter, 91
Stillingfleet, Edward, 13, 17–20
Stopp, F. J., 183
Strabo, 119
Strazel, Jean, 28
Suda, 56, 66, 112, 154
Suetonius, 55, 58, 60, 66
Sulpizio, Giovanni, 188

Tacitus, 39, 56–57, 81, 87
Tarrutius, L., 92, 132–133, 141
Tatian, 100, 105
Temporarius, Ioannes, 174, 303n64
Theocritus, 69, 70
Theopompus, 100
Thucydides, 70
Tibullus, 227
Timpanaro, Sebastiano, 9, 28, 55, 59–60

Tirimbocchi, Gaspare de', 49
Tory, Geoffroy, 83
Tourreil, Jacques de, 221–222
Trithemius, Joannes, 93
Tryphiodorus, 53, 155
Turnèbe, Adrien, 148, 154, 174

Ullman, B. L., 27
Ulpian, 232
Ursus, N. R., 195, 199–200

Valckenaer, L. C., 229–230
Valerius Flaccus, 62
Valerius Maximus, 53
Valla, Lorenzo, 7–8, 27, 42, 53, 61, 78, 162, 197–198, 261n26
Varnhagen von Ense, K., 242
Varro, 37, 106, 109
Vasoli, Cesare, 85
Velleius Paterculus, 197–198
Vergicius, Angelus, 169
Vergil, Polydore, 110
Vettori, Pier, 28
Vico, G. B., 137, 224
Villoison, J. B. G. d'Ansse de, 226, 229, 231, 232–233, 234, 315n81
Vinet, Elie, 108, 109, 112, 118
Virgil, 37–38, 40, 42–43, 47, 49, 58, 69, 73, 74, 172
Vives, Juan Luis, 93, 172
Volsco, Antonio, 28
Vossius, G. J., 156–157, 277n16
Vossius, I., 157–158, 208

Wacker, Johann, 179–181, 185, 202
Walton, Brian, 237
Waquet, F., 247n3
Warburton, William, 159
Weiss, Roberto, 27
Wettstein, J. R., 228
Wieland, C. M., 243
Witsius, Herman, 157
Wolf, Friedrich August, 44, 214–243
Wolf, Hieronymus, 28

Wood, Robert, 219, 224, 318n145
Worm, Ole, 208
Wotton, Henry, 179–180
Wowerius, Janus, 165
Wyttenbach, D., 314n72

Xenophon, 86, 87
Xerxes, 191–192

Zarlino, Gioseffo, 139
Zenodotus, 224, 231, 234, 237
Zoroaster, 158, 167–168, 209

DATE DUE

AUG 29 '95			

HIGHSMITH   # 45220